Mechanical Engineer's Data Handbook

To my daughters, Helen and Sarah

Mechanical Engineer's Data Handbook

J. Carvill

CRC Press, Inc.
Boca Raton, Florida

Direct all enquiries to CRC Press, Inc.,
2000 Corporate Blvd., N.W., Boca Raton,
Florida, 33431.

This edition published in the
United States of America by CRC Press, Inc. 1993
and in the UK by Butterworth-Heinemann Ltd 1993

Library of Congress Cataloging-in-Publication Data
CIP data available from the Library of Congress

All Rights Reserved
International Standard Book No. 0-8493-7780-3

Printed and bound in Great Britain by Bath Press

Contents

Preface

There are several good mechanical engineering data books on the market but these tend to be very bulky and expensive, and are usually only available in libraries as reference books.

The Mechnical Engineer's Data Handbook has been compiled with the express intention of providing a compact but comprehensive source of information of particular value to the engineer whether in the design office, drawing office, research and development department or on site. It should also prove to be of use to production, chemical, mining, mineral, electrical and building services engineers, and lecturers and students in universities, polytechnics and colleges. Although intended as a personal handbook it should also find its way into the libraries of engineering establishments and teaching institutions.

The Mechanical Engineer's Data Handbook covers the main disciplines of mechanical engineering and incorporates basic principles, formulae for easy substitution, tables of physical properties and much descriptive matter backed by numerous illustrations. It also contains a comprehensive glossary of technical terms and a full index for easy cross-reference.

I would like to thank my colleagues at the University of Northumbria, at Newcastle, for their constructive suggestions and useful criticisms, and my wife Anne for her assistance and patience in helping me to prepare this book.

J. Carvill

Symbols used in text

a	Acceleration	j	Operator $\sqrt{-1}$
A	Area	J	Polar second moment of area
$\mathscr{A}$	Anergy	k	Radius of gyration; coefficient of thermal conductivity; pipe roughness
b	Breadth		
b.p.	Boiling point	K	Bulk modulus; stress concentration factor
B	Breadth, flux density		
c	Clearance, depth of cut; specific heat capacity	KE	Kinetic energy
		K_w	Wahl factor for spring
C	Couple; Spring coil index; velocity (thermodynamics); heat capacity	l	Length
		L	Length
C_d	Drag coefficient, discharge coefficient	m	Mass; mass per unit length; module of gear
COP	Coefficient of performance		
C_p	Specific heat at constant pressure	$\dot{m}$	Mass flow rate
C_v	Specific heat at constant volume; velocity coefficient	m.p.	Melting point
		M	Mass; moment; bending moment; molecular weight
CV	Calorific value		
d	Depth; depth of cut; diameter; deceleration	MA	Mechanical advantage
		n	Index of expansion; index; number of; rotational speed
D	Depth; diameter; flexural rigidity		
e	Strain; coefficient of restitution; emissivity	N	Rotational speed; number of
		N_s	Specific speed
E	Young's Modulus; energy; luminance; effort	N_u	Nusselt number
		p	Pressure; pitch
EL	Elastic limit; endurance limit	P	Power; force; perimeter
ELONG%	Percentage elongation	P_r	Prandtl number
$\mathscr{E}$	Exergy	PE	Potential energy
f	Frequency; friction factor; feed	PS	Proof stress
F	Force; luminous flux	Q	Heat quantity; volume flow rate; metal removal rate
F_g	Strain gauge factor		
FL	Fatigue limit	r	Radius; pressure or volume ratio
FS	Factor of safety	R	Radius; electric resistance; reaction, thermal resistance; gas constant
g	Acceleration due to gravity		
G	Shear modulus; Gravitational constant	R_e	Reynolds number
G_r	Grashof number	RE	Refrigeration effect
h	Height; thickness; specific enthalpy; shear, heat transfer coefficient	R_o	Universal gas constant
		s	Specific entropy; stiffness
h.t.c.	Heat transfer coefficient	S	Entropy, shear force, thermoelectric sensitivity
H	Enthalpy; height, magnetic field strength		
i	Slope; operator $\sqrt{-1}$	SE	Strain energy
I	Moment of inertia; Second moment of area; luminous intensity, electric current	S_t	Stanton number
		t	Temperature; thickness; time

T	Time; temperature; torque; tension; thrust; number of gear teeth	α	Angle; coefficient of linear expansion; angular acceleration; thermal diffusivity; Resistance temperature coefficient
TS	Tensile strength		
u	Velocity; specific strain energy; specific internal energy	β	Angle; coefficient of superficial expansion
U	Internal energy; strain energy; overall heat transfer coefficient	γ	Angle; coefficient of volumetric expansion; ratio of specific heats
		δ	Angle
UTS	Ultimate tensile stress	ε	Permittivity
v	Velocity; specific volume	η	Efficiency
V	Velocity; voltage, volume	θ	Angle; temperature
VR	Velocity ratio	λ	Wavelength
w	Weight; weight per unit length	μ	Absolute viscosity; coefficient of friction
W	Weight; load; work; power (watts)	v	Poisson's ratio; kinematic viscosity
x	Distance (along beam); dryness fraction	ρ	Density; resistivity; velocity ratio
X	Parameter (fluid machines)	ρ_o	Resistivity
y	Deflection	σ	Stress; Stefan–Boltzmann constant
YP	Yield point	τ	Shear stress
YS	Yield stress	ϕ	Friction angle; phase angle; shear strain; pressure angle of gear tooth
Z	Bending modulus; impedance; number of		
Z_p	Polar modulus	ω	Angular velocity

Strengths of materials

1.1 Types of stress

Engineering design involves the correct determination of the sizes of components to withstand the maximum stress due to combinations of direct, bending and shear loads. The following deals with the different types of stress and their combinations. Only the case of two-dimensional stress is dealt with, although many cases of three-dimensional stress combinations occur. The theory is applied to the special case of shafts under both torsion and bending.

1.1.1 *Direct, shear and bending stress*

Tensile and compressive stress (direct stresses)

Stress $\sigma = \dfrac{\text{load}}{\text{area}} = \dfrac{P}{A}$

Strain $e = \dfrac{\text{extension}}{\text{original length}} = \dfrac{x}{L}$

$\dfrac{\text{Stress}}{\text{Strain}} = \dfrac{\sigma}{e} = \text{Young's modulus, } E.$ Thus $E = \dfrac{PL}{Ax}$

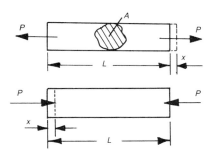

Poisson's ratio

Poisson's ratio $v = \dfrac{\text{strain in direction of load}}{\text{strain at right angles to load}}$

$\qquad = \dfrac{\delta B/B}{\delta L/L} = \dfrac{e_B}{e_L}$

Note: if e_L is positive, e_B is negative.

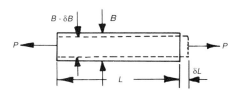

Shear stress

Shear stress $\tau = \dfrac{P}{A}$

Shear strain $\phi = \dfrac{\tau}{G}$, where $G = $ Shear modulus.

$$\phi = \dfrac{x}{L}$$

$$G = \dfrac{PL}{Ax}$$

Note: A is parallel to the direction of P.

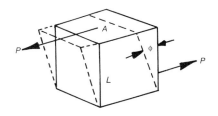

Bending stress

Bending stress $\sigma = \dfrac{My}{I}$

where:
M = bending moment
I = second moment of area of section
y = distance from centroid to the point considered

Maximum stress $\sigma_m = \dfrac{My_m}{I}$

where y_m = maximum value of y for tensile and compressive stress.

Radius of curvature $R = \dfrac{EI}{M}$

Bending modulus $Z = I/y_m$ and $\sigma_m = M/Z$

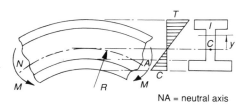

NA = neutral axis

Combined bending and direct stresses

$\sigma_c = P/A \pm M/Z$ where $Z = \dfrac{I}{y_m}$

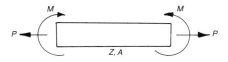

Hydrostatic (three-dimensional) stress

Volumetric strain $e_v = \dfrac{\sigma V}{V}$

Bulk modulus $K = p/e_v$
where p = pressure and V = volume.

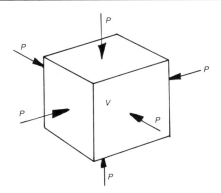

Relationship between elastic constants

$$K = \frac{E}{3(1-2v)}; \quad G = \frac{E}{2(1+v)}; \quad E = \frac{9GK}{(G+3K)}$$

Compound stress

For normal stresses σ_x and σ_y with shear stress τ:
Maximum principal stress $\sigma_1 = (\sigma_x + \sigma_y)/2 + \tau_{max}$
Minimum principal stress $\sigma_2 = (\sigma_x + \sigma_y)/2 - \tau_{max}$

where: maximum shear stress $\tau_{max} = \sqrt{\left(\dfrac{\sigma_x - \sigma_y}{2}\right)^2 + \tau^2}$

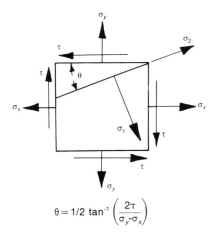

$$\theta = 1/2 \, \tan^{-1}\left(\frac{2\tau}{\sigma_y - \sigma_x}\right)$$

Combined bending and torsion

For solid and hollow circular shafts the following can be derived from the theory for two-dimensional (Compound) stress. If the shaft is subject to bending moment

M and torque T, the maximum direct and shear stresses, σ_m and τ_m are equal to those produced by 'equivalent' moments M_e and T_e where

$\tau_m = T_e/Z_p$ and $\sigma_m = M_e/Z$

where Z_p = polar modulus

$T_e = \sqrt{M^2 + T^2}$ and $M_e = (M + T_e)/2$

$Z = \dfrac{\pi D^3}{32}$ (solid shaft) or $\dfrac{\pi}{32} \dfrac{(D^4 - d^4)}{D}$ (hollow shaft)

$Z_p = \dfrac{\pi D^3}{16}$ (solid shaft) or $\dfrac{\pi}{16} \dfrac{(D^4 - d^4)}{D}$ (hollow shaft)

See section 1.1.7.

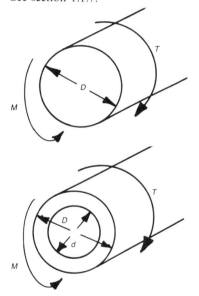

1.1.2 Impact stress

In many components the load may be suddenly applied to give stresses much higher than the steady stress. An example of stress due to a falling mass is given.

Maximum tensile stress in bar

$\sigma_m = \sigma_s [1 + \sqrt{1 + (2h/x_s)}]$

where:
σ_s = steady stress = mg/A
x_s = steady extension = mgL/AE
h = height fallen by mass m.

Stress due to a 'suddenly applied' load ($h = 0$)

$\sigma_m = 2\sigma_s$

Stress due to a mass M moving at velocity v

$\sigma_m = v\sqrt{\dfrac{mE}{AL}}$

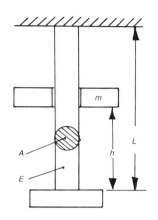

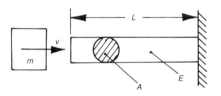

1.1.3 Compound bar in tension

A compound bar is one composed of two or more bars of different materials rigidly joined. The stress when loaded depends on the cross-sectional areas (A_a and A_b) areas and Young's moduli (E_a and E_b) of the components.

Stresses

$\sigma_a = \dfrac{F}{A_a + (E_b/E_a)A_b}$

$\sigma_b = \dfrac{F}{A_b + (E_a/E_b)A_a}$

Strains

$e_a = \sigma_a/E_a$; $e_b = \sigma_b/E_b$ (note that $e_a = e_b$)

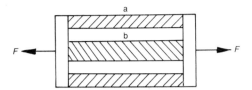

1.1.4 *Stresses in knuckle joint*

The knuckle joint is a good example of the application of simple stress calculations. The various stresses which occur are given.

Symbols used:
P = load
σ_t = tensile stress
σ_b = bending stress
σ_c = crushing stress
τ = shear stress
D = rod diameter
D_p = pin diameter
D_o = eye outer diameter
a = thickness of the fork
b = the thickness of the eye

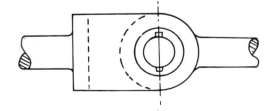

Failure may be due to any one of the following stresses.

(1) Tensile in rod $\sigma_t = 4P/\pi D^2$

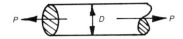

(2) Tensile in eye $\sigma_t = P/(D_o - D_p)b$

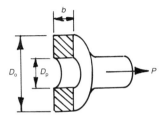

(3) Shear in eye $\tau = P/(D_o - D_p)b$

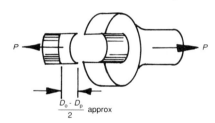

$\dfrac{D_o - D_p}{2}$ approx

(4) Tensile in fork $\sigma_t = P/(D_o - D_p)2a$

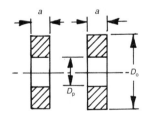

(5) Shear in fork $\tau = P/(D_o - D_p)2a$

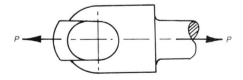

(6) Crushing in eye $\sigma_c = P/bD_p$

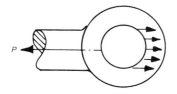

(7) Crushing in fork $\sigma_c = P/2D_p a$

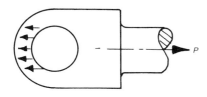

(8) Shear in pin $\tau = 2P/\pi D_p^2$

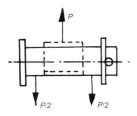

(9) Bending in pin $\sigma_b = \dfrac{4P(a+b)}{\pi D_p^3}$

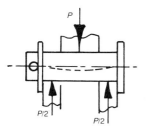

(10) Crushing in pin due to eye $\sigma_c = P/bD_p$

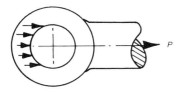

(11) Crushing in pin due to fork $\sigma_c = P/2aD_p$

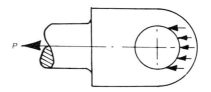

1.1.5 Theories of failure

For one-dimensional stress the factor of safety (FS) based on the elastic limit is simply given by

$$FS = \frac{\text{Elastic limit}}{\text{Actual stress}}.$$

When a two- or three-dimensional stress system exists, determination of FS is more complicated and depends on the type of failure assumed and on the material used.

Symbols used:

σ_{el} = elastic limit in simple tension
σ_1, σ_2, σ_3 = maximum principal stresses in a three-dimensional system
FS = factor of safety based on σ_{el}
v = Poisson's ratio

Maximum principal stress theory (used for brittle metals)

FS = smallest of σ_{el}/σ_1, σ_{el}/σ_2 and σ_{el}/σ_3

Maximum shear stress theory (used for ductile metals)

FS = smallest of $\sigma_{el}/(\sigma_1 - \sigma_2)$, $\sigma_{el}/(\sigma_1 - \sigma_3)$ and $\sigma_{el}/(\sigma_2 - \sigma_3)$

Strain energy theory (used for ductile metals)

$$FS = \sigma_{el}/\sqrt{\sigma_1^2 + \sigma_2^2 + \sigma_3^2 - 2v(\sigma_1\sigma_2 + \sigma_2\sigma_3 + \sigma_1\sigma_3)}$$

Shear strain energy theory (best theory for ductile metals)

$$FS = \sigma_{el}/\sqrt{[(\sigma_1 - \sigma_2)^2 + (\sigma_2 - \sigma_3)^2 + (\sigma_3 - \sigma_1)^2]/2}$$

Maximum principal strain theory (used for special cases)

FS = smallest of $\sigma_{el}/(\sigma_1 - v\sigma_2 - v\sigma_3)$, $\sigma_{el}/(\sigma_2 - v\sigma_1 - v\sigma_3)$ and $\sigma_{el}/(\sigma_3 - v\sigma_2 - v\sigma_1)$

Example

In a three-dimensional stress system, the stresses are $\sigma_1 = 40\,MN\,m^{-2}$, $\sigma_2 = 20\,MN\,m^{-2}$ and $\sigma_3 = -10\,MN\,m^{-2}$. $\sigma_{el} = 200\,MN\,m^{-2}$ and $v = 0.3$. Calculate the factors of safety for each theory.

Answer: (a) 5.0; (b) 4.0; (c) 4.5; (d) 4.6; (e) 5.4.

1.1.6 *Strain energy (Resilience)*

Strain energy U is the energy stored in the material of a component due to the application of a load. Resilience u is the strain energy per unit volume of material.

Tension and compression

Strain energy $U = \dfrac{Fx}{2} = \dfrac{\sigma^2 AL}{2E}$

Resilience $U = \dfrac{\sigma^2}{2E}$

Shear

Resilience $U = \dfrac{\tau^2}{2G}$

The units for U and u are joules and joules per cubic metre.

1.1.7 *Torsion of various sections*

Formulae are given for stress and angle of twist for a solid or hollow circular shaft, a rectangular bar, a thin tubular section, and a thin open section. The hollow shaft size equivalent in strength to a solid shaft is given for various ratios of bore to outside diameter.

Solid circular shaft

Maximum shear stress $\tau_m = \dfrac{16T}{\pi D^3}$

where: D = diameter, T = torque.

Torque capacity $T = \dfrac{\pi D^3 \tau_m}{16}$

Power capacity $P = \dfrac{\pi^2 ND^3}{8}\tau_m$

where: N = the number of revolutions per second.

Angle of twist $\theta = \dfrac{32TL}{\pi GD^4}$ rad

where: G = shear modulus, L = length

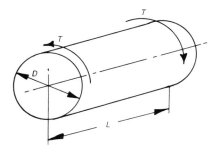

Hollow circular shaft

$\tau_m = \dfrac{16TD}{\pi(D^4 - d^4)}$; $T = \dfrac{\pi(D^4 - d^4)}{16D}\tau_m$

where: D = outer diameter, d = inner diameter.

$P = \dfrac{\pi^2 N(D^4 - d^4)\tau_m}{8D}$; $\theta = \dfrac{32TL}{\pi G(D^4 - d^4)}$

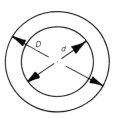

Rectangular section bar

For $d > b$:

$\tau_m = \dfrac{(1.8b + 3d)T}{b^2 d^2}$ (at middle of side d)

$\theta = \dfrac{7TL(b^2 + d^2)}{2Gb^3 d^3}$

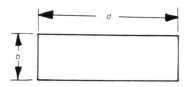

Thin tubular section

$\tau_{\mathrm{m}} = T/2tA$; $\theta = TpL/4A^2tG$

where:
t = thickness
A = area enclosed by mean perimeter
p = mean perimeter

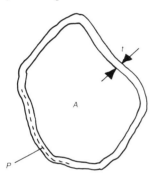

Thin rectangular bar and thin open section

$\tau_{\mathrm{m}} = 3T/dt^2$; $\theta = 3TL/Gdt^3$ (rectangle)
$\tau_{\mathrm{m}} = 3T/\Sigma dt^2$; $\theta = 3TL/G\Sigma dt^3$ (general case)
$\Sigma dt^2 = (d_1t_1^2 + d_2t_2^2 + \cdots \Sigma dt^3 = (d_1t_1^3 + d_2t_2^3 + \cdots)$

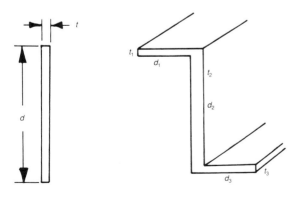

Strain energy in torsion

Strain energy $U = \frac{1}{2}T\theta$

for solid circular shaft $u = \dfrac{\tau_{\mathrm{m}}^2}{4G}$

for hollow circular shaft $u = \dfrac{\tau_{\mathrm{m}}^2}{4G}\left(\dfrac{D^2 + d^2}{D^2}\right)$

where $U = u\,\dfrac{\pi D^2 L}{4}$ *solid shaft*

$ = u\,\dfrac{\pi(D^2 - d^2)L}{4}$ *hollow shaft*

Torsion of hollow shaft

For a hollow shaft to have the same strength as an equivalent solid shaft:

$D_{\mathrm{o}}/D_{\mathrm{s}} = \sqrt[3]{\dfrac{1}{1 - k^4}}$; $W_{\mathrm{h}}/W_{\mathrm{s}} = \dfrac{1 - k^2}{\sqrt[3]{(1 - k^4)^2}}$

$\theta_{\mathrm{h}}/\theta_{\mathrm{s}} = \sqrt[3]{(1 - k^4)}$

$k = D_{\mathrm{i}}/D_{\mathrm{o}}$

where:
D_{s}, D_{o}, D_{i} = solid, outer and inner diameters
W_{h}, W_{s} = weights of hollow and solid shafts
θ_{h}, θ_{s} = angles of twist of hollow and solid shafts

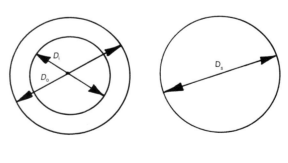

k	0.5	0.6	0.7	0.8	0.9
$D_{\mathrm{o}}/D_{\mathrm{s}}$	1.02	1.047	1.095	1.192	1.427
$W_{\mathrm{h}}/W_{\mathrm{s}}$	0.783	0.702	0.613	0.516	0.387
$\theta_{\mathrm{h}}/\theta_{\mathrm{s}}$	0.979	0.955	0.913	0.839	0.701

1.2 Strength of fasteners

1.2.1 Bolts and bolted joints

Bolts, usually in conjunction with nuts, are the most widely used non-permanent fastening. The bolt head is usually hexagonal but may be square or round. The shank is screwed with a vee thread for all or part of its length.

In the UK, metric (ISOM) threads have replaced Whitworth (BSW) and British Standard Fine (BSF) threads. British Association BA threads are used for small sizes and British Standard Pipe BSP threads for pipes and pipe fittings. In the USA the most common threads are designated 'unified fine' (UNF) and 'unified coarse' (UNC).

Materials

Most bolts are made of low or medium carbon steel by forging or machining and the threads are formed by cutting or rolling. Forged bolts are called 'black' and machined bolts are called 'bright'. They are also made in high tensile steel (HT bolts), alloy steel, stainless steel, brass and other metals.

Nuts are usually hexagonal and may be bright or black. Typical proportions and several methods of locking nuts are shown.

Bolted joints

A bolted joint may use a 'through bolt', a 'tap bolt' or a 'stud'.

Socket head bolts

Many types of bolt with a hexagonal socket head are used. They are made of high tensile steel and require a special wrench.

Symbols used:
D = outside or major diameter of thread
L = Length of shank
T = Length of thread
H = height of head
F = distance across flats
C = distance across corners
R = radius of fillet under head
B = bearing diameter

Extract from table of metric bolt sizes (mm)

Nominal size	D	H	F	Thread pitch Coarse	Fine
M10	10	7	17	1.5	1.25
M12	12	8	19	1.75	1.25
M16	16	10	24	2.0	1.5
M20	20	13	30	2.5	1.5

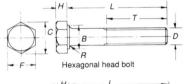

Hexagonal head bolt

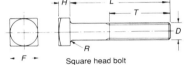

Square head bolt

Types of bolt

Bolted joint (through bolt) application

Tap bolt application

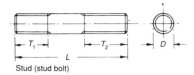

Stud (stud bolt)

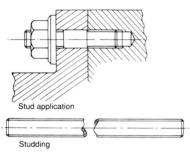

Stud application

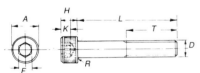

Studding

Stud and application

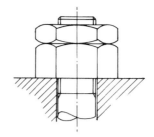

Locked nuts (jam nuts)

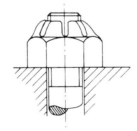

Slotted nut   Castle nut

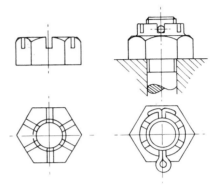

Typical metric sizes (mm):

D = 10.0 R = 0.6
A = 16.0 F = 8.0
H = 10.0 K = 5.5

L/T according to application

Hexagon socket head screw

Spring lock nut (compression stop nut)

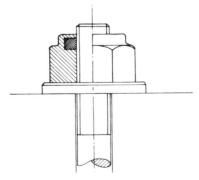

Elastic stop nut (Nyloc nut)

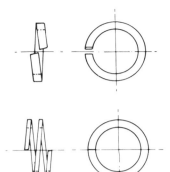

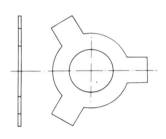

Helical spring lock washer and
two-coil spring lock washer

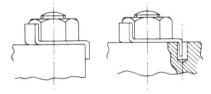

Tab washer and application

Approximate dimensions of bolt heads and nuts (ISO metric precision)

Exact sizes are obtained from tables.

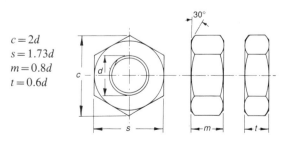

$c = 2d$
$s = 1.73d$
$m = 0.8d$
$t = 0.6d$

Bolted joint in tension

The bolt shown is under tensile load plus an initial tightening load. Three members are shown bolted together but the method can be applied to any number of members.

Symbols used:
P_e = external load
P_t = tightening load
P = total load
A = area of a member (A_1, A_2, etc.)
A_b = bolt cross-sectional area
t = thickness of a member (t_1, t_2, etc.)
L = length of bolt
E = Youngs modulus (E_b, E_1, etc.)
x = deflection of member per unit load
x_b = deflection of bolt per unit load
D = bolt diameter
D_r = bolt thread root diameter
A_r = area at thread root
T = bolt tightening torque

$$x_b = \frac{L}{A_b E_b}; \quad x_1 = \frac{t_1}{A_1 E_1}; \quad x_2 = \frac{t_2}{A_2 E_2}; \text{ etc.}$$

$$P = P_t + P_e \cdot \frac{\Sigma x}{\Sigma x + x_b}$$

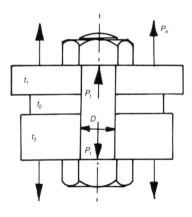

Tightening load

(a) Hand tightening:

$P_t = kD$

where:
$k = 1500$ to 3000; P_t is in newtons and D is in millimetres.

(b) Torque-wrench tightening:

$$P_t = T/0.2D$$

Shear stress in bolt

$$\tau_{max} = \sqrt{\left(\frac{P}{2A_r}\right)^2 + \left(\frac{16T}{\pi D^3}\right)^2}$$

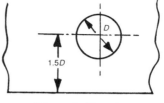

Distance of bolt from edge

1.2.2 Bolted or riveted brackets - stress in bolts

Bracket in torsion

Force on a bolt at r_1 from centroid of bolt group
$$P_1 = Par_1/(r_1^2 + r_2^2 + r_3^2 + \ldots)$$
Vertical force on each bolt $P_v = P/n$
where: n = number of bolts.
Total force on a bolt $P_t = $ vector sum of P_1 and P_v
Shear stress in bolt $\tau = P_t/A$
where: $A = $ bolt area. This is repeated for each bolt and the greatest value of τ is noted.

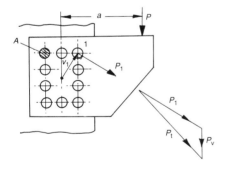

Bracket under bending moment

(a) Vertical load:
Tensile force on bolt at a_1 from pivot point

$$P_1 = Pda_1/(a_1^2 + a_2^2 + a_3^2 + \ldots)$$

Tensile stress $\sigma_1 = P_1/A$

where: $A = $ bolt area.

and similarly $\sigma_2 = \dfrac{P_2}{A}$, etc.

Shear stress $\tau = P/(nA)$

where: $n = $ number of bolts.

Maximum tensile stress in bolt at a_1, $\sigma_m = \dfrac{\sigma_1}{2} + \dfrac{1}{2}\sqrt{\sigma_1^2 + 4\tau^2}$

(b) Horizontal load:

Maximum tensile stress $\sigma_m = \sigma_1 + P/(nA)$ for bolt at a_1

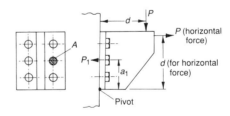

1.2.3 Bolts in shear

This deals with bolts in single and double shear. The crushing stress is also important.

Single shear

Shear stress $\tau = 4P/\pi D^2$

Double shear

Shear stress $\tau = 2P/\pi D^2$

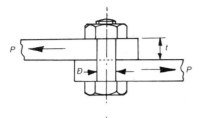

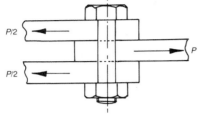

Crushing stress

$$\sigma_c = P/Dt$$

1.2.4 Rivets and riveted joints in shear

Lap joint

Symbols used:
  t = plate thickness
  D = diameter of rivets
  L = distance from rivet centre to edge of plate
  p = pitch of rivets
  σ_p = allowable tensile stress in plate
  σ_b = allowable bearing pressure on rivet
  τ_r = allowable shear stress in rivet
  τ_p = allowable shear stress in plate
  P = load

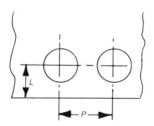

Allowable load per rivet:
Shearing of rivet $P_1 = \tau_r \pi D^2 / 4$
Shearing of plate $P_2 = \tau_p 2Lt$
Tearing of plate $P_3 = \sigma_p (p - D)t$
Crushing of rivet $P_4 = \sigma_b Dt$

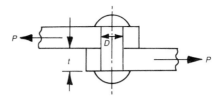

Efficiency of joint:

$$\eta_j = \frac{\text{least of } P_1 P_2 P_3 P_4}{\sigma_p pt} \times 100\%$$

Butt joint

The rivet is in 'double shear', therefore $P_1 = \tau_r \pi D^2 / 2$ per row.

In practice, P_1 is nearer to $\tau_r \pi \dfrac{3D^2}{8}$.

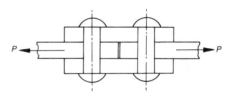

Several rows of rivets

The load which can be taken is proportional to the number of rows.

1.2.5 Strength of welds

A well-made 'butt weld' has a strength at least equal to that of the plates joined. In the case of a 'fillet weld' in shear the weld cross section is assumed to be a 45° right-angle triangle with the shear area at 45° to the plates. For transverse loading an angle of 67.5° is assumed as shown.

For brackets it is assumed that the weld area is flattened and behaves like a thin section in bending. For ease of computation the welds are treated as thin lines. Section 1.2.6 gives the properties of typical weld groups.

Since fillet welds result in discontinuities and hence stress concentration, it is necessary to use stress concentration factors when fluctuating stress is present.

Butt weld

The strength of the weld is assumed equal to that of the plates themselves.

Fillet weld

Parallel loading:

Shear stress $\tau = F/tL$
Weld throat $t \doteq 0.7w$
where w = weld leg size.

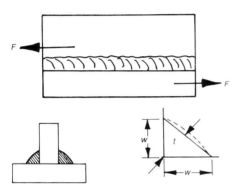

Transverse loading:

Shear stress $\tau = F/tL$
Throat $t = 0.77w$

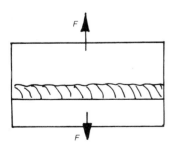

Welded bracket with bending moment

Symbols used:
I = second moment of area of weld group (treated as lines) = constant $\times t$
$Z = I/y_{max}$ = bending modulus

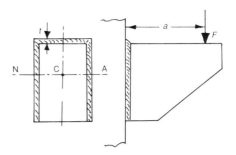

Maximum shear stress due to moment $\tau_b = M/Z$ (*an assumption*)
where: M = bending moment.

Direct shear stress $\tau_d = F/A$
where: A = total area of weld at throat, F = load.

Resultant stress $\tau_r = \sqrt{\tau_b^2 + \tau_d^2}$
from which t is found.

Welded bracket subject to torsion

Maximum shear stress due to torque (T) $\tau_t = Tr/J$ $(T = Fa)$
Polar second moment of area $J = I_x + I_y$
where: r = distance from centroid of weld group to any point on weld.

Direct shear stress $\tau_d = F/A$

Resultant stress (τ_r) is the vector sum of τ_d and τ_t; r is chosen to give highest value of τ_r. From τ_r the value of t is found, and hence w.

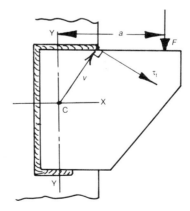

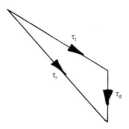

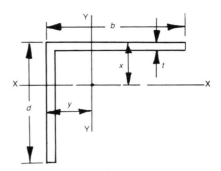

1.2.6 Properties of weld groups - welds treated as lines

Symbols used:
Z = bending modulus about axis XX
J = polar second moment of area
t = weld throat size

(1) $Z = d^2 t/3$; $J = dt(3b^2 + d^2)/6$

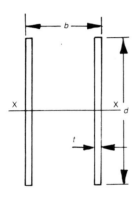

(4) $Z = (bd + d^2/6)t$; $J = \left[\dfrac{(2b+d)^3}{12} - \dfrac{b^2(b+d)^2)}{(2b+d)} \right] t$

$$y = \frac{b^2}{2b+d}$$

(2) $Z = bdt$; $J = bt(3d^2 + b^2)/6$

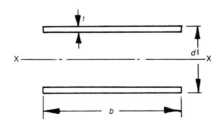

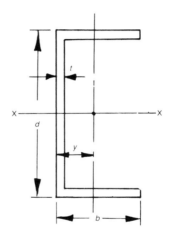

(5) $Z = (2bd + d^2)t/3$ (at top); $J = \left[\dfrac{(b+2d)^3}{12} - \dfrac{d^2(b+d)^2}{(b+2d)} \right] t$

$Z = \dfrac{d^2(2b+d)t}{3(b+d)}$ (at bottom); $x = \dfrac{d^2}{b+2d}$

(3) $Z = (4bd + d^2)t/6$ (at top); $J = \dfrac{[(b+d)^4 - 6b^2 d^2]}{12(b+d)} t$

$Z = \dfrac{(4bd^2 + d^3)t}{6(2b+d)}$ (at bottom); $x = \dfrac{d^2}{2(b+d)}$; $y = \dfrac{b^2}{2(b+d)}$

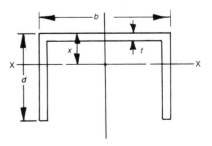

(6) $Z = (bd + d^2/3)t$; $J = t(b+d)^3/6$

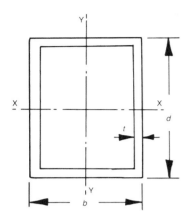

(7) $Z = (4bd + d^2)t/3$ (at top); $J = \left[\dfrac{d^3(4b+d)}{6(b+d)} + \dfrac{b^3}{6}\right]t$

$Z = \dfrac{(4bd^2 + d^3)t}{(6b + 3d)}$ (at bottom); $x = \dfrac{d^2}{2(b+d)}$

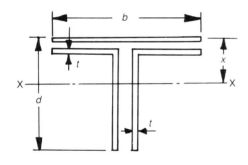

(8) $Z = (2bd + d^2/3)t$; $J = (2b^3 + 6bd^2 + d^3)t/6$

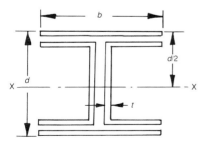

(9) $Z = \pi D^2 t/4$; $J = \pi D^3 t/4$

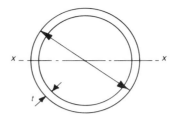

1.2.7 *Stresses due to rotation*

Flywheels are used to store large amounts of energy and are therefore usually very highly stressed. It is necessary to be able to calculate the stresses accurately. Formulae are given for the thin ring, solid disk, annular wheel and spoked wheel, and also the rotating thick cylinder.

Thin ring

Symbols used:
ρ = density
r = mean radius
v = tangential velocity = $r\omega$

Tangential stress $\sigma_t = \rho v^2 = \rho r^2 \omega^2$

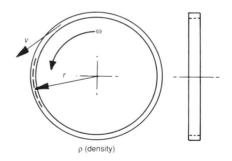

ρ (density)

Solid disk

Maximum tangential and radial stress (σ_r)

$\sigma_t = \sigma_r = \rho v^2 (3+v)/8$ at $r = 0$

where: v = Poisson's ratio, $v = r\omega$.

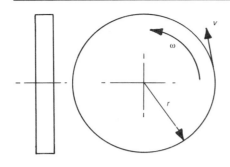

No. of spokes	Value of constant c
4	$0.073\left(\dfrac{r}{t}\right)^2 + 0.643 + A_r/A_s$
6	$0.0203\left(\dfrac{r}{t}\right)^2 + 0.957 + A_r/A_s$
8	$0.0091\left(\dfrac{r}{t}\right)^2 + 1.274 + A_r/A_s$

Annular wheel

For axial length assumed 'small':

$$\sigma_{t\,max} = \rho v^2 \frac{(3+v)}{4}\left(1 + \frac{(1-v)}{(3+v)}\left(\frac{r_1}{r_2}\right)^2\right)\text{(at }r_1)$$

$$\sigma_{r\,max} = \rho v^2 \frac{(3+v)}{8}\left(1 - \left(\frac{r_1}{r_2}\right)^2\right)\text{(at }r = \sqrt{r_1 r_2})$$

where: $v = r_2 \omega$

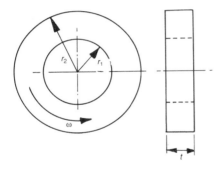

Spoked wheel

Greatest tangential stress $\sigma_t = \rho v^2$

$$\left[1 - \frac{\cos\theta}{3c\sin\alpha} \pm \frac{2r}{ct}\left(1/\alpha - \frac{\cos\theta}{\sin\alpha}\right)\right]\text{ at angle }\theta$$

where: r = mean radius of rim.

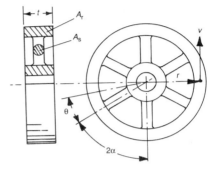

Tensile stress in spokes $\sigma_s = \dfrac{2Ar}{3cA_s}\cdot\rho v^2$

Long thick cylinder

Maximum tangential stress

$$\sigma_t = \frac{\rho v^2}{4(1-v)}\left[(1-2v) + (3-2v)\left(\frac{r_1}{r_2}\right)^2\right]\text{(at }r_1)$$

Maximum radial stress $\sigma_r = \dfrac{\rho v^2(3-2v)}{8(1-v)}\left(1 - \dfrac{r_1}{r_2}\right)^2$

$$\text{(at }r = \sqrt{r_1 r_2})$$

Maximum axial stress $\sigma_a = \dfrac{\rho v^2}{4(1-v)}\dfrac{v}{}\left[1 - \left(\dfrac{r_1}{r_2}\right)^2\right]$

(tensile at r_1, compressive at r_2)

1.3 Fatigue and stress concentration

In most cases failure of machine parts is caused by fatigue, usually at a point of high 'stress concentration', due to fluctuating stress. Failure occurs suddenly as a result of crack propagation without plastic deformation at a stress well below the elastic limit. The stress may be 'alternating', 'repeated', or a combination of these. Test specimens are subjected to a very large number of stress reversals to determine the 'endurance limit'. Typical values are given.

At a discontinuity such as a notch, hole or step, the stress is much higher than the average value by a factor K, which is known as the 'stress concentration factor'. The Soderberg diagram shows the alternating and steady stress components, the former being multiplied by K, in relation to a safe working line and a factor of safety.

1.3.1 *Fluctuating stress*

Alternating stress

The stress varies from σ_r compressive to σ_r tensile.

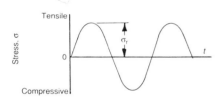

Repeated stress

The stress varies from zero to a maximum tensile or compressive stress, of magnitude $2\sigma_r$.

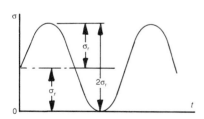

Combined steady and alternating stress

The average value is σ_m with a superimposed alternating stress of range σ_r.

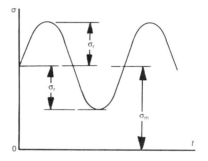

SN curves - endurance limit

The number of cycles N of alternating stress to cause failure and the magnitude of the stress σ_f are plotted. At $N = 0$, failure occurs at σ_u, the ultimate tensile strength. At a lower stress σ_e, known as the 'endurance limit', failure occurs, in the case of steel, as N approaches infinity. In the case of non-ferrous metals, alloys and plastics, the curve does not flatten out and a 'fatigue stress' σ_{FS} for a finite number of stress reversals N^1 is specified.

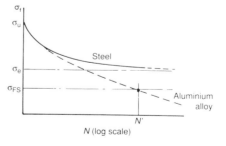

Soderberg diagram (for steel)

Alternating stress is plotted against steady stress. Actual failures occur above the line PQ joining σ_e to σ_u. PQ is taken as a failure line. For practical purposes the yield stress σ_y is taken instead of σ_u and a safety factor FS is applied to give a working line AB. A typical point on the line is C, where the steady stress component is σ_m and the alternating component is $K\sigma_r$, where K is a 'stress concentration coefficient' which allows for discontinuities such as notches, holes, shoulders, etc. From the figure:

$$FS = \frac{\sigma_y}{\sigma_m + (\sigma_y/\sigma_e)K\sigma_r}$$

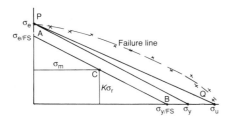

1.3.2 Endurance limit and fatigue stress for various materials

Steel

Most steels have an endurance limit which is about half the tensile strength. An approximation often used is as follows:

Endurance limit $= 0.5$ tensile strength up to a tensile strength of $1400\,\text{N mm}^{-2}$
Endurance limit $= 700\,\text{N mm}^{-2}$ above a tensile strength of $1400\,\text{N mm}^{-2}$

Cast iron and cast steel

Approximately:
Endurance limit $= 0.45 \times$ tensile strength up to a tensile strength of $600\,\text{N mm}^{-2}$
Endurance limit $= 275\,\text{N mm}^{-2}$ above a tensile strength of $600\,\text{N mm}^{-2}$.

Non-ferrous metals and alloys

There is no endurance limit and the fatigue stress is taken at a definite value of stress reversals, e.g. 5×10^7. Some typical values are given.

Endurance limit for some steels

Steel	Condition	Tensile strength, σ_u ($N\,mm^{-2}$)	Endurance limit, σ_e ($N\,mm^{-2}$)	σ_e/σ_u
0.4% carbon (080M40)	Normalized	540	270	0.50
	Hardened and tempered	700	340	0.49
Carbon, manganese (150M19)	Normalized	540	250	0.46
	Hardened and tempered	700	325	0.53
3% Chrome molybdenum (709M40)	Hardened and tempered	1000	480	0.48
Spring steel (735A50)	Hardened and tempered	1500	650	0.43
18,8 Stainless	Cold rolled	1200	490	0.41

Wrought aluminium alloys

Material	Tensile strength, σ_u (N mm^{-2})	Fatigue stress, σ_{FS} (N mm^{-2}), (5×10^7 cycles)	σ_u/σ_{FS}
N3 non-heat-treated	110	48	0.44
	130	55	0.42
	175	70	0.40
H9 heat treated	155	80	0.52
	240	85	0.35

Plastics

Plastics are very subject to fatigue failure, but the data on fatigue stress are complex. A working value varies between 0.18 and 0.43 times the tensile strength. Curves are given for some plastics.

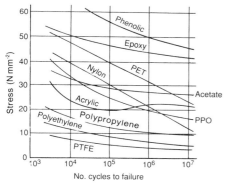

Room temperature fatigue characteristics of engineering plastics

Effect of surface finish on endurance limit

The values of endurance limits and fatigue stress given are based on tests on highly polished small specimens. For other types of surface the endurance limit must be multiplied by a suitable factor which varies with tensile strength. Values are given for a tensile strength of 1400 N mm^{-2}.

Surface	Surface factor
Polished	1.0
Ground	0.90
Machined, cold drawn	0.65
Hot rolled	0.37
As-forged	0.25

There are also factors which depend upon size, temperature, etc.

1.3.3 Causes of fatigue failure in welds

Under fatigue loading, discontinuities lead to stress concentration and possible failure. Great care must be taken in welds subject to fluctuating loads to prevent unnecessary stress concentration. Some examples are given below of bad cases.

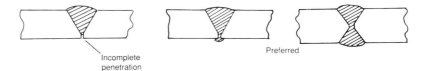

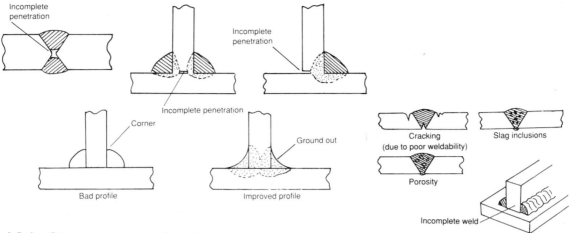

1.3.4 *Stress concentration factors*

Stress concentration factors are given for various common discontinuities; for example, it can be seen that for a 'wide plate' with a hole the highest stress is 3 times the nominal stress. General values are also given for keyways, gear teeth, screw threads and welds.

Stress concentration factor is defined as:

$$K = \frac{\text{Highest value of stress at a discontinuity}}{\text{Nominal stress at the minimum cross-section}}$$

Plate with hole at centre of width

$K = \sigma_{max}/\sigma$; $\sigma = P/wh$
σ_{max} occurs at A and B.

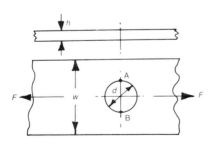

d/w	0.00	0.10	0.20	0.30	0.40	0.50	0.55
K	3.00	3.03	3.14	3.36	3.74	4.32	4.70

Note: In this case the area of maximum cross-section is used.

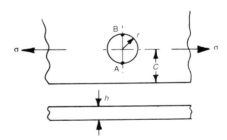

Semi-infinite plate with hole near edge

σ_a = stress at A
σ_b = stress at B
σ = stress away from hole
$K_a = \sigma_a/\sigma$; $K_b = \sigma_b/\sigma$

r/c	0.00	0.10	0.20	0.30	0.40	0.50	0.60	0.70	0.80	0.85
K_a	3.00	3.05	3.15	3.25	3.40	3.70	4.12	4.85	6.12	7.15
K_b	3.00	3.03	3.07	3.10	3.15	3.18	3.25	3.32	3.42	3.50

Bending of stepped flat bar with fillets (values of **K**)

$$K = \frac{\sigma_{max}}{6M/hd^2}$$

D/d	r/d							
	0.01	0.01	0.04	0.06	0.10	0.15	0.20	0.30
1.01	1.64	1.44	1.32	1.28	1.24	—	—	—
1.02	1.94	1.66	1.46	1.38	1.32	—	—	—
1.05	2.42	2.04	1.74	1.60	1.48	1.40	1.34	1.29
1.10	2.80	2.34	1.96	1.78	1.60	1.49	1.40	1.31
1.20	3.30	2.68	2.21	1.96	1.70	1.55	1.44	1.34
1.50	3.80	2.98	2.38	2.08	1.78	1.59	1.48	1.36
2.00	—	3.14	2.52	2.20	1.86	1.64	1.51	1.37
3.00	—	3.30	2.68	2.34	1.93	1.67	1.53	1.38

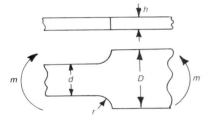

Tension of stepped bar with fillets (values of **K**)

$$K = \frac{\sigma_{max}}{P/hd}$$

D/d	R/d								
	0.01	0.02	0.04	0.06	0.10	0.15	0.20	0.25	0.30
1.01	1.68	1.48	1.34	1.26	1.20	—	—	—	—
1.02	2.00	1.70	1.49	1.39	1.30	—	—	—	—
1.05	2.50	2.08	1.74	1.60	1.45	—	—	—	—
1.10	2.96	2.43	1.98	1.78	1.60	1.50	1.43	1.39	1.36
1.20	3.74	2.98	2.38	2.14	1.89	1.72	1.62	1.56	1.53
1.30	4.27	3.40	2.67	2.38	2.06	1.86	1.73	1.64	1.59
1.50	4.80	3.76	3.00	2.64	2.24	1.99	1.84	1.74	1.67
2.00	—	—	3.30	2.90	2.44	2.13	1.95	1.84	1.76

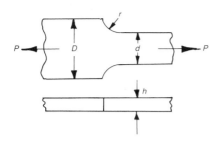

	Bending			r/d			
D/d	0.04	0.06	0.10	0.15	0.20	0.25	0.30
1.05	2.33	2.04	1.76	1.60	1.50	1.42	1.36
1.10	2.52	2.19	1.89	1.69	1.56	1.46	1.39
1.20	2.75	2.36	1.98	1.75	1.60	1.49	1.41
1.30	2.96	2.52	2.02	1.78	1.62	1.51	1.42
1.50	—	2.60	2.07	1.81	1.64	1.53	1.43
2.00	—	2.67	2.10	1.83	1.67	1.55	1.45

Bending of grooved shaft (values of **K**)

$$K = \frac{\sigma_{max}}{32M/\pi d^3}$$

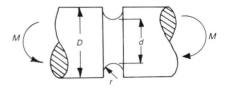

Torsion of grooved shaft (values of **K**)

$$K = \frac{\tau_{max}}{16T/\pi d^3}$$

	Torsion				r/d			
D/d	0.02	0.03	0.04	0.06	0.10	0.15	0.20	0.30
1.05	2.01	1.80	1.65	1.52	1.38	1.30	1.25	1.20
1.10	2.20	1.95	1.81	1.63	1.45	1.35	1.29	1.22
1.20	2.43	2.12	1.94	1.72	1.51	1.39	1.32	1.24
1.30	2.58	2.20	2.00	1.76	1.54	1.41	1.33	1.24
1.50	2.69	2.25	2.03	1.79	1.56	1.42	1.34	1.25
2.00	2.80	2.30	2.05	1.80	1.57	1.43	1.34	1.25

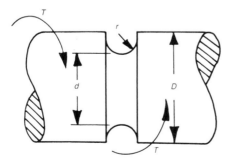

Bending of stepped shaft (values of **K**)

$$K = \frac{\sigma_{max}}{32M/\pi d^3}$$

D/d	r/d									
	0.01	0.02	0.03	0.04	0.05	0.08	0.10	0.15	0.20	0.25
1.01	1.65	1.44	1.36	1.32	1.29	1.25	1.24	—	—	—
1.02	1.96	1.64	1.54	1.46	1.41	1.34	1.32	—	—	—
1.05	2.41	2.04	1.84	1.73	1.65	1.52	1.48	—	—	—
1.10	2.85	2.34	2.08	1.94	1.84	1.66	1.60	—	—	—
1.20	3.40	2.62	2.32	2.14	2.00	1.75	1.65	1.50	1.42	1.30
1.50	3.73	2.90	2.52	2.30	2.13	1.84	1.72	1.54	1.43	1.35
2.00	—	—	2.70	2.42	2.25	1.92	1.78	1.58	1.46	1.36
3.00	—	—	—	2.60	2.42	2.04	1.88	1.61	1.48	1.38

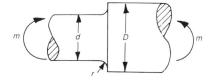

Torsion of stepped shaft (values of **K**)

$$K = \frac{\tau_{max}}{16T/\pi d^3}$$

D/d	r/d							
	0.02	0.03	0.05	0.07	0.10	0.15	0.20	0.30
1.05	1.60	1.48	1.33	1.25	1.20	1.16	1.13	1.09
1.10	1.75	1.60	1.44	1.35	1.28	1.21	1.17	1.12
1.20	1.85	1.72	1.59	1.43	1.33	1.25	1.19	1.14
1.30	—	1.78	1.59	1.47	1.36	1.27	1.21	1.14
1.50	—	—	—	1.50	1.39	1.28	1.22	1.15
1.75	—	—	—	1.51	1.40	1.29	1.24	1.16
2.00	—	—	—	—	1.41	1.31	1.24	1.16
2.50	—	—	—	—	1.42	1.31	1.25	1.16

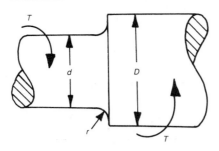

Welds

Reinforced butt weld, $K = 1.2$

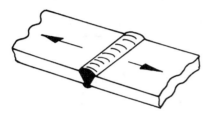

Toe of transverse fillet weld, $K = 1.5$

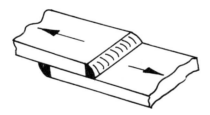

End of parallel fillet weld, $K = 2.7$

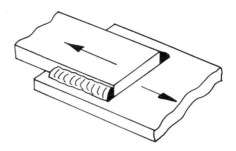

Tee butt joint sharp corner, $K = 2.0$

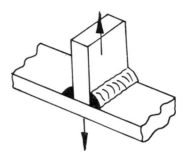

Typical stress concentration factors for various features

Component	K
Keyways	1.36–2.0
Gear teeth	1.5–2.2
Screw threads	2.2–3.8

1.4 Bending of beams

Beams generally have higher stresses than axially loaded members and most engineering problems involve bending. Examples of beams include structural members, shafts, axles, levers, and gear teeth.

To simplify the analysis, beams are usually regarded as being either 'simply supported' at the ends or 'built in'. In practice, the situation often lies between the two.

1.4.1 *Beams - basic theory*

Symbols used:
 x = distance along beam
 y = deflection normal to x
 i = slope of beam = dy/dx
 R = radius of curvature
 S = shear force
 M = bending moment
 w = load per unit length
 W = concentrated load
 I = second moment of area of beam
 E = Young's modulus

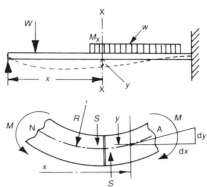

$$\frac{w}{EI}=\frac{d^4y}{dx^4};\ \frac{S}{EI}=\frac{d^3y}{dx^3};\ \frac{M}{EI}=\frac{d^2y}{dx^2};\ i=\frac{dy}{dx};\ y=f(x);\ \frac{1}{R}=\frac{d^2y}{dx^2}\ \text{(approx.)}$$

Principle of superposition

For a beam with several loads, the shear force, bending moment, slope and deflection can be found at any point by adding those quantities due to each load acting separately.

Example For a cantilever with an end load W and a distributed load w, per unit length.

Due to W only: $S_a = W$; $M_a = WL$; $y_b = WL^3/3EI$

Due to w only: $S_a = wL$; $M_a = wL^2/2$; $y_b = wL^4/8EI$

For both W and w: $S_a = W + wL$; $M_a = WL + wL^2/2$;
$y_b = WL^3/3EI + wL^4/8EI$

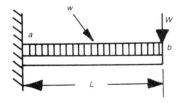

Bending stress

Bending stress at y from neutral axis $\sigma = \dfrac{My}{I}$

Maximum tensile stress ${}_t\sigma_m = \dfrac{M_t y_m}{I}$

where: ${}_t y_m$ = greatest y on tensile side.

Maximum compressive stress ${}_c\sigma_m = \dfrac{M_c y_m}{I}$

where: ${}_c y_m$ = greatest y on compressive side.

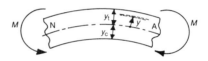

Values of I *for some sections*

Rectangular section $B \times D$
$I = BD^3/12$ about axis parallel to B.
Hollow rectangular section, hole $b \times d$
$I = (BD^3 - bd^3)/12$ about axis parallel to B.
Circular section, diameter D
$I = \pi D^4/64$ about diameter.
Hollow circular section, hole diameter d
$I = \pi(D^4 - d^4)/64$ about diameter.
I section, $B \times D$, flange T, web t
$I = [BD^3 - (B-t)(D-2T)^3]/12$ about axis parallel to B.

1.4.2 *Standard cases of beams*

The table gives maximum values of the bending moment, slope and deflection for a number of standard cases. Many complex arrangements may be analysed by using the principle of superimposition in conjunction with these.

Symbols used:

L = length of beam
I = second moment of area
w = load per unit length
W = total load = wL for distributed loads
E = Young's modulus

Maximum bending moment $M_m = k_1 WL$
Maximum slope $i_m = k_2 WL^2/EI$
Maximum deflection $y_m = k_3 WL^3/EI$

Type of beam	Moment coefficient, k_1	Slope coefficient, k_2	Deflection coefficient, k_3
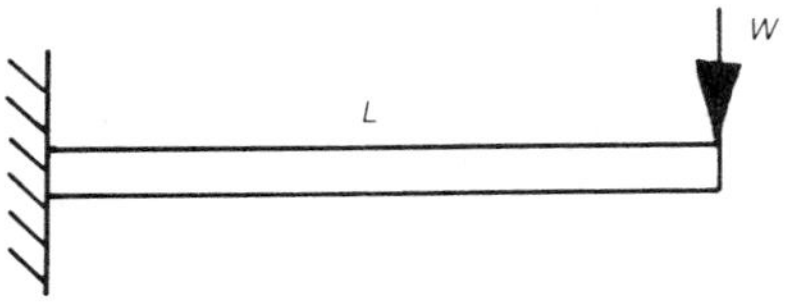	1 at wall	$\frac{1}{2}$ at load	$\frac{1}{3}$ at load
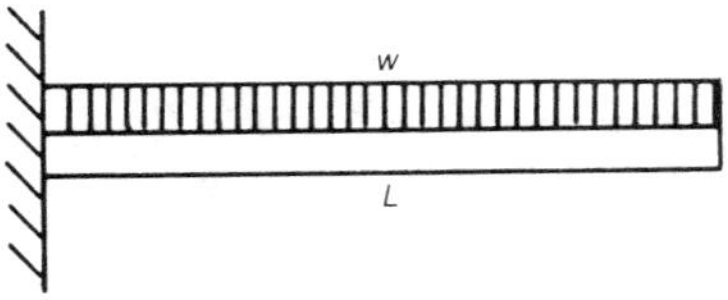	$\frac{1}{2}$ at wall	$\frac{1}{6}$ at free end	$\frac{1}{8}$ at free end
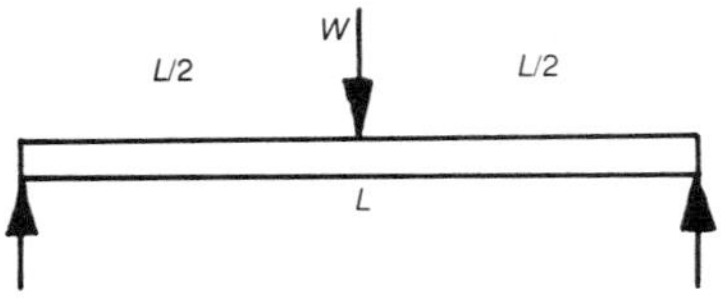	$\frac{1}{4}$ at load	$\frac{1}{16}$ at ends	$\frac{1}{48}$ at load
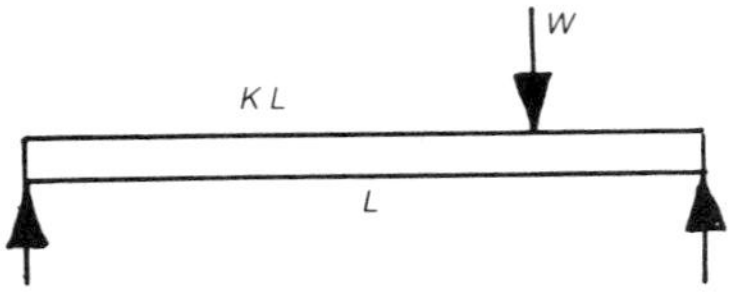	$K(1-K)$ at load	$K(1-K^2)/6$ at right-hand end for $K > \frac{1}{2}$	$K^2(1-K)^2/3$ at load (not maximum)
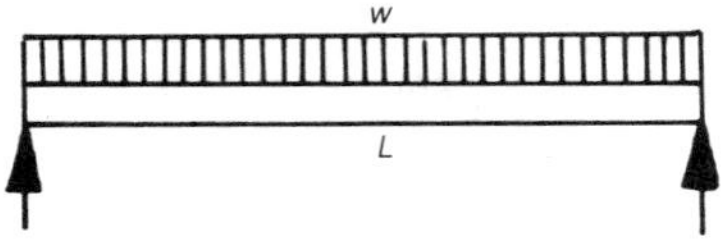	$\frac{1}{8}$ at centre	$\frac{1}{24}$ at ends	$\frac{5}{384}$ at centre
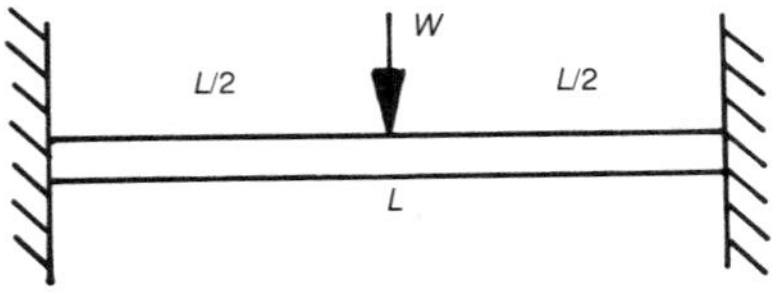	$\frac{1}{8}$ at centre and ends	$\frac{1}{64}$ at ends	$\frac{1}{192}$ at centre
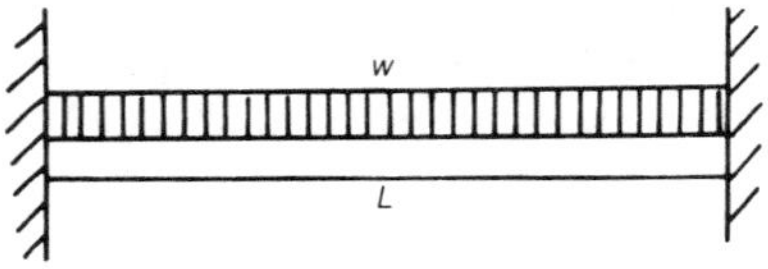	$\frac{1}{12}$ at ends	0.00803 at 0.211L from each end	$\frac{1}{384}$ at centre

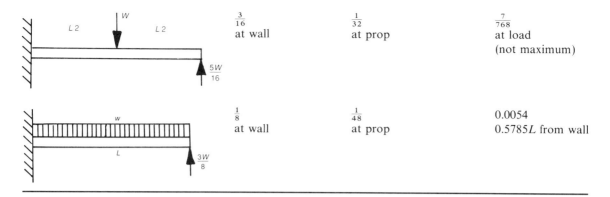

$\frac{3}{16}$ at wall	$\frac{1}{32}$ at prop	$\frac{7}{768}$ at load (not maximum)
$\frac{1}{8}$ at wall	$\frac{1}{48}$ at prop	0.0054 0.5785L from wall

1.4.3 Continuous beams

Most beam problems are concerned with a single span. Where there are two or more spans the solution is more complicated and the following method is used. This uses the so-called 'equation of three moments' (or Clapeyron's equation), which is applied to two spans at a time.

Clapeyron's equation of three moments

Symbols used:
M = bending moment
L = span
I = second moment of area
A = area of 'free' bending moment diagram treating span as simply supported
$\bar{x}$ = distance from support to centroid C of A
y = deflections of supports due to loading

(1) General case:
$$M_1 L_1/I_1 + 2M_2(L_1/I_1 + L_2/I_2) + M_3 L_2/I_2 =$$
$$6(A_1\bar{x}_1/L_1 I_1 + A_2\bar{x}_2/L_2 I_2) + 6E[y_2/L_1 + (y_2 - y_3)/L_2]$$

(2) Supports at same level, same I:
$y_1 = y_2 = y_3 = 0$
$$M_1 L_1 + 2M_2(L_1 + L_2) + M_3 L_2 = 6(A_1\bar{x}_1/L_1 + A_2\bar{x}_2/L_2)$$
(usual case)

(3) Free ends, $M_1 = M_3 = 0$:
$$M_2(L_1 + L_2) = 3(A_1\bar{x}_1/L_1 + A_2\bar{x}_2/L_2)$$

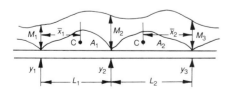

n Spans:
Apply to each group of three supports to obtain $(n-2)$ simultaneous equations which can be solved to give the $(n-2)$ unknown bending moments.

Solution:
For cases (2) and (3). If M_1 and M_3 are known (these are either zero or due to an overhanging load), then M_2 can be found. See example.

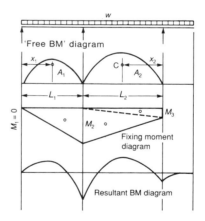

1.4.4 Bending of thick curved bars

In these the calculation of maximum bending stress is more complex, involving the quantity h^2 which is given for several geometrical shapes. The method is used for loaded rings and the crane hook.

Bending of thick curved bars, rings and crane hooks

If M acts as shown:

Stress on inside of curve $\sigma_2 = \dfrac{M}{AR}\left(1 - \dfrac{y_2}{R - y_2} \cdot \dfrac{R^2}{h^2}\right)$

Stress on outside of curve $\sigma_1 = \dfrac{M}{AR}\left(1 + \dfrac{y_1}{R + y_1} \cdot \dfrac{R^2}{h^2}\right)$

where values of h^2 are as given below.

General: $h^2 = \dfrac{R^3}{A}\displaystyle\int \dfrac{dA}{R + y} - R^2$

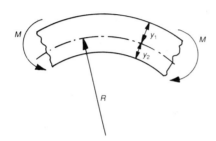

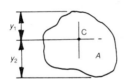

Rectangle: $h^2 = \dfrac{R^3}{D} \ln\left(\dfrac{2R + D}{2D - D}\right) - R^2$

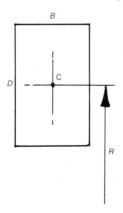

Trapezoid: $h^2 = \dfrac{R^3}{A}\left(\left[C + \dfrac{(B - C)(R + F)}{(E + F)}\right]\right.$

$\left. \ln\left(\dfrac{R + F}{R - E}\right) - (B - C)\right) - R^2$

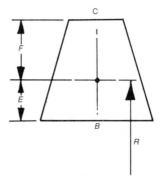

Circle: $h^2 = \dfrac{2R^3}{(R + \sqrt{(R^2 - r^2)})} - R^2$

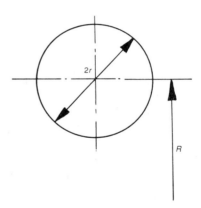

I section: $h^2 = \dfrac{R^3}{A}(B_3 \ln R_4/R_3 + B_2 \ln R_3/R_2$
$+ B_1 \ln R_2/R_1) - R^2$

where: R = radius at centroid, A = total area.
This method can be used for any shape made up of rectangles.

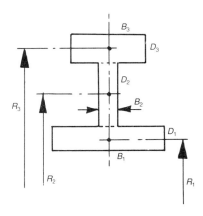

Maximum stresses (at A and B):

Outside, tensile $\sigma_t = \dfrac{W}{\pi A} \dfrac{R^2}{(R^2 + h^2)} \left[1 + \dfrac{R^2}{h^2} \dfrac{y_1}{(y_1 + R)}\right]$

Inside, compressive $\sigma_c = \dfrac{W}{\pi A} \dfrac{R^2}{(R^2 + h^2)} \left[\dfrac{R^2}{h^2} \dfrac{y_2}{(R - y_2)} - 1\right]$

where: A = area of cross-section, R = radius at centroid C. Use appropriate h^2 for the section.

Stresses in a crane hook

There is a bending stress due to moment Wa and a direct tensile stress of W/A at P.

Inside, tensile stress $\sigma_t = \dfrac{Wa}{AR} \left[\dfrac{y_2}{(R - y_2)} \dfrac{R^2}{h^2} - 1\right] + W/A$

Outside, compressive stress $\sigma_c = \dfrac{Wa}{AR} \left[1 + \dfrac{y_1}{(y_1 + R)} \dfrac{R^2}{h^2}\right] - W/A$

Use appropriate h^2 for the section.

Stresses in a loaded thick ring

Maximum bending moment (at A and B):

$M_{max} = \dfrac{WR}{2} \dfrac{R^2}{R^2 + h^2} \dfrac{2}{\pi}$

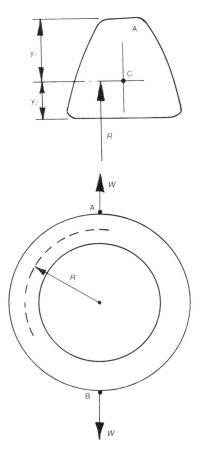

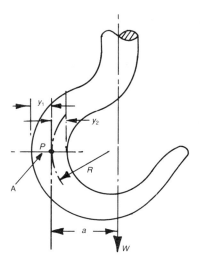

1.4.5 *Bending of thin curved bars and rings*

Stresses and deflections for a loaded thin ring

Maximum bending moment $M_{max} = \dfrac{WR}{\pi}$ (at A)

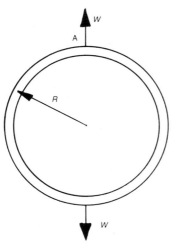

Maximum bending stresses $\sigma_t = \dfrac{M_{max}y_1}{I}$ (tensile on outside)

$$\sigma_c = \frac{M_{max}y_2}{I} \text{ (compressive on inside)}$$

Deflection in direction of load $\delta_w = \dfrac{WR^3}{4EI}\left(\dfrac{\pi^2-8}{\pi}\right)$

Deflection in direction normal to load $\delta_n = -\dfrac{WR^3}{2EI}\cdot\left(\dfrac{4-\pi}{\pi}\right)$ (reduces diameter)

Stresses and deflections in thin curved bars

Case I: $M_{max} = WR$ (at A)

Maximum bending stresses $\sigma_t = \dfrac{M_{max}y_1}{I}$ (tensile on outside)

$$\sigma_c = \frac{M_{max}y_2}{I} \text{ (compressive on inside)}$$

Deflection in direction of load $\delta_w = \dfrac{\pi WR^3}{4EI}$

Deflection in direction normal to load $\delta_n = \dfrac{WR^3}{2EI}$

Case II: $M_{max} = 2WR$ (at A)
Stresses as for case I.

Deflection in direction of load $\delta_w = \dfrac{3\pi WR^3}{2EI}$

Deflection in direction normal to load $\delta_n = \dfrac{2WR^3}{EI}$

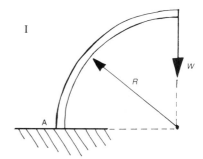

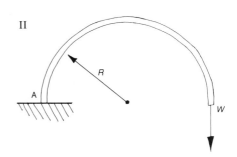

Case III: $M_{max} = WR$ (A to B)
Stresses as for case I.

Deflection in direction of load $\delta_w = \dfrac{WR^2}{EI}\left(\dfrac{\pi R}{4} + L\right)$

Deflection in direction normal

to load $\delta_n = \dfrac{WR}{EI}\left(\dfrac{R^2}{2} + RL + \dfrac{L^2}{2}\right)$

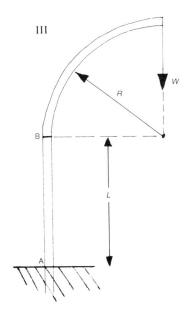

1.4.6 *Transverse vibration of beams*

Formulae are given for the fundamental frequency of transverse vibrations of beams due to the beam's own mass and due to concentrated masses.

Uniform cantilever, beam mass only

Frequency of vibration $f = \dfrac{0.56}{L^2}\sqrt{EI/m}$

where: m = mass per unit length of beam, I = second moment of area, L = length of beam.

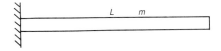

Simply supported beam, beam mass only

$f = \dfrac{1.57}{L^2}\sqrt{EI/m}$

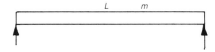

Built-in beam, mass of beam only

$f = \dfrac{3.57}{L^2}\sqrt{EI/m}$

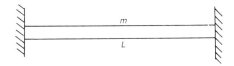

Concentrated mass: for all cases with a single mass

$f = \dfrac{1}{2\pi}\sqrt{g/y}$

where: y = static deflection at load, g = acceleration due to gravity.

For cantilever mass at end $f = 1/2\pi\sqrt{3EI/mL^3}$

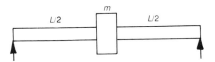

Simply supported beam, central mass

$f = 1/2\pi\sqrt{48EI/mL^3}$

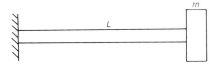

Simply supported beam, non-central mass

$f = 1/2\pi\sqrt{3EI/ma^2b^2}$

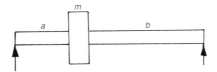

Built-in beam, central mass

$$f = 1/2\pi\sqrt{192EI/mL^3}$$

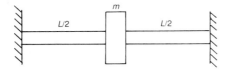

Combined loading (Dunkerley's method)

$$1/f^2 = 1/f_b^2 + 1/f_1^2 + 1/f_2^2 + \dots$$

where: f_b = frequency for beam only, $f_1, f_2, \dots$, are frequencies for each mass.

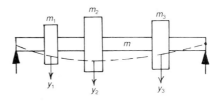

Energy method

If y is the static deflection under a mass m, then

$$f = 1/2\pi\sqrt{\frac{g\Sigma my}{\Sigma my^2}}$$

1.5 Springs

Springs are used extensively in engineering to control movement, apply forces, limit impact forces, reduce vibration and for force measurement.

1.5.1 *Helical torsion and spiral springs*

Close-coiled helical spring

This consists of a wire of circular or rectangular cross-section, wrapped around an imaginary cylinder to form a helix. Springs may be 'compression', with flat ends, or 'tension' with loading hooks. Helical springs may also be used as 'torsion' springs. Formulae are given for stress and deflection as well as frequency of vibration.

Close-coiled helical compression spring

Symbols used:
 D = mean diameter
 d = wire diameter
 c = clearance between coils
 L = free length

p = pitch of coils
n = number of active coils
n_t = total number of coils
y = deflection
E = Young's modulus
W = load
 s = stiffness
C = coil ratio or index = D/d
G = shear modulus
 τ = allowable shear stress
 ρ = density of spring material

K_w = Wahl factor

 (stress concentration factor) = $K_w = \dfrac{4C-1}{4C-4} + \dfrac{0.615}{C}$

Load $W = \pi\tau d^2/8CK_w$
Wire diameter $d = \sqrt{8WCK_w/\pi\tau}$
Stiffness $s = Gd/8nC^3$
Deflection $y = W/s$
Total number of coils $n_t = n + 1.5$ (for ground, flattened ends)
Free length $L = (n+1)d + nc$
Ratio $L/D =$ about 2 to 3 for stability
'Close-coiled' length $L_c = (n+1)d$

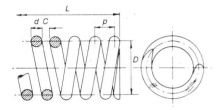

Helical tension spring

The formulae for load and stiffness are the same. There is usually no initial clearance between coils, and there is an initial 'built-in' compression. Various types of end hooks are used.

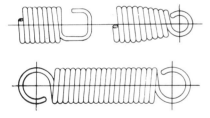

Helical torsion spring

Angle of twist (for torque T) $\theta = 64TDn/Ed^4$
Maximum bending stress $\sigma_m = 32T/\pi d^3$

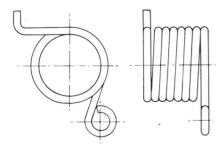

Vibration of helical spring

Axial vibration under own mass:

Frequency of vibration $f = \dfrac{1}{2\pi dCn}\sqrt{G/2\rho}$

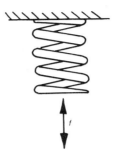

Torsional vibration under end inertia I:

Frequency of vibration $f = \dfrac{1}{2\pi}\sqrt{Ed^4/64DnI}$

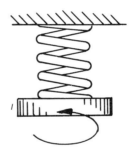

Compression helical spring of rectangular section

Section is $b \times d$, where $b =$ major dimension.
Maximum shear stress (side b) $\tau_b = (1.8d + 3b)WDK/2b^2d^2$
Maximum shear stress (side d) $\tau_d = (1.8b + 3d)WDK/2b^2d^2$
Direct shear stress $\tau = 1.5W/bd$

where: $K = \dfrac{4C-1}{4C-4}$ and $C = D/d$ for case 1 and D/b for case 2.

Case 1 ($d =$ radial dimension): Maximum stress $\tau_{max} = \tau_b + \tau$
Case 2 ($b =$ radial dimension):
Maximum stress $\tau_{max} = \tau_b$ or $\tau_d + \tau$ whichever is the greater.

Stiffness $s = W/y = \dfrac{8}{7n} \cdot \dfrac{Gb^3d^3}{(b^2+d^2)nD^3}$

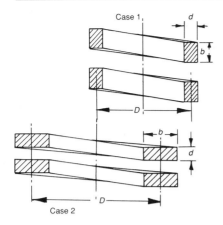

Spiral spring

A spiral spring consists of a strip or wire wound in a flat spiral subjected to a torque to give an angular deflection. The clock spring is an example.

Equation of spiral $D = D_i + p\alpha/\pi$

where:
D = diameter
D_i = minimum diameter
α = angle around spiral (in radians)
p = radial pitch
D_o = maximum diameter

Torque $T = Fa$, where $a = D_o/2$.
Angle of twist $\theta = 1.25\ TL/EI$
Maximum bending stress $\sigma_m = My/I$ where $M = 2T$
Length of strip or wire $= \pi n(D_o + D_i)/2$, where
n = number of turns.
Second moment of area $I = bt^3/12$ (strip) or $\pi d^4/64$ (wire)
Dimension $y = t/2$ (strip) or $d/2$ (wire)

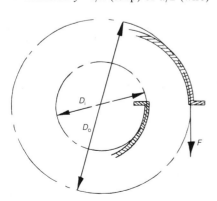

Conical helical compression spring

This is a helical spring in which the coils progressively change in diameter to give increasing stiffness with increasing load. It has the advantage that the compressed height is small. This type of spring is used for upholstery.

Conical helical spring

Symbols used:
D_1 = smaller diameter
D_2 = larger diameter
d = wire diameter
n = number of active coils

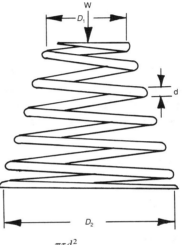

Load $W = \dfrac{\pi \tau d^2}{8CK}$

where: $C = D_2/d$; $K = \left(\dfrac{4C-1}{4C-4}\right) + \dfrac{0.615}{C}$

Stiffness $s = W/y = \dfrac{Gd^4}{2n(D_1 + D_2)(D_1^2 + D_2^2)}$

Allowable working stress (MPa) for helical springs (grade 060A96)

Spring	Wire diameter (mm)		
	1–3.9	4–7.9	8–12
Light duty	590	510	450
Medium duty	470	410	360
Heavy duty	400	340	300

1.5.2 *Leaf and laminated leaf springs*

Leaf springs

A leaf spring consists basically of a beam, usually of flat strip, e.g. a cantilever or simply supported beam, subjected to a load to give a desired deflection proportional to the load.

The laminated leaf spring, or 'carriage spring', is used for vehicle suspensions and is made up of several flat strips of steel of various lengths clamped together. The spring is effectively a diamond-shaped plate cut into strips. Analysis shows that the maximum bending stress is constant.

The quarter-elliptic spring is, in effect, half of the so-called 'semi-elliptic' spring.

Beam leaf springs

Maximum stress $\sigma = k_1 WL/bd^2$
Stiffness $s = W/y = k_2 EI/L^3$

Spring type	k_1	k_2
	6.0	3
	1.5	48
	0.75	192

Laminated leaf springs

Symbols used:
 L = span
 b = width of leaves
 t = thickness of leaves
W = load
 y = deflection
σ_m = maximum bending stress
 n = number of leaves
 E = Young's modulus
 s = stiffness = W/y

Semi-elliptic spring:
Maximum bending stress $\sigma_m = 3WL/2nbt^2$
Stiffness $s = 8Enbt^3/3L^3$

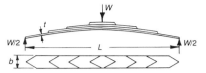

Quarter-elliptic spring:
Maximum bending stress $\sigma_m = 6WL/nbt^2$
Stiffness $s = Enbt^3/6L^3$

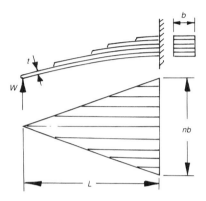

1.5.3 *Torsion bar spring*

The torsion bar is a solid or hollow circular bar clamped at one end with a lever attached to the other. The load is applied to the end of the lever and twists the bar elastically.

Symbols used:
R = lever radius
D = bar diameter
L = bar length
G = torsional modulus
τ = allowable shear stress

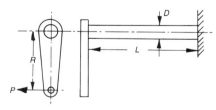

For a hollow shaft of bore d use:
$(D^4 - d^4)$ instead of D^4
$\left(\dfrac{D^4 - d^4}{D}\right)$ instead of D^3

Deflection $y = \dfrac{32PR^2L}{\pi GD^4}$

Stiffness $s = \dfrac{P}{y} = \dfrac{\pi GD^4}{32R^2L}$

Maximum load $P_{max} = \dfrac{\pi D^3 \tau}{16R}$

1.5.4 Belleville washer spring (disk or diaphragm spring)

This is an annular dished steel ring which deflects axially under load. Several springs may be used in series or parallel arrangements to give lower or higher stiffness, respectively. The spring is space saving and its non-linear characteristics can be altered considerably by varying the proportions.

Symbols used:

D_o = outer diameter
D_i = inner diameter
t = thickness
h = height

y = deflection
E = Young's modulus
v = Poisson's ratio
k_1, k_2, k_3 = constants
σ_m = maximum stress
W = load

$$W = \frac{Ey}{(1-v^2)k_1 D_o^2}\left[(h-y)\left(h-\frac{y}{2}\right)t + t^3\right] \text{ (may be negative)}$$

$$\sigma_m = \frac{Ey}{(1-v^2)k_1 D_o^2}\left[k_2\left(h-\frac{y}{2}\right) \pm k_3 t\right]$$

(positive for A, negative for B. Stress is positive or negative depending on the value of y)

D_o/D_i	k_1	k_2	k_3
1.4	0.46	1.07	1.14
1.8	0.64	1.18	1.30
2.2	0.73	1.27	1.46
2.6	0.76	1.35	1.60
3.0	0.78	1.43	1.74
3.4	0.80	1.50	1.88
3.8	0.80	1.57	2.00
4.2	0.80	1.64	2.14
4.6	0.80	1.71	2.26
5.0	0.79	1.77	2.38

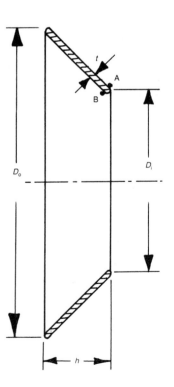

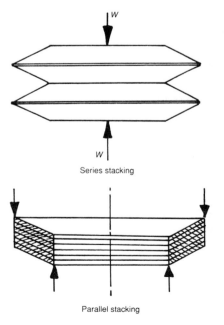

Series stacking

Parallel stacking

1.5.5 *Rubber springs*

Springs of rubber bonded to metal are made in a wide variety of configurations. The rubber is usually in shear and, because of the high internal damping, such springs are used for limiting vibrations.

Two-block shear spring – load P

Shear stress $\tau = P/2A$
Deflection $y = Ph/2AG$
where G = shear modulus.

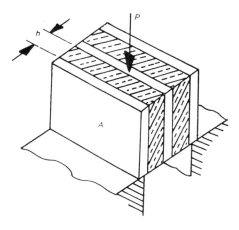

Cylindrical shear spring, load P

Maximum shear stress $\tau_{m} = P/\pi h D_{i}$

Deflection $y = \dfrac{P}{2\pi hG} \ln (D_{o}/D_{i})$

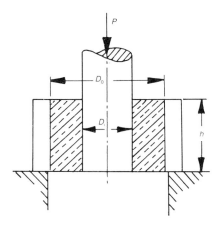

Cylindrical torsion spring, torque T

Maximum shear stress $\tau_{m} = 2T/\pi LD_{i}^{2}$

Angle of twist $\theta = \dfrac{T}{\pi LG}(1/D_{i}^{2} - 1/D_{o}^{2})$

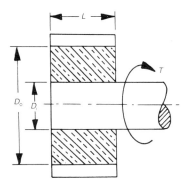

Modulus and strength of rubber

$G = 0.3$ to 1.2 MPa
$E = 0.9$ to 3.6 MPa
Allowable shear stress $= 0.2$ to 0.4 MPa
Deflection limited to 10% to 20% of free height.

1.5.6 *Form factors for springs*

The table gives form factors giving the amount of strain energy stored in different types of spring relative to a bar with uniform direct stress.

Strain energy $u = C_f \sigma_{max}^2/2E$ or $C_f \tau_{max}^2/2G$ per unit volume

Type of spring	Modulus	C_f
Bar in tension or compression	E	1.0
Beam, uniform bending moment rectangular section	E	0.33
Clock spring	E	0.33
Uniformly tapered cantilever rectangular section	E	0.33
Straight cantilever rectangular section	E	0.11
Torsion spring	E	0.25
Belleville washer	E	0.05 to 0.20
Torsion bar	G	0.50
Torsion tube	G	$\frac{1}{2}[1-(d/D)^2] \simeq 0.8$ to 0.9
Compression spring	G	0.50/Wahl factor

1.6   Shafts

Rotating or semirotating shafts are invariably subject to both torsion and bending due to forces on levers, cranks, gears, etc. These forces may act in several planes parallel to the shaft, producing bending moments which may be resolved into two perpendicular planes. In addition, there will be a torque which varies along the length of the shaft. The following shows how the resultant bending moments and bearing reactions can be determined.

In the case of gears, the contact force is resolved into a tangential force and a separating force.

1.6.1   Resultant bending moment diagram

Forces P and Q may be resolved into vertical and horizontal components:

$P_v = P \sin \theta_p$, $Q_v = Q \sin \theta_q$,
$P_h = P \cos \theta_p$, $Q_h = Q \cos \theta_q$

Assuming the bearings act as simple supports, the bending moment (BM) diagram is drawn. From BM diagrams for each plane, moments M_v and M_h may be found and also reactions $_vR_a$, $_vR_b$, $_hR_a$ and $_hR_b$.

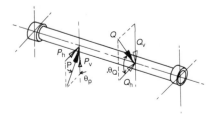

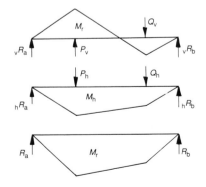

Resultant bending moments, M_r:
At any point $M_r = \sqrt{M_v^2 + M_h^2}$
and the bending stress $= M_r/Z$; $Z =$ modulus

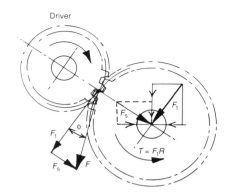

Resultant reactions, R_a and R_b (bearing loads):

$$R_a = \sqrt{{}_vR_a^2 + {}_hR_a^2}$$
$$R_b = \sqrt{{}_vR_b^2 + {}_hR_b^2}$$

A torque diagram is also drawn and the torque and resultant bending moment can be found at any point. The equivalent torque and equivalent bending moment are found as follows:

$$T_e = \sqrt{M_r^2 + T^2}; \quad M_e = (M_r + T_e)/2$$

The shaft diameter is:

$$d = \sqrt[3]{\frac{16T_e}{\pi\tau}} \text{ or } d = \sqrt[3]{\frac{32M_e}{\pi\sigma}} \text{ (whichever is the greater)}$$

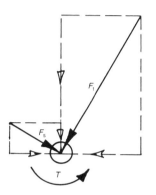

where: τ and $\sigma =$ the allowable shear and bending stresses.

Note: bearings are assumed to act as simple supports.

1.6.2 *Shafts with gears and levers*

Shafts with levers

A force such as P acting at radius R, can be replaced by a force P acting at the shaft centre and a torque PR. P is resolved into components P_v and P_h as before.

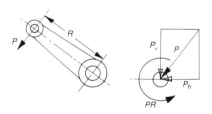

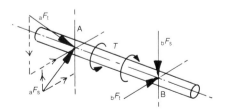

Shafts with gears

The tangential force on the gear teeth is $F_t = P/2\pi NR$
where: $P =$ power, $N =$ speed, $R =$ gear radius.

The 'separating force' is $F_s = F_t \tan\phi$
where: $\phi =$ the pressure angle. F_t and F_s can be assumed to act at the gear centre if a torque F_tR is introduced. F_t and F_s can be resolved into vertical and

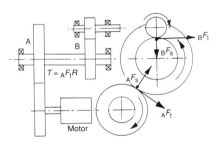

horizontal components, as before. The forces are shown for a shaft AB with two gears.

1.6.3 *Strength of keys and splines*

A key is used to prevent a machine part from moving relative to another part. In the case of a shaft, the key must be strong enough to transmit a high torque and is often made of alloy or high tensile steel. The fit may be either 'close' or 'free' if sliding is desired. The 'keyway' in the shaft and hub is usually produced by milling.

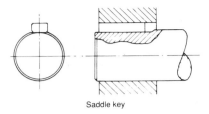

Saddle key

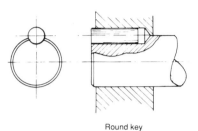

Round key

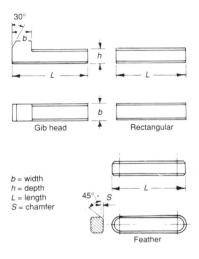

Gib head   Rectangular

b = width
h = depth
L = length
S = chamfer

Feather

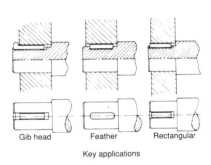

Gib head   Feather   Rectangular

Key applications

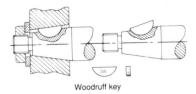

Woodruff key

Splines are a means of keying a hub to a shaft where separate keys are not required. They consist of mating grooves in hub and shaft of rectangular, triangular or involute form. The grooves are designed to allow axial sliding.

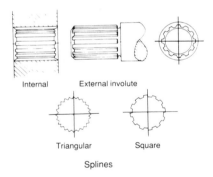

Internal   External involute

Triangular   Square

Splines

Types of key

The main types of key are the 'rectangular' where the keyways are half the key depth, the 'feather' where the keyway is closed at each end, the 'Gib-head' used always at the end of a shaft and with a head so that it can be tapped into place, the 'Woodruff key' which is segmental and for use on tapered shafts, and the inexpensive 'saddle' and 'round' keys.

Torque capacity

d = depth of spline or half depth of key
r = mean radius of spline or shaft radius for key
n = number of splines
L = length of spline or key
b = breadth of key
T = limiting torque
σ_c = allowable crushing stress
τ = allowable shear stress

Keys:
$T = \tau b L r$ (based on shear)
$T = \sigma_c d L r$ (based on crushing)

Splines:
$T = \sigma_c n d L r$ (based on crushing)
(σ_c = about 7 MPa for steel)

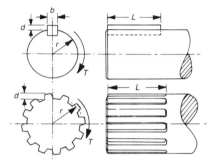

1.6.4 Shaft couplings

Shaft couplings may be 'solid' or 'flexible'. Solid couplings may consist simply of a sleeve joining the shafts, the drive being taken by pins or keys. For large powers, bolted flanges are used to give either a solid or flexible coupling.

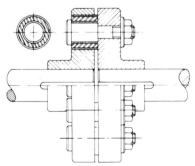

Rubber-bushed pin-type flexible coupling

A large variety of flexible couplings are used to accommodate angular, parallel or axial misalignment. Several types are shown.

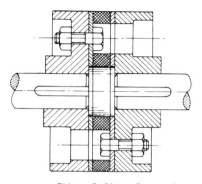

Disk-type flexible coupling

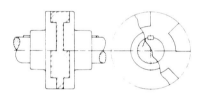

Moulded rubber insert coupling

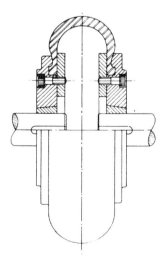

Rubber-tyre-type flexible coupling

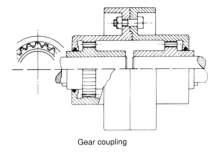

Gear coupling

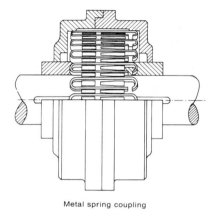

Metal spring coupling

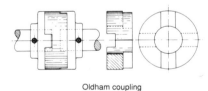

Oldham coupling

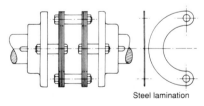

Steel lamination

Metaflex coupling

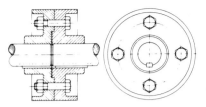

Solid bolted flanged coupling

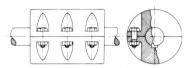

Muff coupling

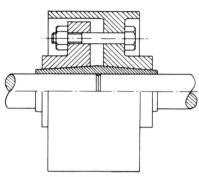

Compression coupling

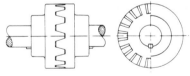

Claw coupling

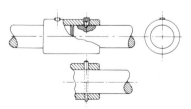

Sleeve coupling

Bonded rubber couplings are simple and cheap and permit large misalignments. Their non-linear characteristics make them useful for detuning purposes. Three annular types are shown and their spring constants given.

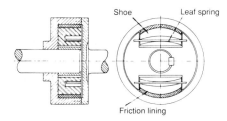

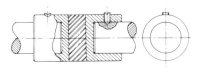

Bonded rubber coupling

Solid bolted shaft coupling

Symbols used:
  D = shaft diameter
  D_p = pitch circle diameter of bolts
  D_b = bolt diameter
  n = number of bolts
  b = width of key
  L = length of key and hub
  P = power transmitted
  N = shaft speed
  FS = factor of safety
  τ_y = shear yield stress

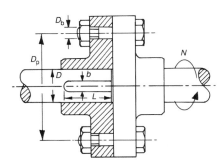

Power capacity $P = \pi^2 N n D_p D_b^2 \tau_y / 4\,\text{FS}$
Key FS $= \pi D N b L \tau_y / P$
Shaft FS $= \pi^2 N D^3 \tau_y / 8P$
If bolts and shaft have same material and FS, then:

Bolt diameter $D_b = \sqrt{D^3 / 2 n D_p}$

Sleeve shaft coupling

Symbols used:
  D = shaft diameter
  D_o = sleeve outer diameter
  T = torque transmitted
  τ = allowable shear stress
  N = speed
  P = power
  b = key width
  L = key length

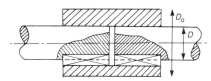

Torque capacity of shaft $T = \pi D^3 \tau / 16$

Torque capacity of key $T = \dfrac{DbL\tau}{2}$

Power capacity of shaft $P = \pi^2 N D^3 \tau / 8$
Torque capacity of sleeve $T = \pi \tau (D_o^4 - D^4) / 16 D_o$
(allowance to be made for keyway)
For equal strength of sleeve and shaft $D_o = 1.22 D$.

Pinned sleeve shaft coupling

Symbols used:
D = shaft diameter
d = pin diameter
Torque capacity of pin $T = \pi d^2 D \tau / 4$

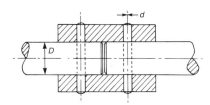

1.6.5 Bonded rubber shaft coupling

Symbols used:
  θ = angle of twist
  T = torque
  G = shear modulus
  s = spring constant $= T / \theta$

Annulus bonded to sleeve:

$$s = \frac{4\pi L G}{1/r_i^2 - 1/r_o^2}$$

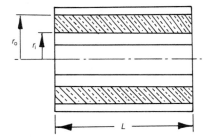

Annulus bonded to disks:

$$s = \frac{\pi G}{2L}(r_o^4 - r_i^4)$$

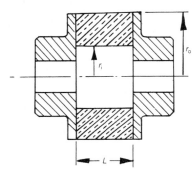

Hyperbolic contour:

$$s = 2\pi G \left(\frac{L_o r_o^2 r_i}{r_o - r_i} \right)$$

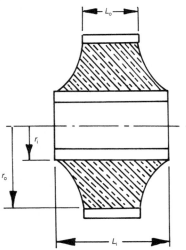

1.6.6 *Critical speed of whirling of shafts*

When a shaft rotates there is a certain speed at which, if there is an initial deflection due to imperfections, the centripetal force is equal to the elastic restoring force. At this point the deflection increases to a large value and the shaft is said to 'whirl'. Above this speed, which depends on the shaft dimensions, the material and the loads carried by the shaft, the shaft whirling decreases. Shafts must be run well below or well above this speed. It can be shown that numerically the critical speed is the same as the frequency of transverse vibrations. Formulae are given for several common cases.

Critical speed for all cases:

$$N_c = \frac{1}{2\pi} \sqrt{g/y}$$

where: g = acceleration due to gravity, y = 'static' deflection at mass.

Cantilevered shaft with disc at end

Mass of shaft neglected.

$$N_c = 1/2\pi \sqrt{3EI/mL^3}$$

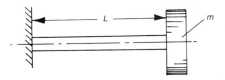

Central disc, 'short' bearings

$$N_c = 1/2\pi \sqrt{48EI/mL^3}$$

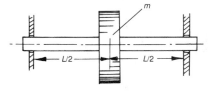

Non-central disc, short bearings

$$N_c = 1/2\pi \sqrt{3EI/ma^2b^2}$$

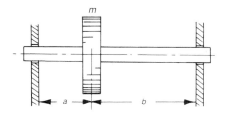

Central disc, 'long' bearings

$$N_c = 1/2\pi\sqrt{192EI/mL^3}$$

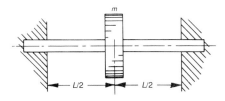

Uniform shaft, one end free

Critical speed $N_c = \dfrac{0.56}{L^2}\sqrt{EI/m}$

where:
$m =$ mass per unit length
$I =$ second moment of area
$E =$ Young's modulus
$L =$ length of shaft

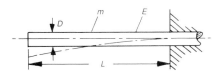

Uniform shaft, in 'short' bearings

$$N_c = \frac{1.57}{L^2}\sqrt{EI/m}$$

where: $m =$ mass per unit length of shaft.

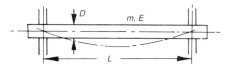

Uniform shaft, 'long' bearings

$$N_c = \frac{3.57}{L^2}\sqrt{EI/m}$$

where: $m =$ mass per unit length of shaft.

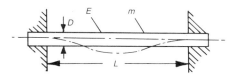

Combined loading on uniform shaft

(1) Dunkerley's method:

$$1/N_c^2 = 1/N_s^2 + 1/N_1^2 + 1/N_2^2 + \ldots$$

where:
$N_c =$ critical speed of system
$N_s =$ critical speed for shaft alone
N_1, N_2, etc. = critical speeds for discs acting alone

(2) Energy method:

$$N_c = \frac{1}{2\pi}\sqrt{\frac{g\Sigma my}{\Sigma my^2}}$$

where: $m =$ any mass of a disc, $y =$ static deflection under the disc.

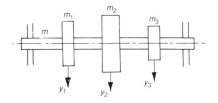

1.6.7 *Torsional vibration of shafts*

For long shafts, e.g. a ship's propeller shaft, torsional vibration may be a problem and the shaft must be designed so that its rotational speed is not numerically near to its natural torsional frequency.

Symbols used:
$f =$ frequency of torsional oscillations (Hz)
$s =$ torsional stiffness $= GJ/L$ (N-m rad^{-1})
$G =$ torsional modulus (N m^{-2})

J = polar second moment of area (m⁴)
D = shaft outer diameter (m)
d = inner diameter
L = length of shaft (m)
I = moment of inertia of disc $= mk^2$ (kg m²)
m = mass of disc (kg)
k = radius of gyration of disc (m)

Single disc on shaft

$$f = \frac{1}{2\pi}\sqrt{s/I}$$

$$J = \frac{\pi D^4}{32} \text{ (for solid shaft); } \frac{\pi}{32}(D^4 - d^4)$$
$$\text{(for hollow shaft)}$$

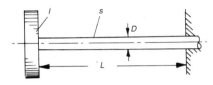

Two discs on uniform shaft

$$f = \frac{1}{2\pi}\sqrt{s(I_1 + I_2)/I_1 I_2}$$

Position of node $L_1 = L/(1 + \frac{I_1}{I_2})$, $L = L_1 + L_2$

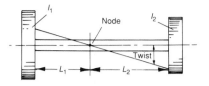

Two discs on stepped solid shaft

$$f = \frac{1}{2\pi}\sqrt{s(I_1 + I_2)/I_1 I_2}$$

$$s = GJ_a/L_e$$

where: $L_e = L_a + L_b(D_a/D_b)^4$ (equivalent length of shaft for uniform diameter D_a) length

$$J_a = \frac{\pi}{32} D_a^4$$

Note: the node must be in length L_a.

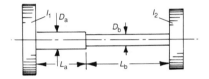

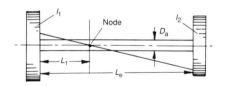

1.7 Struts

A component subject to compression is known as a 'strut' if it is relatively long and prone to 'buckling'. A short column fails due to shearing when the compressive stress is too high, a strut fails when a critical load called the 'buckling' or 'crippling' load causes sudden bending. The resistance to buckling is determined by the 'flexural rigidity' EI or EAk^2, where k is the *least* radius of gyration.

The important criterion is the 'slenderness ratio' L/k, where L is the length of the strut.

The Euler theory is the simplest to use but the much more involved Perry–Robertson formula (BS 449) is regarded as the most reliable.

1.7.1 Euler theory

Buckling load $P = K\pi^2 EI/L^2$

where:

$I = least$ second moment of area $= Ak^2$
$K =$ factor dependent on 'end conditions'
$k = least$ radius of gyration $= \sqrt{I/A}$
$A =$ cross-sectional area
$L =$ length
$E =$ Young's modulus

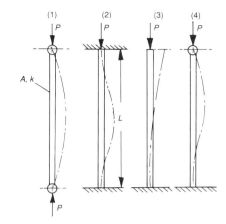

	(1)	(2)	(3) Fixed at one end, free at other	(4) Fixed at one end, pinned at other
End condition	Pinned ends	Fixed ends		
K	1	4	0.25	2.05

1.7.2 Rankine–Gordon formula

Buckling load $P = \sigma A = \dfrac{\sigma_c A}{1 + a\left(\dfrac{L}{k}\right)^2}$

where:

$\sigma =$ failure stress
$\sigma_c =$ elastic limit in compression
$a =$ constant
$A =$ cross-sectional area

		a	
Material	σ_c MPa	Pinned ends	Fixed ends
Mild steel	320	1/7500	1/30000
Wrought iron	250	1/9000	1/36000
Cast iron	550	1/1600	1/6400
Wood	35	1/3000	1/12000

1.7.3 Johnson's parabolic formula

Buckling load $P = \sigma_c A[1 - b(L/k)^2]$
$\sigma_c = 290$ MPa for mild steel
$b = 0.00003$ (pinned ends) or 0.00002 (fixed ends)

1.7.4 Straight-line formula

Buckling load $P = \sigma_c A[1 - K(L/k)]$
$\sigma_c = 110$ MPa (mild steel) or 140 (structural steel)
$K = 0.005$ (pinned ends) or 0.004 (fixed ends)

1.7.5 Perry–Robertson formula

Buckling load $P = A\left[\dfrac{\sigma_c + (K+1)\sigma_e}{2} - \sqrt{\left(\dfrac{\sigma_c + (K+1)\sigma_e}{2}\right)^2 - \sigma_c \sigma_e}\right]$

where:

$K = 0.3\left(\dfrac{L_e}{100k}\right)^2$

L_e = actual length of pinned end strut
   = 0.7 × actual length of fixed ends strut
   = 2.0 × actual length of strut with one end fixed, one end free
   = 0.85 × actual length with one end pinned and one end fixed

σ_e = Euler buckling stress = $\dfrac{\pi^2 E}{(L_e/k)^2}$

σ_c = Yield stress in compression

1.7.6 *Pinned strut with uniformly distributed lateral load*

Maximum bending moment $M_m = \dfrac{wEI}{P}\left(\sec\dfrac{\alpha L}{2} - 1\right)$

Maximum compressive stress $\sigma_m = \dfrac{My}{I} + \dfrac{P}{A}$

Maximum deflection $y_m = -\dfrac{M_m}{P} + \dfrac{wL^2}{8P}$

where: $\alpha = \sqrt{\dfrac{P}{EI}}$

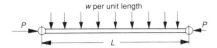

1.8 Cylinders and hollow spheres

In engineering there are many examples of hollow cylindrical and spherical vessels subject to internal or external pressure. The formulae given are based on Lamé's equations. In the case of external pressure, failure may be due to buckling. In the following, p is the difference between the internal and external pressures.

1.8.1

Thin cylinder, internal pressure

Hoop stress $\sigma_h = pD/2t$
Longitudinal stress $\sigma_L = pD/4t$

Radial displacement $x_r = \dfrac{D}{2E}(\sigma_h - v\sigma_L)$

where: v = Poisson's ratio.
For external pressure, use $-p$.

Buckling of thin cylinder due to external pressure

(1) Long tube, free ends:

$$L > 2.45D\sqrt{\dfrac{D}{2t}}; \quad p_b = \dfrac{E}{4(1 - v^2)}\left(\dfrac{2t}{D}\right)^3$$

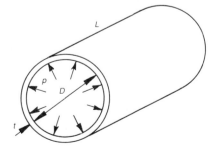

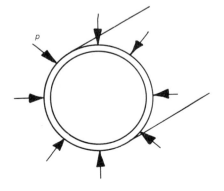

(2) Short tube, ends held circular:

$$p_b = \frac{1.61 E t^2}{LD} \sqrt[4]{\frac{1}{(1-v^2)^3} \cdot \frac{4t^2}{D^2}}$$

Thin spherical vessel, internal pressure

$$\sigma_h = \sigma_L = pD/4t; \quad x_r = \frac{D\sigma_h}{2E}(1-v)$$

For external pressure use $-p$.

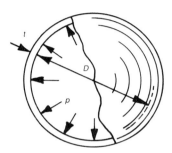

Thin cylinder with hemispherical ends

For equal maximum stress $t_e = 0.5 t_c$
For no distortion $t_e \simeq 0.4 t_c$

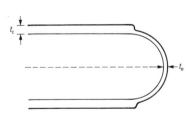

Thick cylinder, internal pressure, no longitudinal pressure

Maximum hoop stress $\sigma_{h\,max} = p \dfrac{(r_b^2 + r_a^2)}{(r_b^2 - r_a^2)}$

(at inner radius); $\sigma_L = 0$

Maximum radial stress $\sigma_{r\,max} = p$
Maximum shear stress $\tau_{max} = p r_b^2 / (r_b^2 - r_a^2)$

(at inner radius)

Change in inner radius $x_a = \dfrac{p r_a}{E}\left(\dfrac{r_b^2 + r_a^2}{r_b^2 - r_a^2} + v\right)$

Change in outer radius $x_b = \dfrac{p r_b}{E}\left(\dfrac{2 r_a^2}{r_b^2 - r_a^2}\right)$

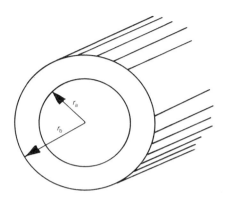

Thick cylinder, internal pressure, all directions

$\sigma_{h\,max}$ and $\sigma_{r\,max}$ as above.

Longitudinal stress $\sigma_L = p\left(\dfrac{r_a^2}{r_b^2 - r_a^2}\right)$

$$x_a = \frac{p r_a}{E}\left[\frac{r_b^2 + r_a^2}{r_b^2 - r_a^2} - v\left(\frac{r_a^2}{r_b^2 - r_a^2} - 1\right)\right]$$

$$x_b = \frac{p r_b}{E}\left[\frac{r_a^2}{r_b^2 - r_a^2}(2 - v)\right]$$

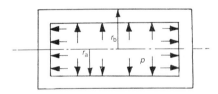

Thick sphere, internal pressure

Symbols used:
  σ = direct stress
  τ = shear stress
  p = pressure
  v = Poisson's ratio
  t = thickness
  D = diameter
  r = radius

x = radial displacement
E = Young's modulus
L = length

$$\sigma_{h\,max} = \frac{p}{2} \frac{(r_b^3 + 2r_a^3)}{(r_b^3 - r_a^3)} \text{ (at inner radius)}$$

$$\sigma_{r\,max} = p \text{ (at inner radius)}$$

$$\tau_{max} = \frac{3p}{4} \left(\frac{r_b^3}{r_b^3 - r_a^3} \right) \text{ (at inner radius)}$$

$$x_a = \frac{pr_a}{E} \left[\frac{(r_b^3 + 2r_a^3)}{2(r_b^3 - r_a^3)} (1 - v) + v \right]$$

$$x_b = \frac{pr_b}{E} \left[\frac{3r_a^3}{2(r_b^3 - r_a^3)} (1 - v) \right]$$

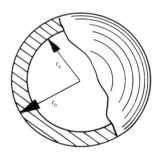

1.8.2 Shrink fit of cylinders

Two hollow cylindrical parts may be connected together by shrinking or press-fitting where a contact pressure is produced. In the case of a hub on a shaft this eliminates the need for a key. Formulae are given for the resulting stresses, axial fitting force and the resulting torque capacity in the case of a shaft.

Symbols used:

r_a = inner radius of inner cylinder ($= 0$ for solid shaft)
r_b = outer radius of inner cylinder
r_b = inner radius of outer cylinder
r_c = outer radius of outer cylinder
x = interference between inner and outer cylinders
L = length of outer cylinder
E_i, E_o = Young's modulus of inner and outer cylinders
v_i, v_o = Poisson's ratio of inner and outer cylinders
p = radial pressure between cylinders
μ = coefficient of friction between cylinders
T = torque capacity of system

P_a = axial force to give interference fit
α = coefficient of linear expansion of inner or outer cylinder
Δt = temperature difference between cylinders

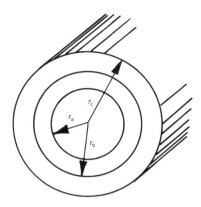

Contact pressure

$$p = \frac{x}{2r_b} \cdot \frac{E_i E_o}{[E_o(K_3 - v_i) + E_i(K_2 + v_o)]}$$

Hoop stresses

Inner cylinder:
$$\sigma_a = -pK_4 \text{ at } r_a$$
$$_i\sigma_b = -pK_3 \text{ at } r_b$$

Outer cylinder:
$$_o\sigma_b = pK_2 \text{ at } r_b$$
$$\sigma_c = pK_1 \text{ at } r_c$$

where: $K_1 = 1/[(r_c/r_b)^2 - 1]$; $K_2 = \dfrac{(r_c/r_b)^2 + 1}{(r_c/r_b)^2 - 1}$;

$K_3 = \dfrac{(r_b/r_a)^2 + 1}{(r_b/r_a)^2 - 1}$; $K_4 = \dfrac{1}{(r_b/r_a)^2 - 1}$;

$$P_a = 2\mu\pi r_b L p; \quad T = P_a r_b.$$

Thermal shrinkage

If the outer cylinder is heated or the inner cylinder is cooled by Δt, then:

$$x = 2\alpha r_b \Delta t$$

1.9 Contact stresses

When a ball is in contact with a flat, concave or convex surface, a small contact area is formed, the size of the area depending on the load and materials. In the case of a roller, a line contact is obtained, giving a rectangular contact area of very small width. The following gives the size of these areas and the maximum stress for several common cases. The theory is of great importance in the design of rolling bearings.

1.9.1 *Contact stresses for balls and rollers*

Symbols used:
E_1, E_2 = Young's moduli
F = load
r_1, r_2 = radii
v_1, v_2 = Poisson's ratio

Two balls in contact

Contact area radius $a = \sqrt[3]{\dfrac{3F((1-v_1^2)/E_1 + (1-v_2^2)/E_2)}{(1/r_1 + 1/r_2)}}$

Contact stress $\sigma_c = 3F/2\pi a^2$

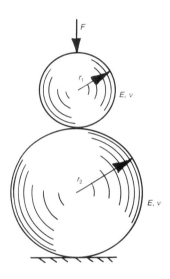

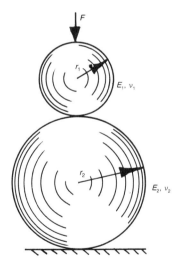

Two balls in contact, same material: $E_1 = E_2$, $v_1 = v_2$

$a = \sqrt[3]{\dfrac{6F(1-v^2)}{E(1/r_1 + 1/r_2)}}$; $\sigma_c = 3F/2\pi a^2$

Ball on flat surface, same material: $r_2 = \infty$, $r_1 = r$

$a = \sqrt[3]{\dfrac{6F(1-v^2)r}{E}}$; $\sigma_c = 3F/2\pi a^2$

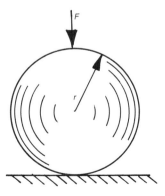

Ball on concave surface, same material:
r_2 negative

$$a = \sqrt[3]{\frac{6F(1-v^2)}{E(1/r_1 - 1/r_2)}}; \quad \sigma_c = 3F/2\pi a^2$$

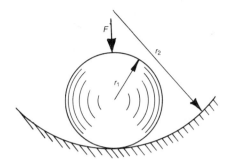

Two rollers in contact

Contact width

$$w = \sqrt{\frac{16F((1-v_1^2)/E_1 + (1-v^2)/E_2)}{\pi L(1/r_1 + 1/r_2)}}; \quad \sigma_c = 4F/\pi wL$$

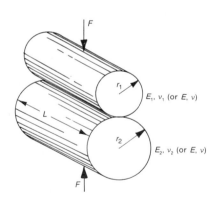

Two rollers in contact, same material

$$w = \sqrt{\frac{32F(1-v^2)}{\pi LE(1/r_1 + 1/r_2)}}; \quad \sigma_c = 4F/\pi wL$$

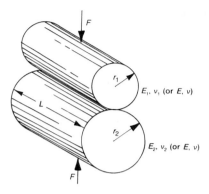

Roller on flat surface, same material: $r_2 = \infty$, $r_1 = r$

$$w = \sqrt{\frac{32F(1-v^2)r}{\pi LE}}; \quad \sigma_c = 4F/wL$$

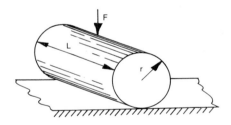

Roller on concave surface, same material: r_2 negative

$$w = \sqrt{\frac{32F(1-v^2)}{\pi LE(1/r_1 - 1/r_2)}}; \quad \sigma_c = 4F/\pi wL$$

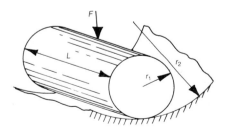

1.10 Loaded flat plates

Formulae are given for the maximum stress and deflection for circular and rectangular flat plates subject to concentrated or distributed loads (pressure) with the edges either clamped or supported. In prac- tice, the edge conditions are usually uncertain and some compromise must be made. The equations are only valid if the deflection is small compared to the plate thickness.

Symbols used:
r = radius of circular plate
a = minor length of rectangular plate
b = major length of rectangular plate
p = uniform pressure loading
P = concentrated load
v = Poisson's ratio (assumed to be 0.3)
E = Young's modulus
t = plate thickness
σ_m = maximum stress
y_m = maximum deflection
D = flexural rigidity = $Et^3/12(1-v^2)$

1.10.1 Stress and deflection of circular flat plates

Circular plate, uniform load, edges simply supported

$$\sigma_m = \frac{3(3+v)pr^2}{8t^2} = \frac{1.238pr^2}{t^2} \text{ (at centre)}$$

$$y_m = \frac{(5+v)pr^4}{64(1+v)D} = \frac{0.696pr^4}{Et^3} \text{ (at centre, } v = 0.3)$$

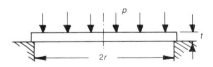

Circular plate, uniform load, clamped edge

$$\sigma_m = \frac{3pr^2}{4t^2} \text{ (at edge)}$$

$$y_m = \frac{pr^4}{64D} = \frac{0.171pr^4}{Et^3} \text{ (at centre)}$$

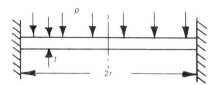

Circular plate, concentrated load at centre, simply supported

$$\sigma_m = \frac{P}{t^2}\left[(1+v)\left(0.485\,ln\frac{r}{t}+0.52\right)+0.48\right]$$

$$= \frac{P}{t^2}\left(0.6305\,ln\frac{r}{t}+1.156\right) \text{ (at centre, lower surface)}$$

$$y_m = \frac{(3+v)Pr^2}{16\pi(1+v)D} = \frac{0.552Pr^2}{Et^3} \text{ (at centre)}$$

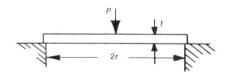

Circular plate, concentrated load at centre, clamped edge

$$\sigma_m = \frac{P}{t^2}(1+v)\left(0.485\,ln\frac{r}{t}+0.52\right)$$

$$= \frac{P}{t^2}\left(0.631\,ln\frac{r}{t}+0.676\right) \text{ (at centre, lower surface)}$$

$$y_m = \frac{Pr^2}{16\pi D} = \frac{0.217Pr^2}{Et^3}$$

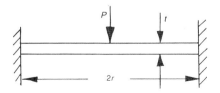

1.10.2 *Stress and deflection of rectangular flat plates*

Rectangular plate, uniform load, simply supported (Empirical)

Since corners tend to rise off the supports, vertical movement must be prevented without restricting rotation.

$$\sigma_m = \frac{0.75pa^2}{t^2[1.61(a/b)^3 + 1]} \text{ (at centre)}$$

$$y_m = \frac{0.142pa^4}{Et^3[2.21(a/b)^3 + 1]} \text{ (at centre)}$$

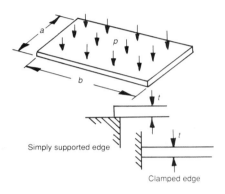

Simply supported edge

Clamped edge

Rectangular plate, uniform load, clamped edges
(empirical)

$$\sigma_m = \frac{pa^2}{2t^2[0.623(a/b)^6 + 1]} \text{ (at middle of edge } b)$$

$$y_m = \frac{0.0284pa^4}{Et^3[1.056(a/b)^5 + 1]} \text{ (at centre)}$$

Rectangular plate, concentrated load at centre, simply supported (empirical)

The load is assumed to act over a small area of radius e.

$$\sigma_m = \frac{1.5P}{\pi t^2}\left[(1+v)\ln\frac{2r}{\pi e} + 1 - k_2\right] \text{ (at centre)}$$

$$y_m = k_1\frac{Pa^2}{Et^3} \text{ (at centre)}$$

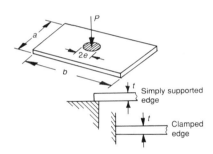

Simply supported edge

Clamped edge

	b/a								
	1.0	1.1	1.2	1.4	1.6	1.8	2.0	3.0	∞
k_1	0.127	0.138	0.148	0.162	0.171	0.177	0.180	0.185	0.185
k_2	0.564	0.445	0.349	0.211	0.124	0.072	0.041	0.003	0.000

Rectangular plate, concentrated load at centre, clamped edges (empirical)

$\sigma_m = k_2P/t^2$ (at middle of edge b)
$y_m = k_1Pa^2/Et^3$ (at centre)

	b/a					
	1.0	1.2	1.4	1.6	1.8	2.0
k_1	0.061	0.071	0.076	0.078	0.079	0.079
k_2	0.754	0.894	0.962	0.991	1.000	1.004

1.10.3 Loaded circular plates with central hole

Symbols used:
 a = outer radius
 b = inner radius
 t = thickness
 P = concentrated load
 p = distributed load
 E = modulus of elasticity

Maximum deflection $y_{max} = k_1 \dfrac{Pa^2}{Et^3}$

or

$$y_{max} = k_1 \frac{pa^4}{Et^3}$$

Maximum stress $\sigma_{max} = k_2 \dfrac{P}{t^2}$

or

$$\sigma_{max} = k_2 \frac{pa^2}{t^2}$$

The following table gives values of k_1 and k_2 for each of the 10 cases shown for various values of a/b. It is assumed that Poisson's ratio $v = 0.3$.

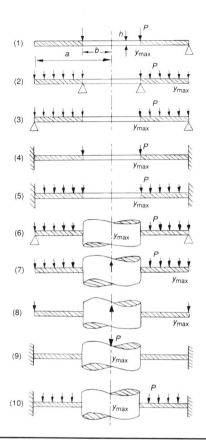

	a/b											
	1.25		1.5		2		3		4		5	
Case	k_1	k_2	k_1	k_2	k_1	k_2	k_1	k_2	k_1	k_2	k_1	k_2
1	0.341	0.100	0.519	1.26	0.672	1.48	0.734	1.880	0.724	2.17	0.704	2.34
2	0.202	0.660	0.491	1.19	0.902	2.04	1.220	3.340	1.300	4.30	1.310	5.100
3	0.184	0.592	0.414	0.976	0.664	1.440	0.824	1.880	0.830	2.08	0.813	2.190
4	0.00504	0.194	0.0242	0.320	0.0810	0.454	0.172	0.673	0.217	1.021	0.238	1.305
5	0.00199	0.105	0.0139	0.259	0.0575	0.480	0.130	0.657	0.162	0.710	0.175	0.730
6	0.00343	0.122	0.0313	0.336	0.1250	0.740	0.221	1.210	0.417	1.450	0.492	1.590
7	0.00231	0.135	0.0183	0.410	0.0938	1.040	0.293	2.150	0.448	2.990	0.564	3.690
8	0.00510	0.227	0.0249	0.428	0.0877	0.753	0.209	1.205	0.293	1.514	0.350	1.745
9	0.00129	0.115	0.0064	0.220	0.0237	0.405	0.062	0.703	0.092	0.933	0.114	1.130
10	0.00077	0.090	0.0062	0.273	0.0329	0.710	0.110	1.540	0.179	2.230	0.234	2.800

2.1 Basic mechanics

2.1.1 *Force*

A force may be represented by an arrow-headed line called a 'vector' which gives 'magnitude', proportional to its length, its 'point of application' and its 'direction'.

Referring to the figure, the magnitude is $20N$, the point of application is $\bigcirc$, and the line of action is XX.

2.1.2 *Triangle of forces*

A force may be resolved into two forces at right angles to one another. The force F shown is at angle θ to axis XX and has components:

$F_x = F \cos \theta$ and $F_y = F \sin \theta$

$$\theta = \tan^{-1}\left(\frac{F_y}{F_x}\right) \text{ and } F = \sqrt{F_x^2 + F_y^2}$$

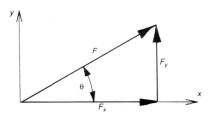

Resultant of several forces

If several forces F_1, F_2, F_3, etc., act on a body, then the resultant force may be found by adding the compo-

nents of these forces in the x and y directions and constructing a triangle of forces.

$F_x = F_1 \cos \theta_1 + F_2 \cos \theta_2 + \ldots$
$F_y = F_1 \sin \theta_1 + F_2 \sin \theta_2 + \ldots$
The resultant force is $F_r = \sqrt{F_x^2 + F_y^2}$

at an angle $\theta_r = \tan^{-1}\left(\frac{F_y}{F_x}\right)$ to the x axis.

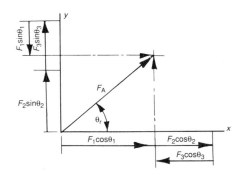

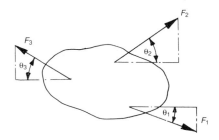

Polygon of forces

The force vectors may be added by drawing a polygon of forces. The line completing the polygon is the resultant (note that its arrow points in the opposite direction), and its angle to a reference direction may be found.

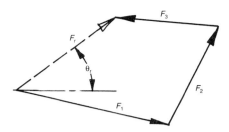

Balance of forces

A system of forces is balanced, i.e. in equilibrium, when the resultant F_r is zero, in which case its components F_x and F_y are each zero.

2.1.3 *Moment of a force, couple*

The moment of a force F about a point O at a perpendicular distance d from its line of action, is equal to Fd.

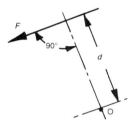

Resultant of several moments

If forces F_1, F_2, etc., act on a body at perpendicular distances d_1, d_2, etc., from a point O, the moments are,
$M_1 = F_1 d_1$, $M_2 = F_2 d_2$, etc. about O
The resultant moment is $M_r = M_1 + M_2 + \ldots$
Clockwise moments are reckoned positive and counterclockwise moments negative. If the moments 'balance' $M_r = 0$ and the system is in equilibrium.

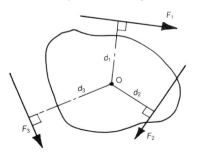

Couple

If two equal and opposite forces have parallel lines of action a distance a apart, the moment about any point O at distance d from one of the lines of action is

$$M = F_d - F(d-a) = Fa$$

This is *independant of d* and the resultant force is zero. Such a moment is called a 'couple'.

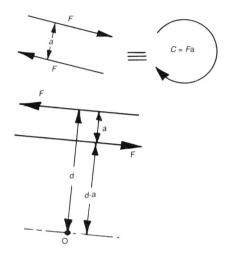

Resolution of a moment into a force and a couple

For a force F at a from point O; if equal and opposite forces are applied at O, then the result is a couple Fa and a net force F.

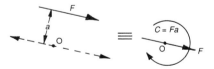

General condition for equilibrium of a body

Complete equilibrium exists when both the forces and the moments balance, i.e. $F_r = 0$ and $M_r = 0$.

2.1.4 *Linear and circular motion*

Relationships for distance travelled, velocity and time of travel are given for a constant linear acceleration.

Similar relationships are given for circular motion with constant angular acceleration. In practice, acceleration may vary with time, in which case analysis is much more difficult.

2.1.5   Acceleration

Linear acceleration

Symbols used:
u = initial velocity
v = final velocity
t = time
a = acceleration
x = distance

Also: $v = \dfrac{\mathrm{d}x}{\mathrm{d}t}$; $a = \dfrac{\mathrm{d}v}{\mathrm{d}t} = v\dfrac{\mathrm{d}v}{\mathrm{d}x}$

And: $x = \displaystyle\int v\,\mathrm{d}t$; $v = \displaystyle\int a\,\mathrm{d}t$

Equations of motion:

$v = u + at$

$x = \dfrac{(u+v)}{2t}$

$v^2 = u^2 + 2ax$
$x = ut + \frac{1}{2}at^2$

Angular acceleration

Let:

ω_1 = initial angular velocity
ω_2 = final angular velocity
t = time
θ = angle of rotation
α = angular acceleration

Also: $\omega = \dfrac{\mathrm{d}\theta}{\mathrm{d}t}$; $\alpha = \dfrac{\mathrm{d}\omega}{\mathrm{d}t} = \omega\dfrac{\mathrm{d}\omega}{\mathrm{d}\theta}$

And: $\theta = \displaystyle\int \omega\,\mathrm{d}t$: $\omega = \displaystyle\int \alpha\,\mathrm{d}t$

Equations of motion:

$\omega_2 = \omega_1 + \alpha t$

$\theta = \dfrac{(\omega_1 + \omega_2)}{2t}$

$\omega_2^2 = \omega_1^2 + 2\alpha\theta$
$\theta = \omega_1 t + \frac{1}{2}\alpha t^2$

2.1.6   Centripetal acceleration

For a mass m rotating at ω rad s^{-1} at radius r:

Tangential velocity $v = r\omega$

Centripetal acceleration $= \dfrac{v^2}{r} = r\omega^2$

Centripetal force $= mr\omega^2$ (acting inwards on m)
Centrifugal force $= mr\omega^2$ (acting outwards on pivot)

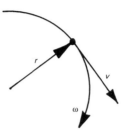

2.1.7   Newton's laws of motion

These state that:

(1) A body remains at rest or continues in a straight line at a constant velocity unless acted upon by an external force.
(2) A force applied to a body accelerates the body by an amount which is proportional to the force.
(3) Every action is opposed by an equal and opposite reaction.

2.1.8   Work, energy and power

Kinetic, potential, strain and rotational kinetic energy are defined and the relationships between work, force and power are given.

Work done $W =$ force × distance $= Fx$ (Nm = J)

Work done by variable force $W = \displaystyle\int F\,\mathrm{d}x$

Work done by torque (T) $W = T\theta$
where: θ = angle of rotation.

Also $W = \displaystyle\int T\,\mathrm{d}\theta$

Kinetic energy $\mathrm{KE} = \dfrac{mv^2}{2}$

Rotational kinetic energy $KE = \dfrac{I\omega^2}{2}$

where: $I =$ moment of inertia of body

Change of kinetic energy $= \dfrac{m}{2}(v^2 - u^2)$

Potential energy $PE = mgh$
where: $g =$ acceleration due to gravity (9.81 m s^{-2}),
        $h =$ height above a datum.

Strain energy $SE = Fx = \dfrac{kx^2}{2}$

where: $x =$ deflection, $k =$ stiffness.

Conversion of potential energy to kinetic energy:

$$mgh = \dfrac{mv^2}{2}$$

Therefore $v = \sqrt{2gh}$ or $h = \dfrac{v^2}{2g}$

Power

Power $P = \dfrac{W}{t} = \dfrac{Fx}{t} = Fv$ ($\text{Nms}^{-1} = \text{J s}^{-1} = \text{W}$)

Rotational power $P =$ torque $\times$ angular velocity

$$= T\omega = \dfrac{T\theta}{t}$$

Also, if $N =$ the number of revolutions per second
$P = 2\pi N T$
where: $2\pi N =$ angular velocity ω.

2.1.9   Impulse and momentum

Impulse.   An impulsive force is one acting for a very short time δt. Impulse is defined as the product of the force and the time, i.e. $= F\delta t$.

Momentum is the product of mass and velocity $= mv$

Change of momentum $= mv - mu$

Angular momentum $= I\omega$

Change of angular momentum $= I(\omega_2 - \omega_1)$

Force $F =$ rate of change of momentum $= \dfrac{d(mv)}{dt}$

If m is constant then force $F = m\dfrac{dv}{dt} = ma$ (mass $\times$ acceleration)

Similarly: Torque $T = \dfrac{d(I\omega)}{dt}$ (rate of change of angular momentum)

If I is constant $T = I\dfrac{d\omega}{dt} = I\alpha$

2.1.10   Impact

The following deals with the impact of elastic and inelastic spheres, although it applies to bodies of any shape.

Consider two spheres rolling on a horizontal plane. Velocities before impact are u_1 and u_2 for spheres of mass m_1 and m_2. After impact their velocities are v_1 and v_2.

Coefficient of restitution

$$e = -\frac{\text{difference in final velocities}}{\text{difference in initial velocities}} = -\frac{(v_1 - v_2)}{(u_1 - u_2)}$$

Note: $e = 1$ for perfectly elastic spheres; $e = 0$ for inelastic spheres.

Velocities after impact (velocities positive to right):

$$v_1 = \frac{[m_1 u_1 + m_2 u_2 - em_2(u_1 - u_2)]}{(m_1 + m_2)}$$

$$v_2 = \frac{[m_1 u_1 + m_2 u_2 + em_1(u_1 - u_2)]}{(m_1 + m_2)}$$

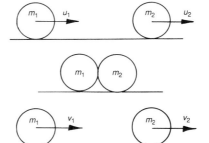

Loss of kinetic energy due to impact $= \dfrac{m_1(u_1^2 - v_1^2) + m_2(u_2^2 - v_2^2)}{2}$

If $e = 1$, KE loss $= 0$.

If $e = 0$, KE loss $= \dfrac{m_1 m_2 (v_1 - v_2)^2}{2(m_1 + m_2)}$.

2.1.11 Centre of percussion

Let:

h = distance from pivot to centre of gravity
p = distance from pivot to centre of percussion
k = radius of gyration of suspended body about centre of gravity

$$p = \frac{h^2 + k^2}{h}$$

Uniform thin rod

$$k^2 = \frac{L^2}{12}, \ h = \frac{L}{2}; \ p = \frac{2}{3}L$$

Cylinder

$$k^2 = \left(\frac{L^2}{12} + \frac{D^2}{16}\right); \ h = \frac{L}{2};$$

$$p = \left(\frac{2}{3}L + \frac{D^2}{8L}\right)$$

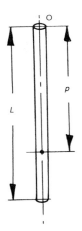

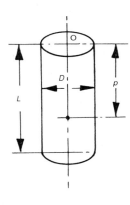

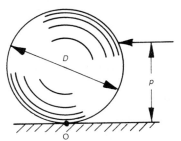

Sphere

$$k^2 = \frac{D^2}{10}; \ h = \frac{D}{2}; \ p = \frac{7}{10}D$$

The physical meaning of centre of percussion is that it is the point where an impact produces no reaction at the pivot point.

2.1.12 Vehicles on curved track

Horizontal curved track

Skidding speed $v_s = \sqrt{gr\mu}$

Overturning speed $v_o = \sqrt{ga\dfrac{r}{h}}$

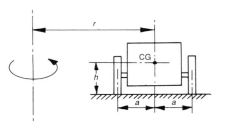

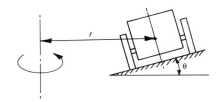

Curved track banked at angle θ

Skidding speed $v_s = \sqrt{\dfrac{g\mu r\left(1 + \tan\dfrac{\theta}{\mu}\right)}{(1 - \mu\tan\theta)}}$

where: μ = coefficient of friction,
h = height of CG above ground.

2.1.13 The gyroscope

The flywheel of moment of inertia $I\ (=mk^2)$ rotates at angular velocity ω_1 about the x axis. An applied couple C about the z axis produces an angular velocity

ω_2 about the y axis. Directions of rotation are as shown in the figure.

Couple $C = I\omega_1\omega_2$

Conversely, if a rotation ω_2 is applied to the wheel bearings, then a couple C is produced.

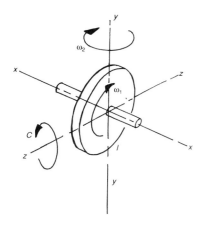

Conical pendulum

Periodic time $t_p = 2\pi\sqrt{\dfrac{h}{g}}$

String tension $T = mL\omega^2$

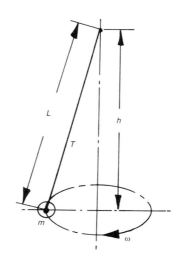

2.1.14   *The pendulum*

Simple pendulum

Periodic time $t_p = 2\pi\sqrt{\dfrac{L}{g}}$

Frequency $f = \dfrac{1}{t_p}$

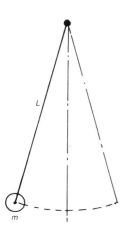

Compound pendulum

Periodic time $t_p = 2\pi\sqrt{\dfrac{(h^2 + k^2)}{gh}}$

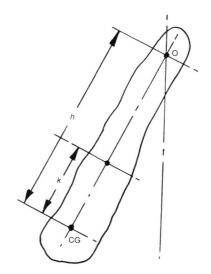

$\dfrac{h^2 + k^2}{h} = L' =$ the length of the equivalent simple pendulum.

(Also equal to the distance to the centre of percussion)
Where: $k =$ radius of gyration about CG, $h =$ distance from pivot to CG.

2.1.15 Gravitation

This deals with the mutual attraction which exists between bodies. The magnitude of the force depends on the masses and the distance between them. For two masses m_1 and m_2 a distance d apart, the force is:

$$F = G \frac{m_1 m_2}{d^2}$$

where: G is the 'gravitational constant' $= 6.67 \times 10^{-11} \, \text{N m}^2 \, \text{kg}^{-2}$

For a body m_2 on the earth's surface $m_1 = 5.97 \times 10^{24} \, \text{kg}$ (earth's mass), $d = 6.37 \times 10^6 \, \text{m}$ (earth's radius). Then

$$F = \frac{6.67 \times 5.97}{6.37^2} \times 10 m_2 = 9.81 m_2 = g m_2$$

Thus: $g = 9.81 \, \text{m s}^{-2}$

Variation of g with height and latitude

If:

$L =$ degrees latitude ($0°$ at equator)
$h =$ height above sea level (km)
$g = 9.806294 - 0.025862 \cos 2L + 0.000058 \cos^2 2L - 0.003086h$

2.1.16 The solar system

The following table gives useful information on the sun, moon and earth.

	Earth	Sun	Moon
Mass (kg)	5.97×10^{24}	2×10^{30}	7.34×10^{22}
Radius (km)	Equatorial 6378	696 000	1738
	Polar 6357		
Average density (kg m^{-3})	5500	1375	3300
Period of revolution			
About axis	23 h 56 min	25 days	27.33 days
orbital	365.26 days		27.33 days
Acceleration due to gravity (m s^{-2})	9.81	2.75×10^8	1.64
Mean orbital radius (km)	149.6×10^6	—	384 400
Miscellaneous information	Tilt of polar axis $23\frac{1}{2}°$	Type G star. Absolute magnitude 5.0. Surface temperature 6000°C. Centre temperature $14 \times 10^6 °\text{C}$	Period between new moons = $29\frac{1}{2}$ days

2.1.17 *Machines*

Mechanical advantage MA (or force ratio) $= \dfrac{\text{Load}}{\text{Effort}}$

Velocity ratio VR (or movement ratio)

$$= \frac{\text{Distance moved by effort}}{\text{Distance moved by load}}$$

Efficiency $\eta = \dfrac{\text{Useful work out}}{\text{Work put in}} = \dfrac{\text{MA}}{\text{VR}}$

2.1.18 *Levers*

The lever is a simple machine consisting of a pivoted beam. An effort E lifts a load W. Referring to the figure, and assuming no friction:

$$W = E\frac{a}{b}; \quad MA = \frac{W}{E}; \quad VR = \frac{a}{b}$$

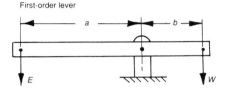

First-order lever

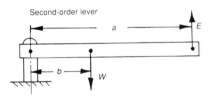

Second-order lever

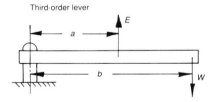

Third-order lever

2.1.19 *Projectiles*

When a projectile is fired at an angle to a horizontal plane under gravity, the trajectory is a parabola if air resistance is neglected. It can be shown that the maximum range is achieved if the projection angle is 45°. The effect of air resistance is to reduce both range and height.

Assuming no air resistance:

Range $R = \dfrac{v^2}{g}\sin 2\theta \ \left(\text{at } \theta = 45°; \ R_{\max} = \dfrac{v^2}{g}\right)$

Height $h = \dfrac{v^2}{g}\sin^2 \theta \ \left(\text{at } \theta = 45°; \ h = \dfrac{v^2}{4g}\right)$

Time of flight $t = 2\sin\theta\dfrac{v}{g} \ \left(\text{for } \theta = 45°; \ t = \sqrt{2}\dfrac{v}{g}\right)$

Projection up a slope (of angle β):

Range $R = \dfrac{v^2}{g}\dfrac{2\sin(\theta - \beta)\cos\theta}{\cos^2\beta};$

$$R_{\max} = \frac{v^2}{g}\frac{1}{(1 + \sin\beta)} \ \left(\text{at } \theta = 45° + \frac{\beta}{2}\right)$$

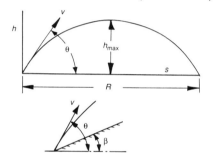

2.1.20 *Rockets*

For a rocket travelling vertically against gravity, the mass of fuel is continually decreasing as the fuel is burnt, i.e. the total mass being lifted decreases uniformly with time.

The following formulae give the velocity and height at any time up to burn-out, and the velocity, height and time expired at burn-out.

Let:
  $V = $ jet velocity (assumed constant)
  $U = $ rocket velocity
  $M_f = $ mass of fuel at blast off
  $M_r = $ mass of rocket with no fuel
  $\dot{m} = $ mass flow rate of fuel
  $t = $ time after blast-off
  $g = $ acceleration due to gravity (assumed to be constant)

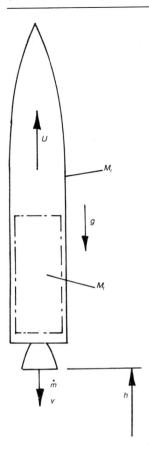

Let: $T = \dfrac{M_f + M_r}{\dot{m}}$

Time to burn-out $t_b = \dfrac{M_f}{\dot{m}}$

Velocity at t: $U = V \ln\left(\dfrac{T}{T-t}\right) - gt$

Velocity at burn-out $U_b = V \ln\left(\dfrac{T}{T-t_b}\right) - gt_b$

Height at t: $h = Vt - \dfrac{gt^2}{2} - V(T-t)\ln\dfrac{T}{(T-t)}$

Height at burn-out: $h_b = Vt_b - \dfrac{gt_b^2}{2} - V(T-t_b)\ln\dfrac{T}{(T-t_b)}$

2.1.21   Satellites

The orbital velocity of a satellite is a maximum at sea level and falls off with height, while the orbital time increases. When the period of rotation is the same as that of the planet, the satellite is said to be 'synchronous', i.e. the satellite appears to be stationary to an observer on earth. This is of great value in radio communications.

Let:
$v =$ velocity
$h =$ height of orbit
$a =$ radius of planet
$r = a + h$
$t =$ time
$g =$ acceleration due to gravity

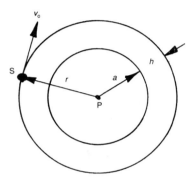

Orbital velocity $v_o = \sqrt{\dfrac{ga^2}{r}}$

Maximum velocity $v_{o\,max} = \sqrt{ga}$ (at sea level)

Periodic time (orbit time) $t_p = 2\pi\sqrt{\dfrac{r^3}{ga^2}}$

Escape velocity $v_e = \sqrt{2ga}$
This is the velocity for a given height when the satellite will leave its orbit and escape the effect of the earth's gravity.

Height of orbit $h = a\left(\sqrt[3]{\dfrac{gt_p^2}{4\pi^2 a}} - 1\right)$

Example

For the earth, $a = 6.37 \times 10^6$ m, $g = 9.81$ m s^{-2}.
Then: $v_{o\,max} = 7.905$ km s^{-1} (at sea level)
     $v_e = 11.18$ km s^{-1} (about 7 miles per second)
Height of synchronous orbit $h_s = 35\,700$ km ($t_p = 24$ h).

2.2   Belt drives

2.2.1   *Flat, vee and timing belt drives*

Formulae are given for the power transmitted by a belt drive and for the tensions in the belt. The effect of centrifugal force is included.

A table of information on timing belt drives is included.

Symbols used:
F_1 = belt tension, tight side
F_2 = belt tension, slack side
r_a = radius of pulley a
r_b = radius of pulley b
N_a = speed of pulley a
N_b = speed of pulley b
m = mass of belt per unit length
P = power transmitted
μ = coefficient of friction between belt and pulley
F_o = initial belt tension
θ_a = arc of belt contact pulley a
θ_b = arc of belt contact pulley b
L = distance between pulley centres
s = percentage slip
v = belt velocity

Speed ratio $\dfrac{N_a}{N_b} = \dfrac{r_b}{r_a}\dfrac{(100-s)}{100}$

(when pulley b is the driver)

Arc of contact ($r_a > r_b$):

$$\theta_a = 180° + 2\sin^{-1}\frac{(r_a - r_b)}{L}$$

$$\theta_b = 180° - 2\sin^{-1}\frac{(r_a - r_b)}{L}$$

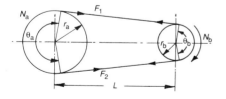

Tension ratio for belt about to slip:

For pulley 'a' $\dfrac{F_1}{F_2} = e^{\mu\theta_a}$

For pulley 'b' $\dfrac{F_1}{F_2} = e^{\mu\theta_b}$

where: e = base of natural logarithms ($= 2.718$).

Power capacity $P = v(F_1 - F_2)$
where: belt velocity $v = 2\pi r_a N_a = 2\pi r_b N_b$ (no slip).

Pulley torque $T_a = r_a(F_1 - F_2)$; $T_b = r_b(F_1 - F_2)$

Initial tension $F_o = \dfrac{(F_1 + F_2)}{2}$

Effect of centrifugal force: the belt tensions are reduced by mv^2 so that

$$\frac{F_1 - mv^2}{F_2 - mv^2} = e^{\mu\theta}$$

Vee belt

The 'wedge' action of the vee belt produces a higher effective coefficient of friction μ'

$$\mu' = \frac{\mu}{\sin\alpha}$$

where: α = the 'half angle' of the vee ($\mu' = 2.9\mu$ for $\alpha = 20°$).

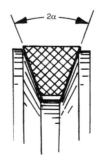

Timing belts

Timing belts have teeth which mate with grooves on the pulleys. They are reinforced with high strength polymer strands to give power capacity up to three times that of conventional belts at three times the speed. There is no slip so a constant ratio is maintained. A large number of speed ratios is available. Belts are made in several strengths and widths.

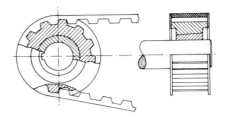

Timing belt sizes (BS 4548: 1970)

Type	Meaning	Pitch (mm)	Widths (mm)	Constant, K
XL	Extra light	5.08	6.4, 7.9, 9.6	—
L	Light	9.53	12.7, 19.1, 25.4	1.53
H	Heavy	12.70	19.1, 25.4, 38.1, 50.8, 76.2	5.19
XH	Extra heavy	22.23	50.8, 76.2, 101.6	12.60
XXH	Double extra heavy	31.75	50.8, 76.2, 101.6, 127.0	—

Service factor

Hours of service per day	< 10	10–16	> 16
% full power	100	72	67

Class	Applications	% full power
1	Typewriters, radar, light domestic	100
2	Centrifugal pumps, fans, woodworking machines, light conveyors	69
3	Punching presses, large fans, printing machines, grain conveyors	63
4	Blowers, paper machines, piston pumps, textile machines	58
5	Brickmaking machines, piston compressors, hoists, crushers, mills	54

Power capacity $P = KNTW \times 10^{-6}$ kilowatts

where:

K = size constant (see table)
N = number of revolutions per minute
T = teeth in smaller pulley
W = width of belt (mm)

Example: Type H belt, $W = 50.8$ mm, $N = 1500$ rev min^{-1}, $T = 20$, for large fan working 12 hours per day. From tables, $K = 5.19$ service factors 72% and 63%.

$P = 5.19 \times 1500 \times 20 \times 50.8 \times 10^{-6} \times 0.63 \times 0.72 = 3.59\,\text{kW}$

Note: at high speeds and with large pulleys the power capacity may be up to 25% less. See manufacturer's tables.

2.2.2   *Winches and pulleys*

Winch

Velocity ratio $VR = \dfrac{R}{r}$

Force to raise load $F = \dfrac{W}{VR} = \eta\,\dfrac{Wr}{R}$

where: $\eta =$ efficiency.

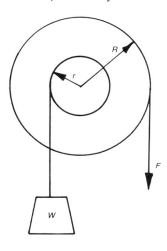

Pulleys

Velocity ratio $VR = 2$

Force to raise load $F = \eta\,\dfrac{W}{2}$

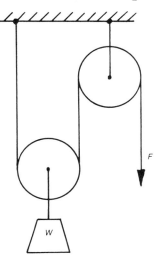

Block and tackle

Velocity ratio $VR = n$
where: $n =$ number of ropes between the sets of
pulleys ($= 5$ in figure).

Force to raise load $F = \eta\,\dfrac{W}{n}$

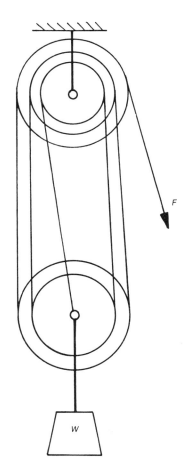

Differential pulley

Velocity ratio $VR = \dfrac{2}{\left(1 - \dfrac{r}{R}\right)}$

Force to raise load $F = \eta\,\dfrac{W}{VR}$

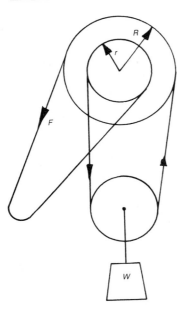

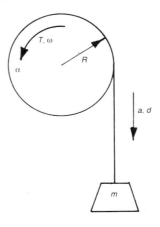

where: $\alpha = \dfrac{a}{R}$

2.2.3 Hoist

Symbols used:
m = mass of load
I = moment of inertia of drum, etc.
R = drum radius
T = torque to drive drum
T_f = friction torque
a = acceleration of load
d = deceleration of load
α = angular acceleration/deceleration

Load being raised and accelerating

Torque $T = T_f + I\alpha + mR(a + g)$

Load rising and coming to rest, no drive

$T_f - I\alpha = mR(d - g)$

Deceleration $d = \dfrac{(mgR + T_f)}{\left(mR + \dfrac{I}{R}\right)}$

Load being lowered and accelerating, no drive

$T_f + I\alpha = mR(g - a)$

Acceleration $a = \dfrac{(mgR - T_f)}{\left(m\dot{R} + \dfrac{I}{R}\right)}$

Load falling and being brought to rest

$T = I\alpha - T_f + mR(g + d)$

2.3 Balancing

2.3.1 Rotating masses

Balancing of rotating components is of extreme importance, especially in the case of high-speed machinery. Lack of balance may be due to a single mass in one plane or masses in two planes some distance apart. The method of balancing is given.

Out of balance due to one mass

For mass m at radius r and angular velocity ω:
Out of balance force $F = mr\omega^2$
This may be balanced by a mass m_b at r_b so that
$m_b r_b = mr$

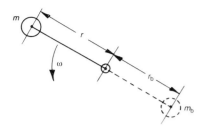

Several out of balance masses in one plane

The forces are: $m_1 r_1 \omega^2$, $m_2 r_2 \omega^2$, etc. These are resolved into vertical and horizontal components:

$F_v = m_1 r_1 \omega^2 \sin \theta_1 + m_2 r_2 \omega^2 \sin \theta_2 + \ldots$
$F_h = m_1 r_1 \omega^2 \cos \theta_1 + m_2 r_2 \omega^2 \cos \theta_2 + \ldots$
Resultant force $F_r = \sqrt{F_v^2 + F_h^2}$

at an angle to horizontal axis $\theta_r = \tan^{-1}\left(\dfrac{F_v}{F_h}\right)$

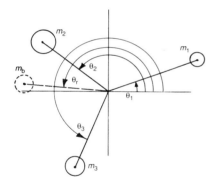

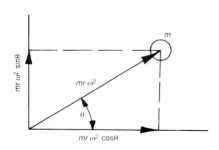

To balance a mass m_b at r_b such that $m_b r_b = \dfrac{F_r}{\omega^2}$

is required at an angle $\theta_r + 180°$.

Dynamic unbalance, forces in several planes

For a force $mr\omega^2$ acting at x from bearing A, the moment of the force about the bearing is $mr\omega^2 x$. This has components:

$mr\omega^2 x \sin \theta$ vertically
$mr\omega^2 x \cos \theta$ horizontally
  For several forces:

Total vertical moment $M_v = m_1 r_1 \omega^2 x_1 \sin \theta_1 + m_2 r_2 \omega^2 x_2 \sin \theta_2 \ldots$
Total horizontal moment $M_h = m_1 r_1 \omega^2 x_1 \cos \theta_1$
$\qquad\qquad\qquad\qquad + m_2 r_2 \omega^2 x_2 \cos \theta_2 \ldots$

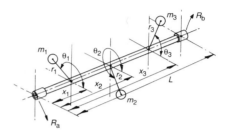

Resultant moment $\quad _a M_r = \sqrt{M_v^2 + M_h^2}$

acting at $\theta_b = \tan^{-1}\left(\dfrac{M_v}{M_h}\right)$.

The reaction at B is: $R_b = \dfrac{_a M_r}{L}$

where: $L =$ span.
The process is repeated, by taking moments about end B, and R_a found.

Method of balancing Complete 'dynamic balance' is achieved by introducing forces equal and opposite to R_a and R_b. In practice, balancing is carried out at planes a short distance from the bearings.

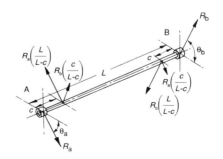

If this distance is c then the balancing forces are

$$R_a\left(\frac{L}{L-c}\right) \text{ and } R_b\left(\frac{L}{L-c}\right)$$

This introduces small errors due to moments

$$R_a\frac{c}{(L-c)} \text{ and } R_b\frac{c}{(L-c)}$$

which can be corrected for as shown in the figure.

A further very small error remains and the process may be repeated until the desired degree of balance is achieved.

2.3.2 Reciprocating masses

For the piston, connecting rod, crank system shown in the figure there exists a piston accelerating force which varies throughout a revolution of the crank. The force can be partially balanced by weights on the crankshaft.

Let:
m = mass of piston
ω = angular velocity of crank
r = radius of crank
L = length of conrod
θ = crank angle

Force to accelerate piston $F = -mr\omega^2\left(\cos\theta + \frac{r}{L}\cos 2\theta\right)$

(approximately, see Section 2.4.1)

Maximum forces $F_1 = mr\omega^2$

(at crankshaft speed, which can be balanced)

$$F_2 = mr\omega^2\frac{r}{L} \text{ (at twice crankshaft speed)}$$

Effect of conrod mass

The conrod mass may be divided approximately between the crankpin and the gudgeon pin. If m_c is the conrod mass:

Effective mass at gudgeon $m_1 = m_c\frac{a}{L}$ added to piston mass.

Effective mass at crankpin $m_2 = m_c\frac{b}{L}$

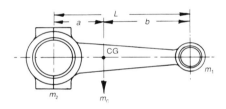

2.4 Miscellaneous machine elements

2.4.1 Simple engine mechanism

Using the same symbols as in the previous section:

Piston displacement $x = r\left(1 - \cos\theta + \frac{K^2}{2}\sin^2\theta + \frac{K^3}{8}\sin^4\theta + \ldots\right)$

Piston velocity $v = r\omega\left[\sin\theta + \frac{K}{2}\sin 2\theta + \frac{K^3}{8}(\sin 2\theta - \frac{1}{2}\sin 4\theta) + \ldots\right]$

Piston acceleration $a = r\omega^2\left[\cos\theta + K\cos 2\theta + \frac{K^3}{4}(\cos 2\theta - \cos 4\theta) + \ldots\right]$

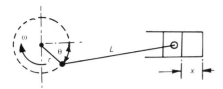

where: $K = \dfrac{r}{L}$.

If K is under about 0.3, it is accurate enough to use only the first two terms containing θ in each formula.

2.4.2 Flywheels

Flywheels are used for the storing of energy in a rotating machine and to limit speed fluctuations. Formulae are given for the calculation of the moment of inertia of flywheels and for speed and energy fluctuation.

Angular velocity $\omega = 2\pi N$

Angular acceleration $\alpha = \dfrac{(\omega_2 - \omega_1)}{t}$

Acceleration torque $T = I\alpha$
where: $I = mk^2$.

Energy stored $E = \dfrac{I\omega^2}{2}$

Calculation of I *for given speed fluctuation*

If P = power,

Energy from engine per revolution $= \dfrac{P}{N}$

Coefficient of speed fluctuation

$$K_N = 2\left(\frac{N_{max} - N_{min}}{N_{max} + N_{min}}\right) = 2\left(\frac{\omega_{max} - \omega_{min}}{\omega_{max} + \omega_{min}}\right)$$

Coefficient of energy fluctuation

$$K_E = \frac{\text{Change in } E}{E} \quad (\text{from } N_{max} \text{ to } N_{min})$$

Required moment of inertia $I = \dfrac{K_E E}{K_N \omega_{mean}^2}$

where: $\omega_{mean} = \dfrac{(\omega_{max} + \omega_{min})}{2}$.

Example The power of an engine is 100 kW at a mean speed of 250 rev min^{-1}. The energy to be absorbed by the flywheel between maximum and minimum speeds is 10% of the work done per revolution.

Calculate the required moment of inertia for the flywheel if the speed fluctuation is not to exceed 2%.

$$K_N = 0.02, \; K_E = 0.1, \; \omega_{mean} = \frac{2\pi \times 250}{60} = 26.2 \,\text{rad s}^{-1}$$

Energy per revolution $E = \dfrac{100\,000 \times 60}{250} = 24\,000 \,\text{J}$

Therefore: $I = \dfrac{0.1 \times 24\,000}{0.02 \times 26.2^2} = 175 \,\text{kg} - \text{m}^2$

Values of I *and* k (*radius of gyration*)

Solid disk:

Mass $m = \rho \pi r^2 b$

Radius of gyration $k = \dfrac{r}{\sqrt{2}}$

Moment of inertia $I = mk^2 = \dfrac{mr^2}{2} = \dfrac{\rho \pi r^4 b}{2}$

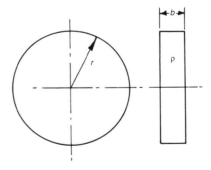

Example For flywheel in previous example ($I = 175$ kg-m². If the flywheel is a solid disc with thickness $\frac{1}{6}$ of the diameter, and the density is 7000 kg m^{-3}, determine the dimensions.

$$I = \frac{\rho \pi r^4 b}{2} \text{ and if } b = \frac{r}{3}, \; I = \frac{\rho \pi r^5}{6} \text{ and } r = \sqrt[5]{\frac{6I}{\pi \rho}}.$$

Therefore $r = \sqrt[5]{\dfrac{6 \times 175}{\pi \times 7000}} = 0.544$ m.

Thus: diameter $D = 1088$ mm, thickness $b = 181$ mm.

Annular ring:

$$m = \rho \pi (r_2^2 - r_1^2) b$$

$$k = \sqrt{\frac{r_2^2 + r_1^2}{2}}$$

$$I = m \frac{(r_2^2 + r_1^2)}{2}$$

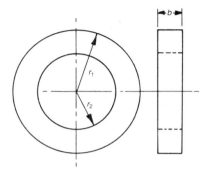

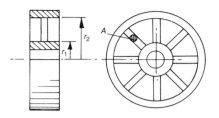

2.4.3 Hooke's joint (cardan joint)

This is a type of flexible shaft coupling used extensively for vehicle drives. They are used in pairs when there is parallel misalignment.

Symbols used:
N_1 = input speed
N_2 = output speed
α = angle of input to output shaft
θ = angle of rotation

Thin ring:

If: r_m = mean radius, A = cross-sectional area.

$$m = 2\pi r_m A \rho$$
$$k = r_m$$
$$I = m r_m^2$$

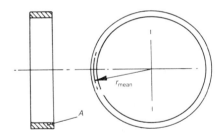

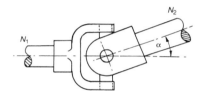

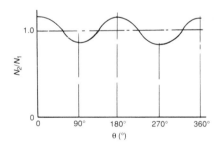

Spokes of uniform cross-section:

$$m = \rho (r_2 - r_1) A$$

$$k = \sqrt{\frac{r_2^2 + r_1 r_2 + r_1^2)}{3}}$$

$$I_s = m k^2$$

Spoked wheel:
The hub and rim are regarded as annular rings.

$$I = I_{hub} + I_{rim} + n I_s$$

where: n = number of spokes.

Speed ratio $\dfrac{N_2}{N_1} = \dfrac{\cos \alpha}{1 - \sin^2 \alpha \cos^2 \alpha}$

Maximum speed ratio $= \dfrac{1}{\cos \alpha}$ (at $\theta = 0°$ or $180°$)

Minimum speed ratio $= \cos \alpha$ (at $\theta = 90°$ or $270°$)

$\dfrac{N_2}{N_1} = 1$, when $\theta = \cos^{-1} \dfrac{\pm 1}{\sqrt{1 + \cos \alpha}}$

2.4.4 *Cams*

A cam is a mechanism which involves sliding contact and which converts one type of motion into another, e.g. rotary to reciprocating. Most cams are of the radial type, but axial rotary cams are also used. Cams may have linear motion. The motion is transmitted through a 'follower' and four types are shown for radial cams.

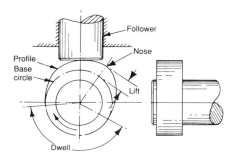

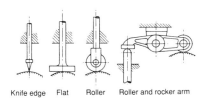

Knife edge   Flat   Roller   Roller and rocker arm

Tangent cam with roller follower

On the flank:
Lift $y = (r_1 + r_0)(\sec \theta - 1)$

where: θ = angle of rotation.

Velocity $v = \omega (r_1 + r_0) \sec \theta \tan \theta$

where: $\omega = \dfrac{\mathrm{d}\theta}{\mathrm{d}t}$ the angular velocity.

Acceleration $a = \omega^2 (r_1 + r_0) \dfrac{(1 + 2 \tan^2 \theta)}{\cos \theta}$

On the nose: the system is equivalent to a con-rod/crank mechanism with crank radius d and conrod length $(r_0 + r_2)$ (see Section 2.4.1).

Maximum lift $y_{max} = d - r_1 + r_2$

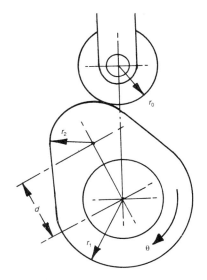

Circular arc cam with flat follower

On flank:
Lift $y = (R - r_1)(1 - \cos \theta)$
Velocity $v = \omega (R - r_1) \sin \theta$
Acceleration $a = \omega^2 (R - r_1) \cos \theta$

On nose:
Lift $y = (r_2 - r_1) + d \cos(\alpha - \theta)$
Velocity $v = \omega d \sin(\alpha - \theta)$
Acceleration $a = -\omega^2 d \cos(\alpha - \theta)$
Maximum lift $y_{max} = d - r_1 + r_2$

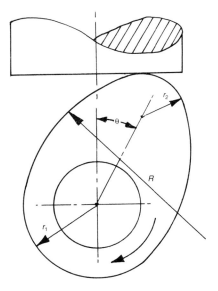

Simple harmonic motion cam

Lift $y = d(1 - \cos\theta)$
where: d = eccentricity.

Velocity $v = \omega d \sin\theta$
Acceleration $a = \omega^2 d \cos\theta$
Maximum lift $y_{max} = d$

The shape of the cam is a circle.

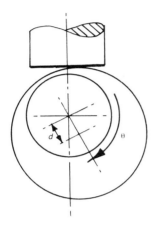

Constant velocity cam, knife-edge follower

Lift $y = y_{max}\left(\dfrac{\theta}{\theta_{max}}\right)$

where: θ_{max} = angle for y_{max}.

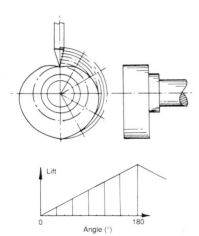

Velocity $v = \left(\dfrac{\omega y_{max}}{\theta_{max}}\right)$

Acceleration $a = 0$ during rise and fall but infinite at direction reversal.

Constant acceleration and deceleration cam, roller follower

The following refers to the motion of the roller centre.

Lift $y = 2y_{max}\left(\dfrac{\theta}{\theta_{max}}\right)^2$ (for first half of lift).

$y = 2y_{max}\left[\dfrac{1}{2} - \left(\dfrac{\theta_{max} - \theta}{\theta_{max}}\right)^2\right]$ (for second half of lift).

Velocity $v = \dfrac{4\omega y_{max}\theta}{\theta_{max}^2}$ (for first half of lift)

$v = 4\omega y_{max}\dfrac{(\theta_{max} - \theta)}{\theta_{max}^2}$ (for second half of lift)

Acceleration and deceleration $a = \dfrac{4\omega^2 y_{max}}{\theta_{max}^2}$ (constant)

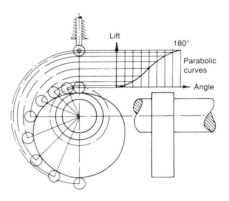

Axial cam (face cam)

The cam profile is on the end of a rotating cylinder and the follower moves parallel to the cylinder axis.

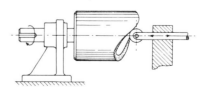

2.4.5 *Governors*

A governor is a device which controls the speed of an engine, a motor or other machine by regulating the fuel or power supply. The controlled speed is called the 'isochronous speed'. Electronic systems are also available.

Watt governor

Isochronous speed $N = \dfrac{1}{2\pi} \sqrt{\dfrac{g}{h}}$

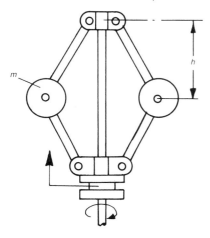

Porter governor

$$N = \dfrac{1}{2\pi} \sqrt{\left(1 + \dfrac{m_1}{m_2}\right) \dfrac{g}{h}}$$

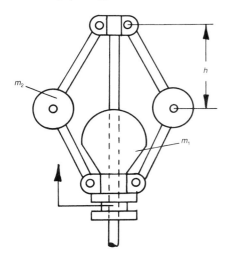

Hartnell governor

$$N = \dfrac{1}{2\pi} \sqrt{\dfrac{k}{2m}\left(\dfrac{b}{a}\right)^2}$$

Where: k = spring stiffness.

Initial spring force $F_o = k\dfrac{bc}{a}$, when $\alpha = 0$.

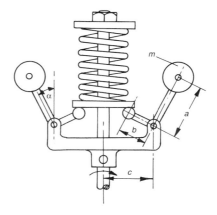

2.4.6 *Screw threads*

Screw threads are used in fasteners such as bolts and screws, and also to provide a linear motion drive which may transmit power. There are several different types of screw thread used for different purposes.

Power transmission (see also Section 2.7.2)

Symbols used:
D = mean diameter of thread
p = pitch of thread

θ = thread angle = $\tan^{-1}\dfrac{p}{\pi D}$

ϕ = friction angle = $\tan^{-1}\mu$
μ = coefficient of friction

Mechanical advantage MA = $\dfrac{1}{\tan(\theta + \phi)}$

Velocity ratio VR = $\dfrac{1}{\tan\theta} = \dfrac{\pi D}{p}$

Efficiency = $\dfrac{\text{MA}}{\text{VR}} = \dfrac{\tan\theta}{\tan(\theta + \phi)}$

Effective coefficient of friction (vee thread) $\mu_o = \mu \sec \beta$ where: β = half angle thread.

Vee thread

The vee thread is used extensively for nuts, bolts and screws. The thread may be produced by machining but rolling is much cheaper.

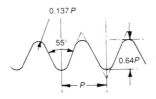

Whitworth thread

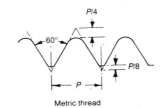

Metric thread

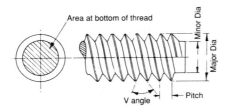

Square thread

Used for power transmission. The friction is low and there is no radial force on the nut.

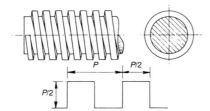

Acme thread

Used for power transmission. Has greater root strength and is easier to machine than the square thread. Used for lathe lead screw.

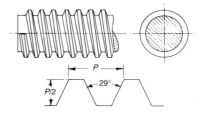

Buttress thread

A power screw with the advantages of both square and Acme threads. It has the greatest strength but takes a large load in one direction only (on the vertical face).

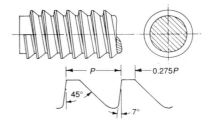

Multi-start thread

This gives a greater pitch with the same thread depth. The nut advance per revolution (lead) is equal to the pitch multiplied by the number of 'starts'.

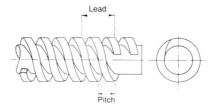

Ball-bearing power screw

The friction is extremely low and hence the efficiency is high. The power is transmitted by balls between the threads on nut and screw. The balls circulate continuously.

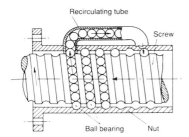

Recirculating tube
Screw
Ball bearing   Nut

2.4.7 *Coefficient of friction for screw threads*

This ranges from 0.12 to 0.20 with an average value of 0.15. It is however much lower for the ball-bearing thread.

2.5 Automobile mechanics

The resistance of a vehicle to motion is made up of 'rolling resistance', 'gradient force' and 'aerodynamic drag'. From the total resistance and a knowledge of the overall efficiency of the drive, the power can be calculated. Additional power is required to accelerate the vehicle. Braking torque is also dealt with.

2.5.1 *Rolling resistance*

Symbols used:
C_r = coefficient of rolling resistance
m = mass of vehicle
v = speed (km h^{-1})
p = tyre pressure (bars)

$$F_r = C_r mg$$

For pneumatic tyres on dry road

$$C_r = 0.005 + \frac{1}{p}\left[0.01 + 0.0095\left(\frac{v}{100}\right)^2\right]$$

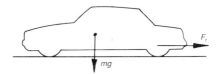

	C_r		C_r
Asphalt or concrete, new	0.01	gravel, rolled, new	0.02
Asphalt or concrete, worn	0.02	gravel, loose, worn	0.04
Cobbles, small, new	0.01	soil, medium hard	0.08
Cobbles, large, worn	0.03	sand	0.1–0.3

2.5.2 *Gradient force*

$$F_g = mg \sin \theta$$

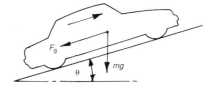

2.5.3 *Aerodynamic drag*

Symbols used:
C_d = drag coefficient
A_f = frontal area (approx. 0.9 bh m^2)
ρ = air density ($\simeq$ 1.2 kg m^{-3})
v = velocity (m s^{-1})
aerodynamic drag force:

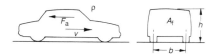

$$F_a = C_d A_f \rho \frac{v^2}{2}$$

Typical values of drag coefficient

	C_d		C_d
Sports car, sloping rear	0.2–0.3	Motorcycle and rider	1.8
Saloon, stepped rear	0.4–0.5	Flat plate normal to flow	1.2
Convertible, open top	0.6–0.7	Sphere	0.47
Bus	0.6–0.8	Long stream-lined body	0.1
Truck	0.8–1.0		

Total force $F_t = F_r + F_g + F_a$

2.5.4 *Tractive effort*

Symbols used:
μ_o = coefficient of adhesion
R_w = load on wheel considered

The horizontal force at which slipping occurs:

$$F_m = \mu_o R_w$$

Coefficient of adhesion for different surfaces

	μ_o		μ_o
Concrete/asphalt, dry	0.8–0.9	Clay, dry	0.5–0.6
Concrete/asphalt, wet	0.4–0.7	Sand, loose	0.3–0.4
Gravel, rolled, dry	0.6–0.7	Ice, dry	0.2
Gravel, rolled, wet	0.3–0.5	Ice, wet	0.1

2.5.5 *Power, torque and efficiency*

Let:
F_t = total resistance
v = velocity
η_0 = overall transmission efficiency
P_e = required engine power
T_e = engine torque

N_e = engine speed
N_w = wheel speed
r = wheel effective radius
F_w = wheel force (4 wheels)

Engine power $P_e = \dfrac{F_t v}{\eta_o}$

Engine torque $T_e = \dfrac{P_e}{2\pi N_e}$

Wheel force (for 4 wheels) $F_w = \dfrac{T_e \eta_o}{r} \dfrac{N_e}{N_w}$

Acceleration power $P_a = m a v_i$
where: $a =$ acceleration, $v_i =$ instantaneous speed.

Transmission efficiency:
Overall efficiency $\eta_o = \eta_c \eta_g \eta_d \eta_a$
Typical values are given in the table.

Clutch efficiency, η_c	0.99
Gearbox efficiency, η_g	0.98 direct drive
	0.95 low gears
Drive shaft, joints	
and bearings, η_d	0.99
Axle efficiency, η_a	0.95
Overall efficiency, η_o	0.90 direct drive
	0.85 low gears

2.5.6 Braking torque

Let:
$I =$ moment of inertia of a *pair* of wheels
$\alpha =$ angular deceleration of wheels
$m =$ mass of vehicle
$\mu =$ coefficient of friction between wheels and road

Front wheels torque $T_f = \mu r m g \dfrac{(b + \mu h)}{L}$

Rear wheels torque $T_r = \mu r m g \dfrac{(a - \mu h)}{L}$

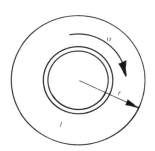

Wheel inertia torque $T_i = I \alpha$

Deceleration $d = \mu g$

Total braking torque (for one wheel):

$$T_{bf} = \dfrac{T_f}{2} + T_i \text{ (front)}$$

$$T_{br} = \dfrac{T_r}{2} + T_i \text{ (rear)}$$

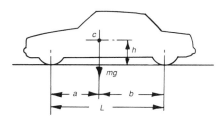

2.6 Vibrations

2.6.1 Simple harmonic motion

Let:
$x =$ displacement
$X =$ maximum displacement
$t =$ time
$f =$ frequency
$t_p =$ periodic time

$m =$ vibrating mass
$k =$ spring stiffness
$\phi =$ phase angle
$\theta =$ angle of rotation

Definition of simple harmonic motion

Referring to the figure, point A rotates with constant angular velocity ω at radius AB. The projection of A on to PQ, i.e. A', moves with simple harmonic motion. If A'B is plotted to a base of the angle of rotation θ, a so-called 'sine curve' is produced. The base of the graph can also represent time. The time for one complete rotation is the 'periodic time' t_p.

If $AB = X$ and $A'B = x$, then $x = X \sin \omega t$, where $\omega = \dfrac{\theta}{t}$.

Periodic time $t_p = \dfrac{2\pi}{\omega}$

Frequency $f = \dfrac{1}{t_p} = \dfrac{\omega}{2\pi}$

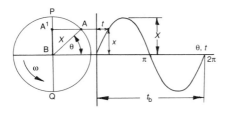

2.6.2 *Free undamped vibration*

Spring mass system

$x = X \cos(\omega_n + \phi)$,

where: $\omega_n = \sqrt{\dfrac{k}{m}}$

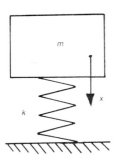

Frequency of vibration $f_n = \dfrac{\omega_n}{2\pi} = \dfrac{1}{2\pi}\sqrt{\dfrac{k}{m}} = \dfrac{1}{2\pi}\sqrt{\dfrac{g}{x_s}}$

where x_s = static deflection

Periodic time $t_p = \dfrac{1}{f_n}$

Torsional vibration

Displacement $\theta = \theta_{max} \cos(\omega_n + \phi)$

Frequency $f_n = \dfrac{\omega_n}{2\pi}$

$\omega_n = \sqrt{\dfrac{T_o}{I}}$. Where: T_o = torque per unit angle of twist, I = moment of inertia of oscillating mass.

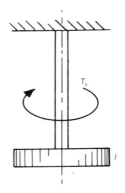

2.6.3 *Free damped vibration*

Critical frequency $\omega_c = \dfrac{c}{2m}$

where: c = damping force per unit velocity

Damping ratio $R = \dfrac{\omega_c}{\omega_n}$

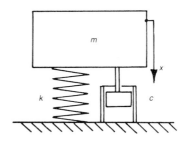

Light damping

Oscillations are produced which decrease in amplitude with time.

$$x = Ce^{-\omega_c t} \cos \omega_d t$$

where: $C = \text{constant}$, $\omega_d = \sqrt{\omega_n^2 - \omega_c^2}$

Periodic time $t_p = \dfrac{2\pi}{\omega_d}$

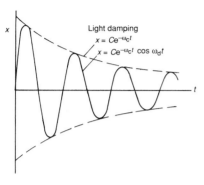

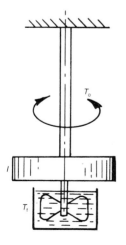

Amplitude ratio $\text{AR} = \dfrac{\text{Initial amplitude}}{\text{Amplitude after } n \text{ cycles}} = e^{n\omega_c t_p}$

AR is a measure of the rate at which the amplitude falls with successive oscillations.

Torsional vibration $\theta = Ce^{-\omega_c t} \cos \omega_d t$

where: $\omega_n = \sqrt{\dfrac{T_o}{I}}$; $\omega_d = \sqrt{\omega_n^2 - \omega_c^2}$, where $\omega_c = \dfrac{T_f}{2I}$;

$T_f = \text{damping torque per unit angular velocity}$.

Critical damping

In this case the damping is just sufficient to allow oscillations to occur: $\omega_c = \omega_n$.

$$x = Ce^{-\omega_c t}$$

where: $C = \text{constant}$.

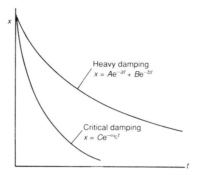

Heavy damping

The damping is heavier than critical and $\omega_c > \omega_n$.

$$x = Ae^{-at} + Be^{-bt}$$

where: A, B, a and b are constants.

2.6.4 *Forced damped vibration*

A simple harmonic force of constant amplitude applied to mass

Let the applied force be $F_a = F \cos \omega t$. When steady conditions are attained the mass will vibrate at the

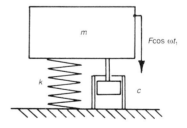

frequency of the applied force. The amplitude varies with frequency as follows:

Magnification factor $Q = \dfrac{\text{Actual amplitude of vibration}}{\text{Amplitude for a static force } F}$

and $\qquad Q = \dfrac{1}{\sqrt{(1-r^2)^2 + 4R^2 r^2}}$

where: $R = \dfrac{\omega_c}{\omega_n}$ and $r = \dfrac{\omega}{\omega_n}$

Phase angle $\alpha = \tan^{-1}\dfrac{2Rr}{(1-r^2)}$

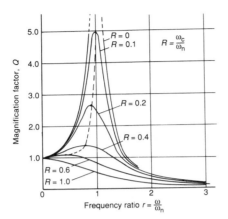

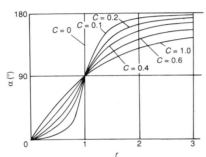

Simple harmonic force of constant amplitude applied to base

$F_a = F\cos\omega t$

$$Q = \sqrt{\dfrac{1+4R^2r^2}{(1-r^2)^2+4R^2r^2}}\,;\ \alpha = \tan^{-1}\dfrac{2Rr}{(1-r^2)}$$

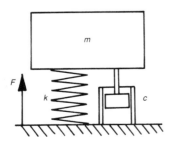

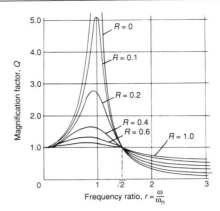

Simple harmonic force applied to mass due to rotary unbalance

$F = m_r a\omega^2 \cos\omega t$ (due to mass m_r rotating at radius a angular velocity ω)

$$Q = \dfrac{r^2}{\sqrt{(1-r^2)^2+4R^2r^2}}\,;\ \alpha = \tan^{-1}\dfrac{2Rr}{(1-r^2)}$$

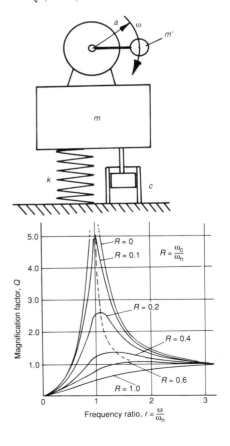

2.6.5 Three mass vibration system

Natural frequency $\omega_n = \sqrt{A \pm B}$ (two values)

$$A = \left(\frac{k_1}{m_1} + \frac{k_1}{m_2} + \frac{k_2}{m_2} + \frac{k_2}{m_3}\right)\bigg/ 2; \quad B = \sqrt{\left[A^2 - k_1 k_2\left(\frac{1}{m_1 m_2} + \frac{1}{m_2 m_3} + \frac{1}{m_1 m_3}\right)\right]}$$

If m_3 is infinite it is equivalent to a wall, hence:

$$A = \left(\frac{k_1}{m_1} + \frac{k_1}{m_2} + \frac{k_2}{m_2}\right)\bigg/ 2; \quad B = \sqrt{\left(A^2 - \frac{k_1 k_2}{m_1 m_2}\right)}$$

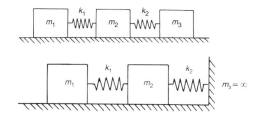

2.7 Friction

2.7.1 Friction laws

For clean dry surfaces the following laws apply approximately. The friction force is proportional to the perpendicular force between contacting surfaces and is independent of the surface area or rubbing speed. This only applies for low pressures and speeds. There are two values of friction coefficient, the 'static' value when motion is about to commence, and the 'dynamic' value, which is smaller, when there is motion.

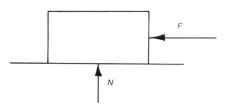

Coefficient of friction $\mu = \dfrac{F}{N}$

2.7.2 Friction on an inclined plane

Force parallel to plane:

$F = W(\mu \cos\theta + \sin\theta)$ (up plane)
$F = W(\mu \cos\theta - \sin\theta)$ (down plane)

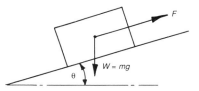

Angle of repose $\phi = \tan^{-1}\mu$; or when $\mu = \tan\phi$
If the angle of the plane is greater than the angle of repose, the body will slide down the plane.

Force horizontal

$F = W\tan(\theta + \phi)$ (up plane)
$F = W\tan(\theta - \phi)$ (down plane)

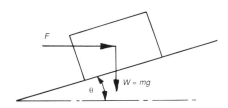

2.7.3 Rolling friction

The force to move a wheeled vehicle $F_r = \mu_r N$
where: μ_r = rolling coefficient of resistance, N = wheel reaction.

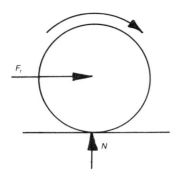

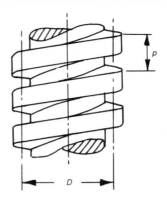

2.7.4 The wedge

Wedge angle $\alpha = \tan^{-1}\left(\dfrac{b}{2h}\right)$

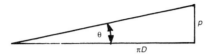

Force Q normal to wedge face $F = 2Q(\mu \cos \alpha + \sin \alpha)$

Force Q normal to force (F), $F = 2Q \tan(\alpha + \phi)$
where: $\mu = \tan \phi$.

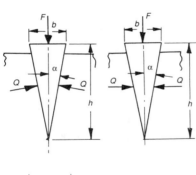

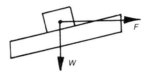

$\theta = \tan^{-1}\left(\dfrac{np}{\pi D}\right)$ (for n starts)

Torque to lower load $T_L = \dfrac{WD}{2}\tan(\theta - \phi)$

Torque to raise load $T_R = \dfrac{WD}{2}\tan(\theta + \phi)$

Efficiency $\eta = \dfrac{\tan \theta}{\tan(\theta + \phi)}$

Maximum efficiency $\eta_{max} = \dfrac{(1 - \sin \phi)}{(1 + \sin \phi)}\left(\text{at } \theta = \left(\dfrac{\pi}{4} - \dfrac{\phi}{2}\right)\right)$

Mechanical advantage $MA = \dfrac{W}{F} = \cot(\theta + \phi)$

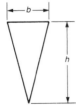

Velocity ratio $VR = \dfrac{\pi D}{p}$

2.7.5 Friction of screw thread

Square section thread

Thread angle $\theta = \tan^{-1}\dfrac{p}{\pi D}$ (for one start)

Vee thread

For a vee thread the 'effective coefficient of friction'
$\mu_e = \mu \sec \beta$

where: $\beta =$ half angle of thread.

Example   For $\beta = 30°$, $\mu_e = 1.155\,\mu$.

2.7.6   *Tables of friction coefficients*

The following tables give coefficients of friction for general combinations of materials, clutch and brake materials, machine tool slides and for rubber on asphalt and concrete.

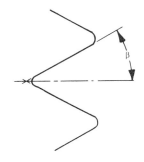

General materials

Materials	Lubrication	Coefficient of friction (low pressure)
Metal on metal	Dry	0.20 average
Bronze on bronze	Dry	0.20
Bronze on cast iron	Dry	0.21
Cast iron on cast iron	Slightly lubricated	0.15
Cast iron on hardwood	Dry	0.49
Cast iron on hardwood	Slightly lubricated	0.19
Metal on hardwood	Dry	0.60 average
Metal on hardwood	Slightly lubricated	0.20 average
Leather on metal	Dry	0.4 average
Rubber on metal	Dry	0.40
Rubber on road	Dry	0.90 average
Nylon on steel	Dry	0.3–0.5
Acrylic on steel	Dry	0.5
Teflon on steel	Dry	0.04
Metal on ice	—	0.02
Cermet on metal	Dry	0.4

Clutches and brakes

Materials	Coefficient of friction		Maximum temperature (0°C)	Maximum pressure (bar)
	Wet	Dry		
Cast iron/cast iron	0.05	0.15–0.2	150	8
Cast iron/steel	0.06	0.15–0.2	250	8–13
Hard steel/hard steel	0.05	—	250	7
Hard steel/chrome-plated hard steel	0.03	—	250	13
Hard drawn phosphor bronze/ hard drawn chrome plated steel	0.03	—	250	10
Powder metal/cast iron or steel	0.05–0.1	0.1–0.4	500	10

Clutches and brakes (*continued*)

Materials	Coefficient of friction		Maximum temperature (0°C)	Maximum pressure (bar)
	Wet	Dry		
Powder metal/chrome plated hard steel	0.05–0.1	0.1–0.3	500	20
Wood/cast iron or steel	0.16	0.2–0.35	150	6
Leather/cast iron or steel	0.12–0.15	0.3–0.5	100	2.5
Cork/cast iron or steel	0.15–0.25	0.3–0.5	100	1
Felt/cast iron or steel	0.18	0.22	140	0.6
Vulcanized paper or fibre/ cast iron or steel	—	0.3–0.5	100	3
Woven asbestos/cast iron or steel	0.1–0.2	0.3–0.6	250	7–14
Moulded asbestos/cast iron or steel	0.08–0.12	0.2–0.5	250	1
Impregnated asbestos/cast iron or steel	0.12	0.32	350	10
Asbestos in rubber/cast iron or steel	—	0.3–0.40	100	6
Carbon graphite/steel	0.05–0.1	0.25	500	20
Moulded phenolic plastic with cloth base/cast iron or steel	0.1–0.15	0.25	150	7

Band brake materials

Material	Lubrication	Coefficient of friction
Leather belt/wood	Well lubricated	0.47
Leather belt/cast iron	Well lubricated	0.12
Leather belt/cast iron	Slightly lubricated	0.28
Leather belt/cast iron	Very slightly lubricated	0.38
Steel band/cast iron	Dry	0.18

Machine tool slides

Materials	Pressure (bars)				
	0.5	1.0	1.5	2.0	4.0
Cast iron/cast iron	0.15	0.20	0.20	0.25	0.30
Cast iron/steel	0.15	0.20	0.25	0.30	0.35
Steel/steel	0.15	0.25	0.30	0.35	0.40

Rubber - sliding

Surface	Wet	Dry
Asphalt	0.25–0.75	0.50–0.80
Concrete	0.45–0.75	0.60–0.85

2.8 Brakes, clutches and dynamometers

2.8.1 *Band brake*

In the simple band brake a force is applied through a lever to a band wrapped part of the way around a drum. This produces tensions in the band and the difference between these multiplied by the drum radius gives the braking torque.

Let:
T = braking torque
P = braking power
F = applied force
p = maximum pressure on friction material
μ = coefficient of friction
N = speed of rotation
a = lever arm
b = belt width
θ = angle of lap of band
r = drum radius
c = distance from belt attachment to fulcrum

Power $P = 2\pi N T$
Torque $T = r(F_1 - F_2)$

$$F_1 = brp; \quad F_2 = F\frac{a}{c}; \quad \frac{F_1}{F_2} = e^{\mu\theta}$$

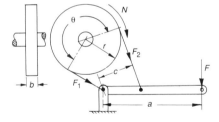

Differential band brake

In this case the dimensions can be chosen so that the brake is 'self-locking', i.e. no force is required, or it can operate in the opposite direction.

$$F = \frac{(F_2 c_2 - F_1 c_1)}{a} = F_2(c_2 - c_1 e^{\mu\theta})$$

If $c_1 e^{\mu\theta}$ is greater than c_2, the brake is self-locking.

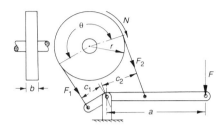

2.8.2 *Block brake*

The friction force is applied through a block made of, or lined with, a friction material. The brake can operate with either direction of rotation, but the friction torque is greater in one direction than the other. As in all friction brakes the limiting factor is the allowable pressure on the friction material.

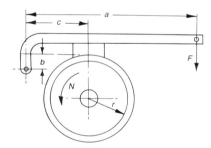

Friction torque $T = \dfrac{Far\mu}{c \pm \mu b}$

Pressure $p = \dfrac{Fa}{(c \pm \mu b)A}$

where: A = block contact area.

Use the positive sign for directions shown in the figure and the negative sign for opposite rotation (greater torque).

Double block brake, spring set

To achieve a greater friction torque, two blocks are used. This also results in zero transverse force on the drum. In this type of brake the force is provided by a spring which normally keeps the brake applied. Further compression is necessary to release the brake. This type of brake is used for lifts, for safety reasons.

Friction torque $T = Far\mu\left[\dfrac{1}{(c+\mu b)} + \dfrac{1}{(c-\mu b)}\right]$

Maximum pressure $p = \dfrac{Fa}{(c-\mu b)A}$

where: F = spring force.

The brake is released by a force greater than F.

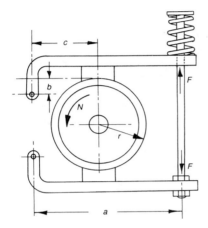

Block brake with long shoe

Here the friction force is applied around a large angle. The torque is increased by a factor K which is a function of the angle of contact. The shoe subtends an angle of 2θ and is pivoted at 'h' where

$h = Kr; \quad K = \dfrac{4\sin\theta}{(2\theta + \sin 2\theta)}$

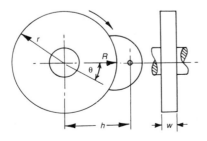

Friction torque $T = K\dfrac{Far}{(c \pm \mu b)}$

Average pressure $p_a = \dfrac{T}{2\mu wr^2 \sin\theta}$

Maximum pressure $p_m = Kp_a$

2.8.3 *Internally expanding shoe brake*

This type is used on vehicles and has two shoes, lined with friction material, which make contact with the inside surface of a hollow drum. For rotation as shown in the figure:

Torque for left-hand shoe $T_L = \dfrac{K\mu Far}{(b + K\mu c)}$

Torque for right-hand shoe $T_R = \dfrac{K\mu Far}{(b - K\mu c)}$

Total torque $T = T_L + T_R$

with K as previously.

Maximum pressure $p_m = \dfrac{T_R}{2\mu wr^2 \sin\theta}$

Average pressure $p_a = \dfrac{p_m}{K}$

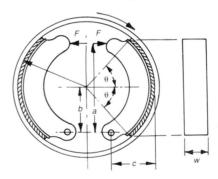

2.8.4 *Disk brake*

Let:
F = force on pad
r = mean radius of pad
A = pad area

Torque capacity (2 pads) $T = 2\mu Fr$

Pad pressure $p = \dfrac{F}{A}$

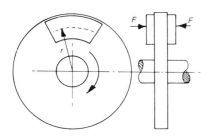

2.8.5 Disk clutch

The simplest type of clutch is the single-plate clutch in which an annular plate with a surface of friction material is forced against a metal disk by means of a spring, or springs, or by other means. There are two theories which give slightly different values of torque capacity.

Uniform-wear theory

Let:
F = spring force
r_o = outer radius of friction material
r_i = inner radius of friction material

Maximum torque capacity $T = F\mu \dfrac{(r_o + r_i)}{2}$

Maximum pressure $p_m = \dfrac{F}{2\pi r_i (r_o - r_i)}$

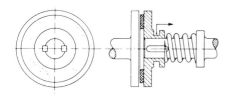

Uniform-pressure theory

$$T = \tfrac{2}{3} F\mu \frac{(r_o^3 - r_i^3)}{(r_o^2 - r_i^2)}$$

$$p = \frac{F}{\pi(r_o^2 - r_i^2)}$$

2.8.6 Cone clutch

By angling the contacting surfaces, the torque capacity is increased; for example, for an angle of 9.6° the capacity is increased by a factor of 6.

θ = cone angle (to the shaft axis, from 8° upwards).

The theory is the same as for the disk clutch but with an effective coefficient of friction

$$\mu' = \frac{\mu}{\sin\theta}$$

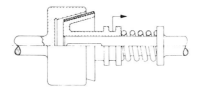

2.8.7 Multi-plate disk clutch

A number of double-sided friction plates may be mounted on splines on one element, and corresponding steel contacting plates on splines on the other element. The assembly is compressed by a spring or springs to give a torque capacity proportional to the number of pairs of contacting surfaces.

Torque capacity $T = n \times$ torque for one plate

where: n = number of pairs of surfaces (6 in the example shown in the figure).

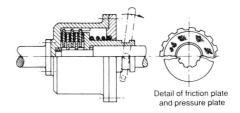

Detail of friction plate
and pressure plate

2.8.8 Centrifugal clutch

Internally expanding friction shoes are held in contact, by the force due to rotation against the force of a light spring. The torque capacity increases as the speed increases.

Let:
m = mass of shoe
k = spring stiffness
x = deflection of spring
μ = coefficient of friction
F = radial force on drum
N = rotational speed
ω = angular velocity

Torque capacity (2 shoes) $T = 2\mu r(mr\omega^2 - kx)$

where: $\omega = 2\pi N$.

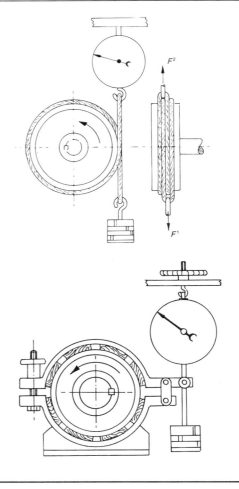

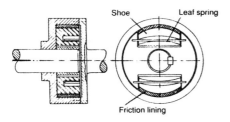

2.8.9 Dynamometers

The power output of a rotary machine may be measured by means of a friction brake. The forces are measured by spring balances or load cells. Other types of dynamometer include fluid brakes and electric generators.

Torque absorbed $T = r(F_1 - F_2)$
Power $P = 2\pi NT$

2.9 Bearings

The full analysis of heavily loaded plain bearings is extremely complex. For so called 'lightly-loaded bearings' the calculation of power loss is simple for both journal and thrust bearings.

Important factors are, load capacity, length to diameter ratio, and allowable pressure on bearing material.

Information is also given on rolling bearings.

2.9.1 Lightly loaded plain bearings

Let:
P = power
L = length
D = diameter
μ = absolute viscosity

t = radial clearance
r_1 = inner radius
r_2 = outer radius
N = rotational speed

Journal bearing:

$$P = \frac{2\pi^3 N^2 D^3 L \mu}{t}$$

Thrust bearing:

$$P = 2\pi^3 N^2 \mu \frac{(r_2^4 - r_1^4)}{t}$$

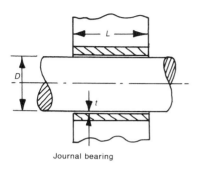

Journal bearing

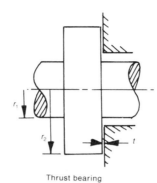

Thrust bearing

2.9.2 *Load capacity for plain bearings*

Machine and bearing	Load capacity, p (MPa)	Length/diameter, L/D
Automobile and aircraft engine main bearings	4–12	0.5–1.75
Automobile and aircraft engine crankpin bearings	4–23	0.5–1.50
Marine steam turbine main bearings	1.5–4	1.0–1.5
Marine steam turbine crankpin bearings	2–4	1.0–1.5
Land steam turbine main bearings	0.5–4	1.0–2.0
Generators and motors	0.3–1.0	1.0–2.5
Machine tools	0.4–2.0	1.5–4.0
Hoisting machinery	0.5–0.7	1.5–2.0
Centrifugal pumps	0.5–0.7	1.0–2.0
Railway axle bearings	2–2.5	1.5–2.0

Load capacity $p = \dfrac{\text{Bearing load}}{\text{Projected area}} = \dfrac{W}{LD}$

This assumes a uniform pressure; actually the maximum pressure is considerably higher.

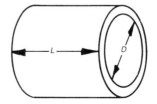

2.9.3 Bearing materials

Metals

Material	Brinell hardness	Thin shaft hardness	Load capacity, p (MPa)	Maximum temperature (°C)
Tin base babbitt	20–30	⩽150	5.5–10.3	150
Lead base babbitt	15–20	⩽150	5.5–8.0	150
Alkali-hardened lead	22–26	200–250	8.0–10.3	260
Cadmium base	30–40	200–250	10.3–15	260
Copper lead	20–30	200	10.3–16.5	175
Tin bronze	60–80	300–400	⩾30	260
Lead bronze	40–70	300	20–30	225
Aluminium alloy	45–50	200–300	⩾30	125
Silver plus overlay	25	300–400	⩾30	260

Porous metals and non-metals

Materials	Load capacity, p (MPa)	Maximum temperature (°C)	Maximum velocity, v (m s^{-1})	Maximum pv (MPa $\times$ m s^{-1})
Porous metals	30	75	7.5	0.7
Rubber	0.35	75	5.0	0.525
Graphite materials	4	350	12.5	5.25 wet, 0.525 dry
Phenolics	35	95	12.5	0.525
Nylon	7	95	2.5	0.875
Teflon	3.5	265	1.2	0.35

2.9.4 Surface finish and clearance for bearings

Type of service	Surface — Journal	Surface — Bearing	Diametral clearance (mm)
Precision spindles $ND < 50 \times 10^3$	Hardened ground steel	Lapped	$0.0175D + 0.0075$
Precision spindles $ND > 50 \times 10^3$	Hardened ground steel	Lapped	$0.02D + 0.01$
Electric motors, generators, etc.	Ground	Broached or reamed	$0.02D + 0.015$
General machinery, continuous running	Turned	Bored or reamed	$0.025D + 0.025$
Rough service machinery	Turned	1.5–3 μm	$0.075D + 0.1$

N = revolutions per minute, D = diameter (mm).

2.9.5 *Rolling bearings*

The term 'rolling bearing' refers to both ball and roller bearings. Ball bearings of the journal type are used for transverse loads but will take a considerable axial load. They may also be used for thrust bearings. Rollers are used for journal bearings but will not take axial load. Taper roller bearings will take axial thrust as well as transverse load.

Advantages of rolling bearings

(1) Coefficient of friction is low compared with plain bearings especially at low speeds. This results in lower power loss.
(2) Wear is negligible if lubrication is correct.
(3) They are much shorter than plain bearings and take up less axial space.

(4) Because of extremely small clearance they permit more accurate location; important for gears for example.
(5) Self-aligning types permit angular deflection of the shaft and misalignment.

Disadvantages of rolling bearings

(1) The outside diameter is large.
(2) The noise is greater than for plain bearings, especially at high speeds.
(3) There is greater need of cleanliness when fitted to achieve correct life.
(4) They cannot always be fitted, e.g. on crankshafts.
(5) They are more expensive for small quantities but relatively cheap when produced in large quantities.
(6) Failure may be catastrophic.

2.9.6 *Types of rolling bearings*

The following table lists the most common types of rolling bearings.

Ball journal	Used for radial load but will take one third load axially. Deep grooved type now used extensively. Light, medium and heavy duty types available.	Light Medium Heavy
Angular contact ball journal	Takes a larger axial load in one direction. Must be used in pairs if load in either direction	
Ball thrust	For axial loads only. Must have at least a minimum thrust	
Self-aligning ball, single row	The outer race has a spherical surface mounted in a ring which allows for a few degrees of shaft misalignment	

Self-aligning ball, double row	Two rows of balls in staggered arrangement. Outer race with spherical surface	
Double row ball journal	Used for larger loads without increase in outer diameter	
Roller journal	For high radial loads but no axial load. Allows axial sliding	
Self-aligning spherical roller	Barrel shaped rollers. High capacity. Self-aligning	
Taper roller	Takes radial and axial loads. Used in pairs for thrust in either direction	
Needle rollers	These run directly on the shaft with or without cages. Occupy small space	
Shields, seals and grooves	Shields on one or both sides prevent ingress of dirt. Seals allow packing with grease for life. A groove allows fitting of a circlip for location in bore.	Shields Shields and seals Circlip groove

2.9.7 *Service factor for rolling bearings*

The bearing load should be multiplied by the following
factor when selecting a bearing.

Type of load	Even	Uneven light shock	Moderate shock	Heavy shock	Very heavy shock
Service factor	1.0	1.2–1.5	1.7–2.0	2.2–2.5	2.7–3.0

2.9.8 *Coefficient of friction for bearings*

Plain bearings – boundary lubrication

	μ
Mixed film (boundary plus hydrodynamic)	0.02–0.08
Thin film	0.08–0.14
Dry (metal to metal)	0.20–0.40

Rolling bearings

	μ
Self-aligning ball	0.0016–0.0066
Rollers	0.0012–0.0060
Thrust ball	0.0013–0.0060
Deep groove ball	0.0015–0.0050
Taper roller	0.0025–0.0083
Spherical roller	0.0029–0.0071
Angular contact	0.0018–0.0019

Plain journal bearings — oil bath lubrication

Lubricant	Velocity $(m\,s^{-1})$	μ Pressure 7 bar	Pressure 30 bar
Mineral grease	1.0	0.0076	0.00016
Mineral grease	2.5	0.0151	0.0027
Mineral oil	1.0	0.0040	0.0012
Mineral oil	2.5	0.007	0.0020

2.10 Gears

Gears are toothed wheels which transmit motion and
power between rotating shafts by means of success-
ively engaging teeth. They give a constant velocity
ratio and different types are available to suit different
relative positions of the axes of the shafts (see table).
Most teeth are of the 'involute' type. The nomen-
clature for spur gears is given in the figures.

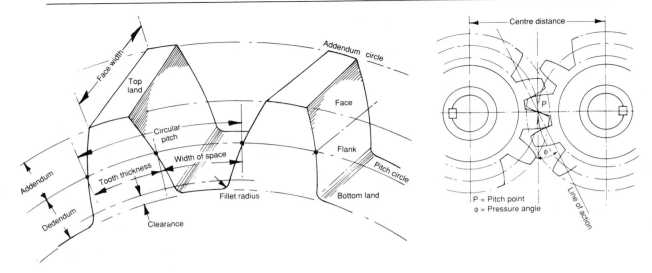

2.10.1 *Classification of gears*

Type of gear	Relation of axes	Pitch surfaces	Elements of teeth
Spur	Parallel	Cylinder	Straight, parallel to axis
Parallel helical	Parallel	Cylinder	Helical
Herringbone	Parallel	Cylinder	Double helical
Straight bevel	Intersecting	Cone	Straight
Spiral bevel	Intersecting	Cone	Spiral
Crossed helical	Crossed but not intersecting	Cylinder	Helical
Worm	Right angle but not intersecting	Cylinder	Helical

2.10.2 *Metric gear teeth*

Metric module $m = \dfrac{D}{T}$ (in millimetres)

where: D = pitch circle diameter, T = number of teeth.
The preferred values of module are: 1, 1.25, 1.5, 2, 2.5, 3, 4, 5, 6, 8, 10, 12, 16, 20, 25, 32, 40 and 50.

Circular pitch $p = \dfrac{\pi D}{T} = \pi m$

Addendum $= m$

Dedendum $= 1.25m$

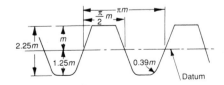

Height of tooth $= 2.25m$

The figure shows the metric tooth form for a 'rack' (i.e. a gear with infinite diameter).

Design of gears

The design of gears is complex and it is recommended
  that British Standards (or other similar sources) be
  consulted.
  See BS 436 for the design of gears and BS 1949 for
  permissible stresses.

2.10.3 *Spur gears*

Symbols used:
 F = tooth force
 F_t = tangential component of tooth force
 F_s = separating component of tooth force
 ϕ = pressure angle of teeth
D_1 = pitch circle diameter of driver gear
D_2 = pitch circle diameter of driven gear
N_1 = speed of driver gear
N_2 = speed of driven gear
n_1 = number of teeth in driver gear
n_2 = number of teeth in driven gear
 P = power
 T = torque
 η = efficiency

Tangential force on gears $F_t = F \cos \phi$
Separating force on gears $F_s = F_t \tan \phi$

Torque on driver gear $T_1 = \dfrac{F_t D_1}{2}$

Torque on driven gear $T_2 = \dfrac{F_t D_2}{2}$

Speed ratio $\dfrac{N_1}{N_2} = \dfrac{D_2}{D_1} = \dfrac{n_2}{n_1}$

Input power $P_i = 2\pi N_1 F_t \dfrac{D_1}{2}$

Output power $P_o = 2\pi N_2 F_t \dfrac{D_2}{2} \eta$

Efficiency $\eta = \dfrac{P_o}{P_i}$

Rack and pinion drive

For a pinion, pitch circle diameter D speed N and
torque T:

Rack velocity $V = \pi D N$

Force on rack $F = \dfrac{2T}{D}$

Rack power $P = F V \eta = 2\pi N T \eta$

where: η = efficiency.

2.10.4 *Helical spur gears*

In this case there is an additional component of force
F_a in the axial direction.

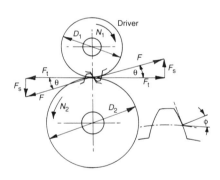

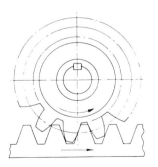

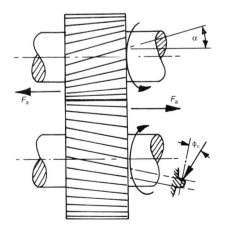

Let:
ϕ_n = pressure angle normal to the tooth
α = helix angle

Separating force $F_s = F_t \dfrac{\tan \phi_n}{\cos \alpha}$

Axial force $F_a = F_t \tan \alpha$

Double helical gears

To eliminate the axial thrust, gears have two sections with helices of opposite hand. These are also called 'herringbone gears'.

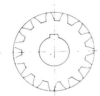

Single helical gear

Double helical gear

2.10.5 Bevel gears

Straight bevel gears

Let:
ϕ = pressure angle of teeth
β = pinion pitch cone angle

Tangential force on gears = F_t
Separating force $F_s = F_t \tan \phi$
Pinion thrust $F_p = F_s \sin \beta$
Gear thrust $F_g = F_s \cos \beta$

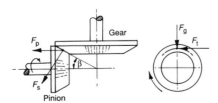

Spiral bevel gear

Let:
α = spiral angle of pinion
ϕ_n = normal pressure angle

Force on pinion $F_p = F_t \left[\dfrac{\tan \phi_n \sin \beta}{\cos \alpha} \pm \tan \alpha \cos \beta \right]$

Force on gear $F_g = F_t \left[\dfrac{\tan \phi_n \cos \beta}{\cos \alpha} \mp \tan \alpha \sin \beta \right]$

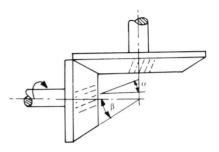

For the diagram shown the signs are '+' for F_p and '−' for F_g. The signs are reversed if the hand of the helix is reversed or the speed is reversed; they remain the same if both are reversed.

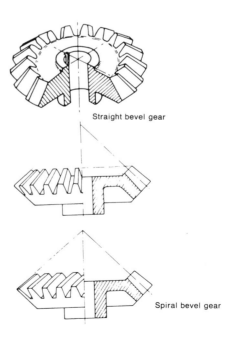

Straight bevel gear

Spiral bevel gear

2.10.6 *Worm gears*

The worm gear is basically a screw (the worm) engaging with a nut (the gear). The gear is, in effect, a partial nut whose length is wrapped around in a circle.

Let:

ϕ_n = normal pressure angle
α = worm helix angle
n_w = number of threads or starts on worm
n_g = number of teeth in gear
D_w = worm pitch circle diameter
D_g = gear pitch circle diameter
L = lead of worm
p = pitch of worm threads and gear teeth
μ = coefficient of friction
η = efficiency
T_w = worm torque
v = velocity of gear teeth
N_w = speed of worm
N_g = speed of gear

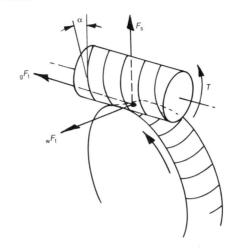

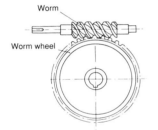

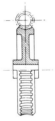

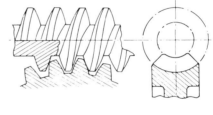

Tangential force on worm $_wF_t$ = axial force on gear $_gF_a = \dfrac{2T_w}{D_w}$

Tangential force on gear $_gF_t$ = axial force on worm $= {}_wF_t\left(\dfrac{\cos\phi_n - \mu\tan\alpha}{\cos\phi_n \tan\alpha + \mu}\right)$

Separating force on each component $F_s = {}_wF_t\left(\dfrac{\sin\phi_n}{\cos\phi_n \sin\alpha + \mu\cos\alpha}\right)$

$\tan\alpha = \dfrac{L}{\pi D_w};\ L = pn_w;\ D_g = pn_g/\pi$

Efficiency $\eta = \left(\dfrac{\cos\phi_n - \mu\tan\alpha}{\cos\phi_n + \mu\cot\alpha}\right)$

Input power $P_w = 2\pi N_w T_w$

Gear tooth velocity $v = \pi D_g N_g$

Coefficient of friction for worm gears

	Velocity (m s^{-1})					
	0.5	1.0	2.0	5.0	10.0	20.0
Hard steel worm/phosphor bronze wheel	0.06	0.05	0.035	0.023	0.017	0.014
Cast iron worm/cast iron wheel	0.08	0.067	0.050	0.037	0.022	0.018

2.10.7 *Epicyclic gears*

The main advantage of an epicyclic gear train is that the input and output shafts are coaxial. The basic type consists of a 'sun gear' several 'planet gears' and a 'ring gear' which has internal teeth. Various ratios can be obtained, depending on which member is held stationary.

Ratio of output to input speed for various types

Let:
$N=$ speed
$n=$ number of teeth
Note that a negative result indicates rotation reversal.

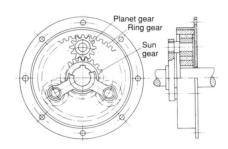

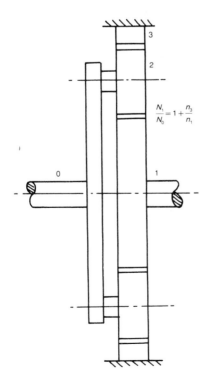

$$\frac{N_1}{N_0}=1+\frac{n_3}{n_1}$$

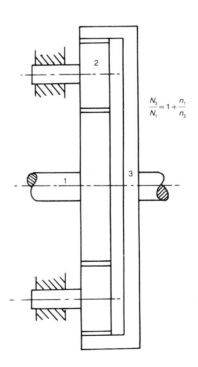

$$\frac{N_3}{N_1}=1+\frac{n_1}{n_3}$$

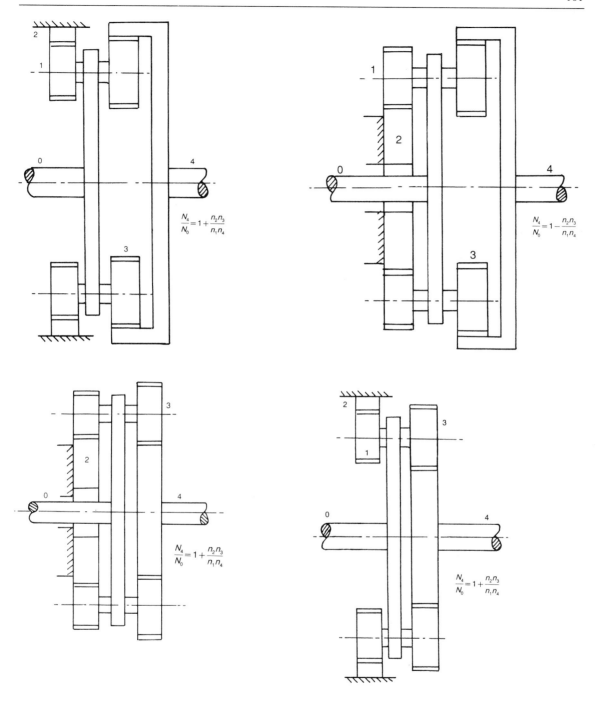

$$\frac{N_4}{N_0} = 1 + \frac{n_2 n_3}{n_1 n_4}$$

$$\frac{N_4}{N_0} = 1 - \frac{n_2 n_3}{n_1 n_4}$$

$$\frac{N_4}{N_0} = 1 + \frac{n_2 n_3}{n_1 n_4}$$

$$\frac{N_4}{N_0} = 1 + \frac{n_2 n_3}{n_1 n_4}$$

3 Thermodynamics and heat transfer

3.1 Heat

3.1.1 Heat capacity

Heat capacity is the amount of heat required to raise the temperature of a body or quantity of substance by 1 K. The symbol is C (units joules per kelvin, $J\,K^{-1}$)
Heat supplied $Q = C(t_2 - t_1)$
where: t_1 and t_2 are the initial and final temperatures.

3.1.2 Specific heat capacity

This is the heat to raise 1 kg of substance by 1 K. The symbol is c (units joules per kilogram per kelvin, $J\,kg^{-1}\,K^{-1}$).
$Q = mc(t_2 - t_1)$
where: m = mass.

3.1.3 Latent heat

This is the quantity of heat required to change the state of 1 kg of substance. For example:
Solid to liquid: specific heat of melting; h_{sf} ($J\,kg^{-1}$)
Liquid to gas: specific heat of evaporation, h_{fg} ($J\,kg^{-1}$)

3.1.4 Mixing of fluids

If m_1 kg of fluid 1 at temperature t_1 is mixed with m_2 kg of fluid 2 at temperature t_2, then

Final mass $m = m_1 + m_2$ at a temperature

$$t = \frac{m_1 c_1 t_1 + m_2 c_2 t_2}{m_1 c_1 + m_2 c_2}$$

3.2 Perfect gases

3.2 Gas laws

For a so-called 'perfect gas':

Boyle's law: pv = constant for a constant temperature T

Charles' law: $\dfrac{V}{T}$ = constant for a constant pressure p

where: p = pressure, V = volume, T = absolute temperature.

Combining the two laws:

$\dfrac{pV}{T}$ = constant = mR

where: m = mass, R = the gas constant

specific volume $v = \dfrac{V}{m}$ $(m^3 kg^{-1})$

so that: $pv = RT$

3.2.2 Universal gas constant

If R is multiplied by M the molecular weight of the gas, then:
Universal gas constant $R_o = MR = 8.3143$ $kJ\,kg^{-1}\,K^{-1}$ (for all perfect gases)

3.2.3 *Specific heat relationships*

There are two particular values of specific heat: that at constant volume c_v, and that at constant pressure c_p.

Ratio of specific heats $\gamma = \dfrac{c_p}{c_v}$

Also $(c_p - c_v) = R$, so that $c_v = \dfrac{R}{(\gamma - 1)}$

3.2.4 *Internal energy*

This is the energy of a gas by virtue of its temperature.
$u = c_v T$ (specific internal energy)
$U = mc_v T$ (total internal energy)
Change in internal energy:
$U_2 - U_1 = mc_v(T_2 - T_1)$
$u_2 - u_1 = c_v(T_2 - T_1)$

3.2.5 *Enthalpy*

Enthalpy is the sum of internal energy and pressure energy pV, i.e.

$h = u + pv$, or $H = U + pV$

where: h = specific enthalpy, H = total enthalpy

and it can be shown that

$h = c_p T$.

Change in enthalpy $h_2 - h_1 = (u_2 - u_1) + p(v_2 - v_1) = c_p(T_2 - T_1)$

$H_2 - H_1 = mc_p(T_2 - T_1)$

3.2.6 *Energy equations*

Non-flow energy equation

Gain in internal energy = Heat supplied − Work done

$u_2 - u_1 = Q - W$

where: $W = \displaystyle\int_1^2 p\,dv$.

Steady flow energy equation

This includes kinetic energy and enthalpy:

$h_2 - h_1 = Q - W - \left(\dfrac{C_2^2 - C_1^2}{2}\right)$

or, if the kinetic energy is small (which is usually the case)

$h_2 - h_1 = Q - W$ (neglecting height differences)

3.2.7 *Entropy*

Entropy, when plotted versus temperature, gives a curve under which the area is heat. The symbol for entropy is s and the units are kilojoules per kilogram per kelvin ($kJ\,kg^{-1}\,K^{-1}$).

$$s_2 - s_1 = \int_1^2 \frac{dQ}{T} \text{ or } Q = \int_1^2 T\,ds$$

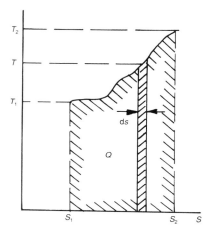

3.2.8 *Exergy and anergy*

In a heat engine process from state 1 with surroundings at state 2 exergy is that part of the total enthalpy drop available for work production.

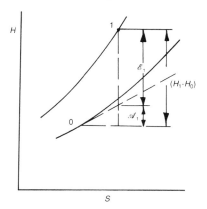

Exergy $\mathscr{E}_1 = (H_1 - H_o) - T_o(S_1 - S_o)$

That part of the total enthalpy not available is called the 'anergy'.

Anergy $\mathscr{A}_1 = T_o(S_1 - S_o)$

3.2.9 Reversible non-flow processes

Constant volume

In this case:

$$\frac{p}{T} = \text{constant}$$

$W = 0.$
$Q = c_v(T_2 - T_1)$

$$(s_2 - s_1) = c_v \ln\left(\frac{T_2}{T_1}\right)$$

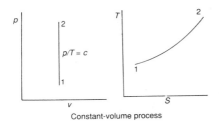

Constant-volume process

Constant pressure

In this case:

$$\frac{v}{T} = \text{constant}$$

$W = p(v_2 - v_1)$
$Q = c_p(T_2 - T_1) = (h_2 - h_1)$

$$(s_2 - s_1) = c_p \ln\left(\frac{T_2}{T_1}\right)$$

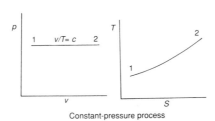

Constant-pressure process

Constant temperature (isothermal)

In this case:

$pv = \text{constant}$

$$W = Q = RT\ln\left(\frac{p_2}{p_1}\right) = RT\ln\left(\frac{v_1}{v_2}\right)$$

$$(s_2 - s_1) = R\ln\left(\frac{p_2}{p_1}\right) = R\ln\left(\frac{v_1}{v_2}\right)$$

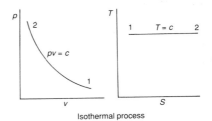

Isothermal process

Constant entropy (isentropic)

In this case:

$$pv^\gamma = \text{constant}, \text{ where } \gamma = \frac{c_p}{c_v}$$

$$W = \frac{p_1 v_1 - p_2 v_2}{\gamma - 1}$$

$Q = 0$
$(s_2 - s_1) = 0$

Also:

$$\frac{p_1}{p_2} = \left(\frac{T_1}{T_2}\right)^{\frac{\gamma}{\gamma-1}}; \quad \frac{v_1}{v_2} = \left(\frac{T_2}{T_1}\right)^{\frac{1}{\gamma-1}}$$

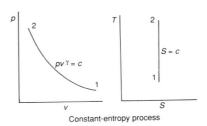

Constant-entropy process

Polytropic process

In this case:
$pv^n = \text{constant}$, where $n = $ any index

$$p_1/p_2 = \left(\frac{T_1}{T_2}\right)^{\frac{n}{n-1}}$$

$$v_1/v_2 = \left(\frac{T_2}{T_1}\right)^{\frac{1}{n-1}}$$

$$W = \frac{p_1 v_1 - p_2 v_2}{n-1}$$

$$Q = W\left(\frac{\gamma - 1}{n - 1}\right)$$

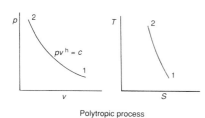

Polytropic process

3.2.10 Irreversible processes

Throttling (constant enthalpy process)

$h_1 = h_2$,
For perfect gas $T_1 = T_2$

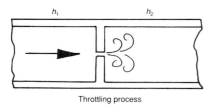

Throttling process

Adiabatic mixing

When two flows of a gas $\dot{m}_1$ and $\dot{m}_2$ at temperatures T_1 and T_2 mix:

Final temperature $T_3 = \dfrac{\dot{m}T_1 + \dot{m}_2 T_2}{\dot{m}_1 + \dot{m}_2}$

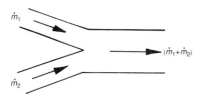

3.2.11 Mixtures of gases

The thermodynamic properties of a mixture of gases can be determined in the same way as for a single gas, the most common example being air for which the properties are well known. Using Dalton's law of partial pressures as a basis, the properties of mixtures can be found as follows.

Symbols used:
m = total mass of mixture
m_A, m_B, etc. = masses of constituent gases
p = pressure of mixture
P_A, P_B, etc. = pressures of constituents
R_A, R_B, etc. = gas constants of constituents
T = temperature of mixture
V = volume of mixture

Dalton's law:
$p = p_A + p_B + p_c + \ldots + p_i$
$m = m_A + m_B + m_C + \ldots + m_i$
where: $p_i = m_i R_i (T/V)$

Apparent gas constant $R = \dfrac{\Sigma (m_i R_i)}{m}$

Apparent molecular weight $M = R_o/R$
where: R_o = universal gas constant.

Internal energy $u = \dfrac{\Sigma (m_i u_i)}{m}$

Enthalpy $h = \dfrac{\Sigma (m_i h_i)}{m}$

Entropy $s = \dfrac{\Sigma (m_i s_i)}{m}$

Specific heats:

$c_p = \dfrac{\Sigma (m_i c_{pi})}{m}$

$c_v = \dfrac{\Sigma (m_i c_{vi})}{m}$

$\gamma = c_p/c_v$

3.3 Vapours

A substance may exist as a solid, liquid, vapour or gas. A mixture of liquid (usually in the form of very small drops) and dry vapour is known as a 'wet vapour'. When all the liquid has just been converted to vapour the substance is referred to as 'saturated vapour' or 'dry saturated vapour'. Further heating produces what is known as 'superheated vapour' and the temperature rise (at constant pressure) required to do this is known as the 'degree of superheat'. The method of determining the properties of vapours is given, and is to be used in conjunction with vapour tables, the most comprehensive of which are for water vapour. Processes are shown on the temperature–entropy and enthalpy–entropy diagrams.

Symbols used:

> p = pressure ($\mathrm{N\,m^{-2}}$ ($\equiv$ pascal); $\mathrm{N\,mm^{-2}}$; bar
> ($\equiv 10^5\,\mathrm{N\,m^{-2}}$); millibar ($\equiv 100\,\mathrm{N\,m^{-2}}$))
> t = temperature (°C)
> t_s = saturation temperature (°C)
> T = absolute temperature (K $\simeq$ °C + 273)
> v = specific volume ($\mathrm{m^3\,kg^{-1}}$)
> v_f = specific volume of liquid ($\mathrm{m^3\,kg^{-1}}$)
> v_g = specific volume of saturated vapour ($\mathrm{m^3\,kg^{-1}}$)
> u = specific internal energy ($\mathrm{kJ\,kg^{-1}}$)
> u_f = specific internal energy of liquid ($\mathrm{kJ\,kg^{-1}}$)
> u_g = specific internal energy of vapour ($\mathrm{kJ\,kg^{-1}}$)
> u_{fg} = specific internal energy change from liquid to vapour ($\mathrm{kJ\,kg^{-1}}$)
> h = specific enthalpy ($\mathrm{kJ\,kg^{-1}}$)
> h_f = specific enthalpy of liquid, kJ/kg
> h_g = specific enthalpy of vapour, kJ/kg
> h_{fg} = specific enthalpy change from liquid to vapour (latent heat) kJ/kg
> s = specific entropy, kJ/kg K
> s_f = specific entropy of liquid, kJ/kg K
> s_g = specific entropy of vapour, kJ/kg K
> s_{fg} = specific entropy change from liquid to vapour, kJ/kg K
> x = dryness fraction

3.3.1 Properties of vapours

Dryness fraction $x = \dfrac{\text{Mass of dry vapour}}{\text{Mass of wet vapour}}$

Specific volume of wet vapour $v_x = v_f(1-x) + xv_g \simeq xv_g$
(since v_f is small)

Specific internal energy of wet vapour
$u_x = u_f + x(u_g - u_f) = u_f + xu_{fg}$

Specific enthalpy of wet vapour
$h_x = h_f + x(h_g - h_f) = h_f + xh_{fg}$

specific entropy of wet vapour
$s_x = s_f + x(s_g - s_f) = s_f + xs_{fg}$

Superheated vapour Tables (e.g. for water) give values of v, u, h, and s for a particular pressure and a range of temperatures above the saturation temperature t_s. For steam above 70 bar use $u = h - pv$.

3.3.2 Temperature–Entropy diagram (T–s diagram)

Various processes are shown for a vapour on the T–s diagram. AB is an isothermal process in which a wet vapour becomes superheated. CD shows an isentropic expansion from the superheat to the wet region. EF is a polytropic process in the superheat region.

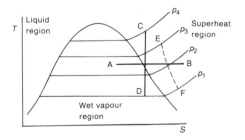

3.3.3 Enthalpy of a vapour

The enthalpy is represented by the area under a constant pressure line on the T–s diagram. Area h_f is the enthalpy of the liquid at saturation temperature, h_{fg} is the enthalpy corresponding to the latent heat,

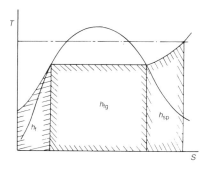

and h_{sp} is the superheat. The total enthalpy is, therefore,

$$h = h_f + h_{fg} + h_{sp}$$

3.3.4 Dryness fraction

The dryness fraction at entropy s is

$$x = \left(\frac{s - s_f}{s_{fg}} \right) \text{ and } h = h_f + xh_{fg}$$

The area xh_{fg} is shown.

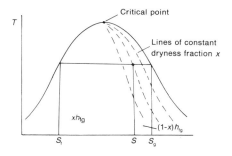

3.3.5 Enthalpy–entropy (h–s) diagram

Lines of constant pressure, temperature, dryness fraction and specific volume are shown on the diagram. AB represents an isentropic process, AC a polytropic process and DE a constant enthalpy process.

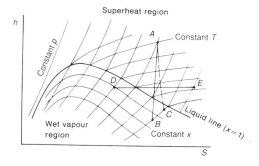

3.4 Data tables

3.4.1 Temperature conversion

Conversion formulae:

$$°C = \frac{°F - 32}{1.8}$$

$$°F = (°C \times 1.8) + 32$$

°C	°F	°C	°F	°C	°F
0	32	160	320	350	662
10	50	170	338	400	752
20	68	180	356	450	842
30	86	190	374	500	932
40	104	200	392	550	1022
50	122	210	410	600	1112
60	140	220	428	650	1202
70	158	230	446	700	1292
80	176	240	464	750	1382
90	194	250	482	800	1472
100	212	260	500	850	1562
110	230	270	518	900	1652
120	248	280	536	950	1742
130	266	290	554	1000	1832
140	284	300	572	1050	1922
150	302	310	590	1100	2012

3.4.2 Latent heats and boiling points

Latent heat of evaporation (kJ kg^{-1}) at atmospheric pressure

Liquid	h_{fg}	Liquid	h_{fg}	Liquid	h_{fg}
Ammonia	1230	Ethanol (ethyl alcohol)	863	Sulphur dioxide	381
		Ether	379		
Bisulphide of carbon	372	Methanol (methyl or wood alcohol)	1119	Turpentine	309
				Water	2248

Latent heat of fusion (kJ kg^{-1}) at atmospheric pressure

Substance	h_{sf}	Substance	h_{sf}	Substance	h_{sf}
Aluminium	387	Paraffin (kerosene)	147.2	Sulphur	39.2
Bismuth	52.9	Phosphorus	21.1	Tin	59.7
Cast iron, grey	96.3	Lead	23.3	Zinc	117.8
Cast iron, white	138.2	Silver	88.2	Ice	334.9
Copper	180	Nickel	309	Magnesium	372

Boiling point (°C) at atmospheric pressure

Substance	b.p.	Substance	b.p.
Ammonia	−33	Methanol (methyl alcohol or wood alcohol)	66
Benzine	80	Naphthalene	220
Bromine	63	Nitric acid	120
Butane	1	Nitrogen	−195
Carbon dioxide	−78.5 (sublimates)	Oxygen	−183
Ethanol (ethyl alcohol)	78	Petrol	200 (approx.)
Ether	38	Propane	−45
Freon 12	−30	Saturated brine	108
Hydrogen	−252.7	Sulphuric acid	310
Kerosine (paraffin)	150–300	Water	100
Mercury	358	Water, sea	100.7 (average)

3.4.3 Properties of air

Analysis of air

Gas	Symbol	Molecular weight	% volume	% mass
Oxygen	O_2	31.999	20.95	23.14
Nitrogen	N_2	28.013	78.09	75.53
Argon	Ar	39.948	0.930	1.28
Carbon dioxide	CO_2	44.010	0.030	0.050

Approximate analysis of air (suitable for calculations)

Gas	Molecular weight	% volume	% mass
Oxygen	32	21	23
Nitrogen	28	79	77

General properties of air (at 300 K, 1 bar)

Mean molecular weight	$M = 28.96$
Specific heat at constant pressure	$c_p = 1.005 \text{ kJ kg}^{-1}$
Specific heat at constant volume	$c_v = 0.718 \text{ kJ kg}^{-1} \text{ K}^{-1}$
Ratio of specific heats	$\gamma = 1.40$
Gas constant	$R = 0.2871 \text{ kJ kg}^{-1} \text{ K}^{-1}$
Density	$\rho = 1.183 \text{ kg m}^{-3}$
Dynamic viscosity	$\mu = 1.853 \times 10^{-5} \text{ Ns m}^{-2}$
Kinematic viscosity	$v = 1.566 \times 10^{-5} \text{ m}^2 \text{ s}^{-1}$
Thermal conductivity	$k = 0.02614 \text{ W m}^{-1} \text{ K}^{-1}$
Thermal diffusivity	$\alpha = 2203 \text{ m}^2 \text{ s}^{-1}$
Prandtl number	$P_r = 0.711$

3.4.4 *Specific heat capacities*

Specific heat capacity of solids and liquids
(kJ kg^{-1} K^{-1})

Aluminium	0.897	Oil, machine	1.676
Aluminium bronze	0.897	Paraffin	2.100
Brass	0.377	Paraffin wax	2.140
Bronze	0.343	Petroleum	2.140
Cadmium	0.235	Phosphorus	0.796
Constantan	0.410	Platinum	0.133
Copper	0.384	Rubber	2.010
Ethanol	2.940	Salt, common	0.880
(ethyl alcohol)		Sand	0.796
Glass: crown	0.670	Seawater	3.940
flint	0.503	Silica	0.800
Pyrex	0.753	Silicon	0.737
Gold	0.129	Silver	0.236
Graphite	0.838	Tin	0.220
Ice	2.100	Titanium	0.523
Iron: cast	0.420	Tungsten	0.142
pure	0.447	Turpentine	1.760
Kerosene	2.100	Uranium	0.116
Lead	0.130	Vanadium	0.482
Magnesia	0.930	Water	4.196
Magnesium	1.030	Water, heavy	4.221
Mercury	0.138	Wood (typical)	2.0 to
Molybdenum	0.272		3.0
Nickel	0.457	Zinc	0.388

Specific heat capacity of gases, gas constant and molecular weight (at normal pressure and temperature)

Gas	Specific heats c_p	(kJ kg^{-1} K^{-1}) c_v	$\gamma = \dfrac{c_p}{c_v}$	Gas constant, R (kJ kg^{-1} K^{-1})	Molecular weight, M
Air	1.005	0.718	1.4	0.2871	28.96
Ammonia	2.191	1.663	1.32	0.528	15.75
Argon	0.5234	0.3136	1.668	0.2081	40
Butane	1.68	1.51	1.11	0.17	58
Carbon dioxide	0.8457	0.6573	1.29	0.1889	44
Carbon monoxide	1.041	0.7449	1.398	0.2968	28
Chlorine	0.511	0.383	1.33	0.128	65
Ethane	1.7668	1.4947	1.18	0.2765	30
Helium	5.234	3.1568	1.659	2.077	4
Hydrogen	14.323	10.1965	1.405	4.124	2
Hydrogen chloride	0.813	0.583	1.40	0.230	36.15
Methane	2.2316	1.7124	1.30	0.5183	16
Nitrogen	1.040	0.7436	1.40	0.2968	28
Nitrous oxide	0.928	0.708	1.31	0.220	37.8
Oxygen	0.9182	0.6586	1.394	0.2598	32
Propane	1.6915	1.507	1.12	0.1886	44
Sulphur dioxide	0.6448	0.5150	1.25	0.1298	64

3.5 Flow through nozzles

Nozzles are used in steam and gas turbines, in rocket motors, in jet engines and in many other applications. Two types of nozzle are considered: the 'convergent nozzle', where the flow is subsonic; and the 'convergent divergent nozzle', for supersonic flow.

Symbols used:

p_1 = inlet pressure
p_2 = outlet pressure
p_c = critical pressure at throat
v_1 = inlet specific volume
v_2 = outlet specific volume
C_2 = outlet velocity
C_c = throat velocity

r = pressure ratio $= \dfrac{p_2}{p_1}$

r_c = critical pressure ratio $= \dfrac{p_c}{p_1}$

A_2 = outlet area
A_c = throat area
n = index of expansion
$\dot{m}$ = mass flow rate

Critical pressure ratio $r_c = \left(\dfrac{2}{n+1}\right)^{\left(\frac{n}{n-1}\right)}$

3.5.1 Convergent nozzle

Outlet pressure p_2 greater than p_c, i.e. $r > r_c$

Outlet velocity $C_2 = \sqrt{\dfrac{2n}{(n-1)} p_1 v_1 \left(1 - r^{\left(\frac{n-1}{n}\right)}\right)}$

Outlet area $A_2 = \dfrac{\dot{m} v_1}{C_2 (r)^{\frac{1}{n}}}$

Outlet pressure p_2 equal to or less than p_c, i.e. $r \leqslant r_c$

Outlet velocity $C_2 = \sqrt{\dfrac{2n}{(n+1)} p_1 v_1}$

Outlet area $A_2 = \dfrac{\dot{m} v_1}{C_2 (r)^{\frac{1}{n}}}$

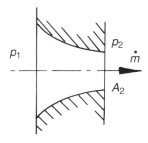

Note that C_2 is independent of p_2 and that the nozzle flow is a maximum. In this case the nozzle is said to be 'choked'.

3.5.2 Convergent–divergent nozzle

In this case:

Throat velocity $C_c = \sqrt{\left(\dfrac{2n}{n+1}\right) p_1 v_1}$

Throat area $A_c = \dfrac{\dot{m} v_1}{C_c (r_c)^{\frac{1}{n}}}$, $r_c = \left(\dfrac{2}{n+1}\right)^{\frac{n}{n-1}}$

Outlet velocity $C_2 = \sqrt{\dfrac{2n}{(n-1)} p_1 v_1 \left(1 - r^{\left(\frac{n-1}{n}\right)}\right)}$

Outlet area $A_2 = \dfrac{\dot{m} v_1}{C_2 (r)^{\frac{1}{n}}}$

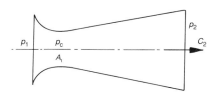

Values of the index n and the critical pressure ratio r_c for different fluids are given in the table.

Fluid	n	r_c
Air ($n = \gamma$)	1.4	0.528
Initially dry saturated steam	1.135	0.577
Initially superheated steam	1.3	0.546

3.6 Steam plant

The simplest steam cycle of practical value is the Rankine cycle with dry saturated steam supplied by a boiler to a power unit, e.g. a turbine, which exhausts to a condenser where the condensed steam is pumped back into the boiler. Formulae are given for work output, heat supplied, efficiency and specific steam consumption. Higher efficiency is obtained if the steam is initially superheated which also reduces specific steam consumption and means smaller plant can be used. If the steam is 'reheated' and passed through a

second turbine the final dryness fraction is increased with beneficial effects (e.g. reduced erosion of turbine blades due to water droplets); in addition, there is a further reduction in specific steam consumption.

In the 'regenerative cycle' efficiency is improved by bleeding off a proportion of the steam at an intermediate pressure and mixing it with feed water pumped to the same pressure in a 'feed heater'. Several feed heaters may be used but these are of the 'closed' variety to avoid the necessity for expensive pumps.

3.6.1 *Rankine cycle — dry saturated steam at turbine inlet*

From the T–s diagram:

$$s_2 = s_1, \quad x_2 = \frac{(s_2 - s_{f2})}{S_{fg2}}$$

$h_2 = h_{f2} + x_2 h_{fg2}$
Work output $W = (h_1 - h_2)$
Heat supplied $Q = (h_1 - h_{f3})$
Cycle efficiency $\eta = W/Q$ (neglecting pump work)
Specific steam consumption
$$SSC = 3600/W \text{ kg kW}^{-1}\text{h}^{-1}$$

Note: if the turbine isentropic efficiency η_i is allowed for: $W = (h_1 - h_2)\eta_i$ and expansion is to point 2^1 on the diagram.

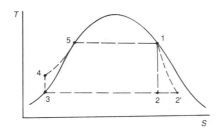

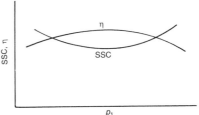

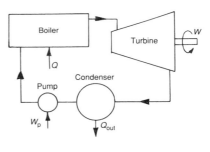

3.6.2 *Rankine cycle — with superheat*

The method is the same as for dry saturated steam. The graph shows the effect of superheat temperature on efficiency and specific steam consumption. In this case h_1 is the enthalpy for superheated steam.

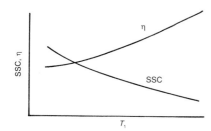

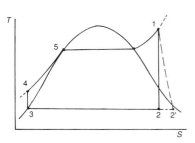

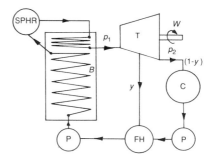

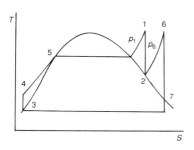

3.6.3 *Rankine cycle with reheat*

At point 2 the steam is reheated to point 6 and passed through a second turbine.

$W = (h_1 - h_2) + (h_6 - h_7)$

$Q = (h_1 - h_3) + (h_6 - h_2)$

The value of p_6 is found using $T_6 = T_1$ (usually) and $s_{g2} = s_1$ from which h_6 is found. The value of h_7 is found using $s_7 = s_6$.

A bleed pressure p_b is selected to correspond to the saturation temperature t_b.

Dryness fractions: $x_b = \dfrac{s_1 - s_{fb}}{s_{fgb}}$; $x_2 = \dfrac{s_1 - s_{f2}}{s_{fg2}}$

Enthalpy: $h_b = h_{fb} + x_b h_{fgb}$; $h_2 = h_{f2} + x_2 h_{fg2}$

Quantity of bled steam $y = \dfrac{h_{fb} - h_{f2}}{h_b - h_2}$ kg/kg total steam

Work done per kg steam $W = (h_1 - h_b) + (1 - y)(h_b - h_2)$

Heat supplied per kg steam $Q = (h_1 - h_{fb})$

Cycle efficiency $\eta = \dfrac{W}{Q}$

Specific steam consumption (SSC) $= \dfrac{3600}{W}$ kg kW^{-1}h^{-1}

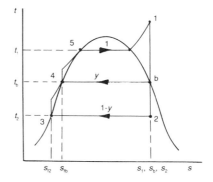

3.6.4 *Regenerative cycle*

Turbine inlet conditions p_1, t_1, h_1
Turbine outlet conditions p_2, t_2, h_2
Bleed steam conditions p_b, t_b, h_b

For maximum efficiency $t_b = \dfrac{(t_1 + t_2)}{2}$

3.7 Steam turbines

This section deals with the two main types of steam turbine, the 'impulse turbine' and the 'impulse-reaction turbine'. The theory is given for a single-stage impulse turbine and velocity compounded impulse turbine.

In the impulse-reaction turbine the fixed and mov-ing blades are of similar form, consisting of converging passages to give a pressure drop in each case. In the case of 50% reaction (Parson's turbine) the enthalpy drop is the same for both fixed and moving blades.

Stage efficiency, overall efficiency and the reheat factor are defined.

3.7.1 *Impulse turbine*

Single-stage impulse turbine

Symbols used:
C = nozzle velocity
C_b = blade velocity
C_a = axial velocity
ρ = ratio of blade to nozzle velocity
β_1 = blade inlet angle
β_2 = blade outlet angle (in this case $\beta_1 = \beta_2$)
α = nozzle angle
$\dot{m}$ = mass flow rate of steam
k = blade friction coefficient = $\dfrac{\text{outlet relative velocity}}{\text{inlet relative velocity}}$
P = stage power
η = stage diagram efficiency
T_a = axial thrust on blades
R_m = mean radius of nozzle arc
v = specific volume of steam at nozzle outlet
θ = nozzle arc angle (degrees)
N = speed of rotation
h = nozzle height
A = nozzle area

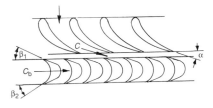

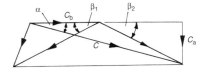

Power $P = \dot{m}C^2\rho(\cos\alpha - \rho)(1 + k)$

where: $\rho = \dfrac{C_b}{C}$ and $C_b = 2\pi R_m N$

Efficiency $\eta = 2\rho(\cos\alpha - \rho)(1 + k)$

Maximum efficiency

$$\eta_{max} = (1 + k)\cos^2\frac{\alpha}{2}\left(\text{at } \rho = \cos\alpha\,\frac{(1 + k)}{2}\right)$$

Axial thrust $T_a = \dot{m}C(1 - k)\sin\alpha$

Mass flow rate $\dot{m} = \dfrac{C_a A}{v}$

Nozzle area $A = \dfrac{\pi R_m \theta h}{180}$

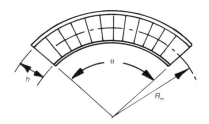

Pressure compounded impulse turbine

The steam pressure is broken down in two or more stages. Each stage may be analysed in the same manner as described above.

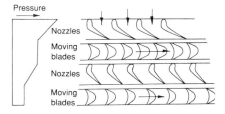

Velocity compounded impulse turbine

One row of nozzles is followed by two or more rows of moving blades with intervening rows of fixed blades of the same type which alter the direction of flow.

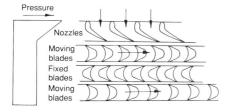

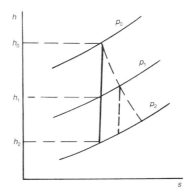

Two-row wheel Assume $\beta_1 = \beta_2$, $k = 1$ and that all blades are symmetrical.

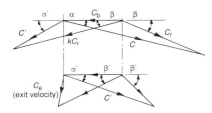

N = nozzles
M = moving blades
F = fixed blades

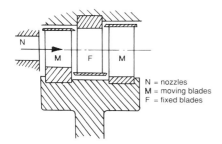

Maximum efficiency $\eta_{max} = \cos^2 \alpha \left(\text{at } \rho = \dfrac{\cos \alpha}{4} \right)$

in which case the steam leaves the last row axially.

3.7.2 *Impulse-reaction turbine*

In this case there is 'full admission', i.e. $\theta = 360°$. Both nozzles and moving blades are similar in shape and have approximately the same enthalpy drop. Referring to the figure:

Enthalpy drop $= (h_0 - h_1)$ (for the fixed blades)
$= (h_1 - h_2)$ (for the moving blades)

Mass flow rate $\dot{m} = \dfrac{C_a A}{v}$

Area of flow $A = 2\pi R_m h$

Degree of reaction $R = \dfrac{(h_1 - h_2)}{(h_0 - h_1) + (h_1 - h_2)} = \dfrac{h_1 - h_2}{h_0 - h_2}$

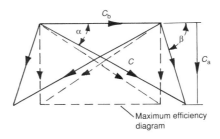

Maximum efficiency diagram

50% reaction (Parson's) turbine

In this case the velocity diagram is symmetrical.

Mass flow rate $\dot{m} = \dfrac{2\pi R_m h C \sin \alpha}{v}$

where: α = blade outlet angle.

Enthalpy drop per stage $\Delta h_s = C^2 \rho (2 \cos \alpha - \rho)$

where: $\rho = \dfrac{C_b}{C}$ and $C_b = 2\pi R_m N$.

Stage power $P_s = \dot{m} \Delta h_s$

Stage efficiency $\eta_s = \dfrac{2\rho(2 \cos \alpha - \rho)}{1 + \rho(2 \cos \alpha - \rho)}$

Maximum efficiency $\eta_{max} = \dfrac{2 \cos^2 \alpha}{(1 + \cos^2 \alpha)}$ (when $\rho = \cos \alpha$)

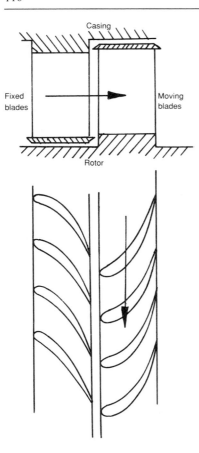

3.7.3 Reheat factor and overall efficiency

Referring to the 'condition curve' on the h–s diagram:

Δh_A = available stage enthalpy drop
Δh_I = isentropic stage enthalpy drop
Δh_{OA} = available overall enthalpy drop
Δh_{OI} = isentropic overall enthalpy drop

Stage efficiency $\eta_s = \dfrac{\Delta h_A}{\Delta h_I}$

Overall efficiency $\eta_o = \dfrac{\Delta h_{OA}}{\Delta h_{OI}}$

Reheat factor $RF = \dfrac{\eta_o}{\eta_s}$

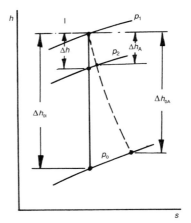

3.8   Gas turbines

The gas turbine unit operates basically on the constant-pressure cycle, particularly in the case of the 'closed cycle'. In the 'open cycle' air is drawn in from the atmosphere, compressed and supplied to a combustion chamber where fuel is burnt with a large amount of 'excess air'. The hot gases drive a turbine which drives the compressor and also provides useful work. The efficiency increases with compression ratio.

The output power increases with both compression ratio and turbine inlet temperature.

The effect of losses and variation in fluid properties is shown on the basic cycle. The efficiency of the basic cycle can be greatly increased by incorporating a heat exchanger between the compressor outlet and the combustion chamber inlet. It uses the exhaust gases from the turbine to preheat the incoming air.

3.8.1 *Simple cycle*

Compression ratio $r = \dfrac{p_2}{p_1} = \dfrac{p_3}{p_4}$

Let: $c = r^{\frac{\gamma-1}{\gamma}} = \dfrac{T_2}{T_1} = \dfrac{T_3}{T_4}$ and $t = \dfrac{T_3}{T_1}$

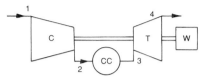

C = compressor
CC = combustion chamber turbine

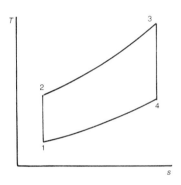

Heat supplied $Q = c_p T_1 (t - c)$ per kg of air
Work done = Turbine work out − Compressor work in

$$W = c_p T_1 \left[t \left(1 - \frac{1}{c} \right) - (c - 1) \right]$$

Efficiency $\eta = 1 - \dfrac{1}{c}$

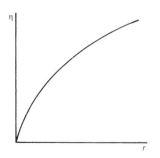

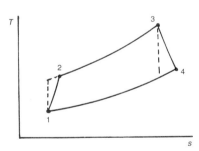

Simple cycle with isentropic efficiencies and variable specific heats

Let:
$_c c_p$ = specific heat for compressor
$_t c_p$ = specific heat for turbine
$_{cc} c_p$ = specific heat for combustion chamber
γ_c = ratio of specific heats for compressor
γ_t = ratio of specific heats for turbine
η_c = isentropic compressor efficiency
η_t = isentropic turbine efficiency

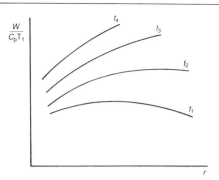

Work done = Turbine work out − Compressor work in.

$$W = {}_t c_p (T_3 - T_4) \eta_t - {}_c c_p \frac{(T_2 - T_1)}{\eta_c} \text{ per kg of air}$$

Heat supplied $Q = {}_{cc} c_p \left[T_3 - T_1 - \dfrac{(T_2 - T_1)}{\eta_c} \right]$ per kg of air

Work ratio $= \dfrac{\text{Net work out}}{\text{Gross work}} = \dfrac{W}{{}_t c_p (T_3 - T_4) \eta_t}$

Efficiency $\eta = \dfrac{W}{Q}$

Note: $r = \dfrac{p_2}{p_1}$; $\dfrac{T_2}{T_1} = r^{\left(\frac{\gamma_c - 1}{\gamma_c} \right)}$; $\dfrac{T_3}{T_4} = r^{\left(\frac{\gamma_t - 1}{\gamma_t} \right)}$

3.8.2 *Simple cycle with heat exchanger*

Heat supplied $Q = c_p T_1 t \left(1 - \dfrac{1}{c} \right)$

Work done $W = c_p T_1 \left[t \left(1 - \dfrac{1}{c} \right) - (c - 1) \right]$

Efficiency $\eta = 1 - \dfrac{c}{t}$

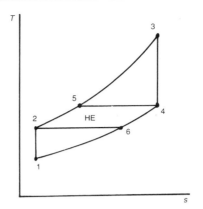

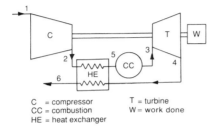

C = compressor   T = turbine
CC = combustion   W = work done
HE = heat exchanger

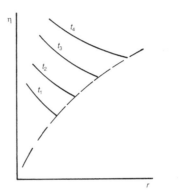

3.9 Heat engine cycles

3.9.1 *Carnot cycle*

The ideal gas cycle is the Carnot cycle and, in practice, only about half of the Carnot cycle efficiency is realized between the same temperature limits.

Efficiency $\eta = 1 - \dfrac{T_2}{T_1}$

$(s_1 - s_4) = R \ln \dfrac{p_4}{p_2} - c_p \ln \dfrac{T_1}{T_2}$

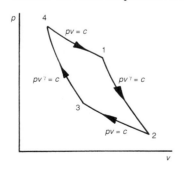

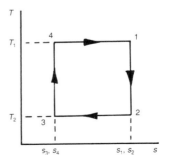

Work done (per kg) $W = (T_1 - T_2)(s_1 - s_4)$

Heat supplied (per kg) $Q = T_1(s_1 - s_4)$

Work ratio $\dfrac{W}{W_{\text{gross}}} = \dfrac{(s_1 - s_4)(T_1 - T_2)}{T_1(s_1 - s_4) + c_v(T_1 - T_2)}$

3.9.2 Constant pressure cycle

In this cycle, heat is supplied and rejected at constant pressure; expansion and compression are assumed to take place at constant entropy. The cycle was once known as the Joule or Brayton cycle and used for hot-air engines. It is now the ideal cycle for the closed gas turbine unit.

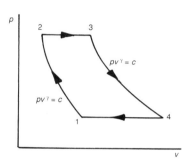

Efficiency $\eta = 1 - \dfrac{1}{r^{\left(\frac{\gamma-1}{\gamma}\right)}}$, where $r = \dfrac{p_2}{p_1}$

$W = c_p(T_3 - T_4) - c_p(T_2 - T_1)$

Work ratio $= 1 - \dfrac{T_1}{T_3} r^{\left(\frac{\gamma-1}{\gamma}\right)}$

$Q = c_p(T_3 - T_2)$

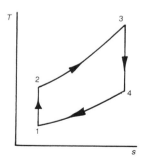

3.9.3 Otto cycle (constant-volume cycle)

This is the basic cycle for the petrol engine, the gas engine and the high-speed oil engine. Heat is supplied and rejected at constant volume, and expansion and compression take place isentropically. The thermal efficiency depends only on the compression ratio.

Efficiency $\eta = 1 - \dfrac{1}{r^{\gamma-1}}$

where: $r = \dfrac{v_1}{v_2}$ and $\dfrac{T_2}{T_1} = \dfrac{T_3}{T_4} = r^{\gamma-1}$

$W = c_v(T_3 - T_2 - T_4 + T_1)$

$Q = c_v(T_3 - T_2)$

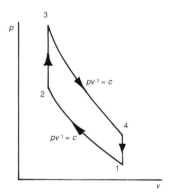

3.9.4 Diesel cycle (constant-pressure combustion)

Although this is called the 'diesel cycle', practical diesel engines do not follow it very closely. In this case heat is added at constant pressure; otherwise the cycle is the same as the Otto cycle.

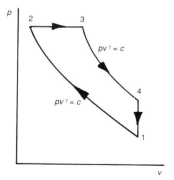

$$\text{Efficiency} = 1 - \frac{(\beta^\gamma - 1)}{(\beta - 1)\gamma r^{\gamma - 1}}$$

where: $r = \dfrac{v_1}{v_2}$ and $\beta = \dfrac{v_3}{v_2}$. ('cut-off' ratio)

$$W = c_p(T_3 - T_2) - c_v(T_4 - T_1)$$

$$Q = c_p(T_3 - T_2)$$

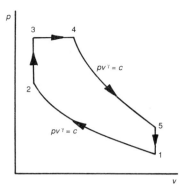

3.9.5 *Dual combustion cycle*

Modern diesel engines follow a similar cycle to this ideal one. In this case combustion takes place partly at constant volume and partly at constant pressure.

$$\text{Efficiency } \eta = 1 - \frac{(k\beta^\gamma - 1)}{[(k - 1) + (\beta - 1)\gamma k]r^{\gamma - 1}}$$

where: $r = \dfrac{v_1}{v_2}$, $k = \dfrac{p_3}{p_2}$ and $\beta = \dfrac{v_4}{v_3}$.

$$W = c_v(T_3 - T_2) + c_p(T_4 - T_3) - c_v(T_5 - T_1)$$

$$Q = c_v(T_3 - T_2) + c_p(T_4 - T_3)$$

3.9.6 *Practical engine cycles*

In actual engines the working substance is air only in the induction and compression strokes. During expansion and exhaust the working substance consists of the products of combustion with different properties to air. In addition, the wide variations in temperature and pressure result in variation in the thermal properties. Another factor is 'dissociation' which results in a lower maximum temperature than is assumed in elementary treatment of the combustion process.

3.10 Reciprocating spark ignition internal combustion engines

3.10.1 *Four-stroke engine*

The charge of air and fuel is induced into the engine cylinder as the piston moves from top dead centre (TDC) to bottom dead centre (BDC). The charge is then compressed and ignited by the sparking plug before TDC producing high pressure and temperature at about TDC. The gas expands and work is produced as the piston moves to BDC. A little before BDC the exhaust valve opens and the gases exhaust. The process is completed during the next stroke. A typical 'timing diagram' (section 3.10.3) and the p–v diagram are shown. Formulae are given for power, mean effective pressure, efficiency and specific fuel consumption.

Pressure–volume (p–v) diagram:
 A = area of power loop
 B = area of pumping loop
 L_d = length of diagram
 K = indicator constant

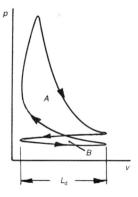

Indicated mean effective pressure

$$p_i = (A - B)\frac{K}{L_d} \;(\text{N mm}^{-2})$$

Indicated power $P_i = p_i A_p L N \dfrac{n}{2}$ (watts)

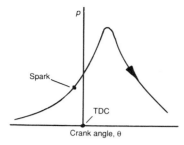

Typical timing diagram

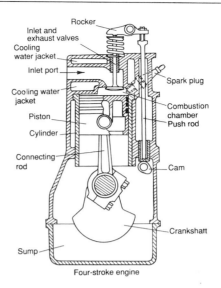

Four-stroke engine

where: N = number of revolutions per second, n = number of cylinders, A_p = piston area (m^2), L = stroke (m)

Torque $T = FR$ (Nm)

where: F = force on brake arm (N), R = brake radius (m).

Brake power $P_b = 2\pi NT$ (watts)

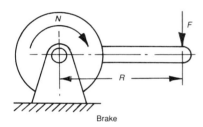

Brake

Friction power $P_f = P_i - P_b$

Mechanical efficiency $\eta_m = \dfrac{P_b}{P_i}$

Brake mean effective pressure (BMEP) $p_b = \dfrac{4\pi T}{ALn}$

$= \text{constant} \times T(\text{N m}^{-2})$

Brake thermal efficiency $\eta_b = \dfrac{P_b}{\dot{m}LCV}$

where: $\dot{m}$ = mass flow rate of fuel (kg s^{-1}), LCV = lower calorific value of fuel (J kg^{-1}).

Specific fuel consumption $\text{SFC} = \dfrac{\dot{m}}{P_b}$ (kg s^{-1} W^{-1})

Volumetric efficiency $\eta_v = \dfrac{\text{Volume of induced air at NTP}}{\text{Swept volume of cylinder}}$

where: NTP = normal temperature and pressure.

3.10.2 Two-stroke engine

In an engine with crankcase compression, the piston draws a new charge into the crankcase through a spring-loaded valve during the compression stroke. Ignition occurs just before TDC after which the working stroke commences. Near the end of the stroke the exhaust port is uncovered and the next charge enters the cylinder. The exhaust port closes shortly after the transfer port, and compression begins. The piston is shaped to minimize mixing of the new charge with the exhaust. (See section 3.10.3)

Pressure–volume (p–v) diagram:
A = area of power loop
B = area of pumping loop

Indicated mean effective pressure (IMEP): $p_i = (A - B)\dfrac{K}{L_d}$

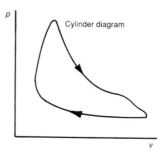

Cylinder diagram

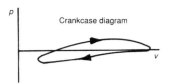

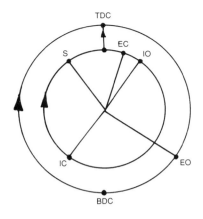

where: K = indicator constant.

Indicated power $P_i = p_i A_p L N n$

Brake mean effective pressure (BMEP) $p_b = \dfrac{2\pi T}{ALn}$

Other quantities are as for the four-stroke engine.

Compression-ignition engines

Both four-stroke and two-stroke engines may have compression ignition instead of spark ignition. The air is compressed to a high pressure and temperature and the fuel injected. The high air temperature causes combustion.

Two-stroke engine

I = inlet angle (approx. $80°$)
E = exhaust angle (approx. $120°$)
T = transfer angle (approx. $100°$)

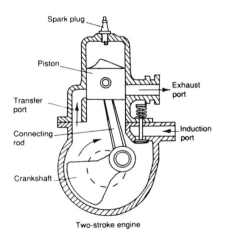

Two-stroke engine

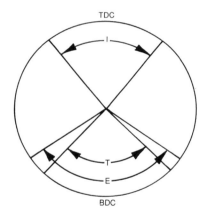

3.10.3   *Timing diagrams*

Four-stroke engine

IO = inlet valve opens
IC = inlet valve closes
 S = spark occurs
EO = exhaust valve opens
EC = exhaust valve closes

3.10.4   *Performance curves for internal combustion engines*

Typical curves are shown for mechanical efficiency versus brake power, BMEP versus torque, and volumetric efficiency versus speed. The effect of mixture strength on the p–v and p–θ diagrams is shown and curves of power and MEP against speed are given. The curve of specific fuel consumption versus brake power, known as the 'consumption loop' shows the effect of mixture strength on fuel consumption.

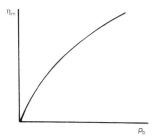

Mechanical efficiency : s brake power

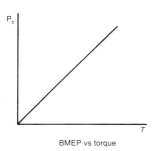

BMEP vs torque

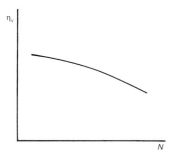

Volumetric efficiency vs speed

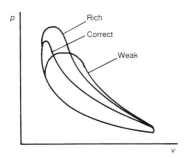

Effect of mixture strength on $p - v$ diagram

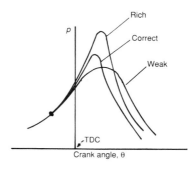

Effect of mixture strength on $p - \theta$ diagram

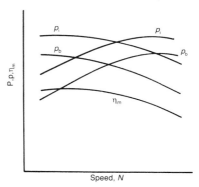

Power, MEP, mechanical efficiency vs speed

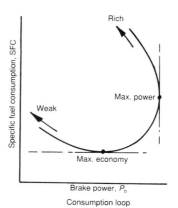

Consumption loop

3.11 Air compressors

The following deals with positive-displacement-type compressors as opposed to rotodynamic types. The reciprocating compressor is the most suitable for high pressures and the Roots blower and vane compressor are most suitable for low pressures.

3.11.1 Reciprocating compressor

This consists of one or more cylinders with cranks, connecting rods and pistons. The inlet and outlet valves are of the automatic spring-loaded type. Large cylinders may be water cooled, but small ones are usually finned.

Air is drawn into the cylinder at slightly below atmospheric pressure, compressed to the required discharge pressure during part of the stroke, and finally discharged at outlet pressure. A small clearance volume is necessary. The cylinders may be single or double acting.

Symbols used:
p = free air pressure (atmospheric conditions)
p_1 = inlet pressure
p_2 = discharge pressure

r = pressure ratio = $\dfrac{p_2}{p_1}$

T = free air temperature
T_1 = inlet air temperature
T_2 = discharge temperature
V_s = swept volume
V_c = clearance volume
$V_a = V_s + V_c$
$V_a - V_d$ = induced volume
R = gas constant for air
n = index of expansion and compression
γ = ratio of specific heats for air
$\dot{m}$ = air mass flow rate
Q = free air volume flow rate
N = number of revolutions per second
Z = number of effective strokes per revolution
   (= 1 for single acting; 2 for double acting)
η = efficiency
W = work done per revolution
P_i = indicated power
S = number of stages

Free air flow $Q = (V_a - V_d)\dfrac{T}{T_1}\dfrac{p_1}{p}NZ$

where: $V_a = (V_s + V_c)$.

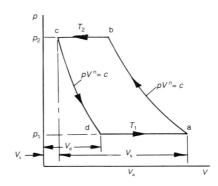

Mass air flow $\dot{m} = \dfrac{Qp}{RT}$

Indicated power $P_i = \dfrac{n}{(n-1)}\dot{m}R(T_2 - T_1)$

where: $T_2 = T_1 r^{\left(\frac{n-1}{n}\right)}$ and $r = \dfrac{p_2}{p_1}$.

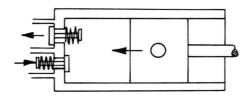

Volumetric efficiency $\eta_v = 1 - \dfrac{V_c}{V_s}(r^{\frac{1}{n}} - 1)$

Clearance ratio $CR = \dfrac{V_c}{V_s}$

Also $\dfrac{V_d}{V_c} = r^{\frac{1}{n}}$

3.11.2 *Multi-stage compressor*

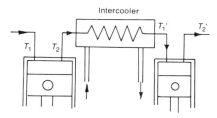

For S stages, the ideal pressure for each stage is:

$$r_i = \left(\frac{p_2}{p_1}\right)^{\frac{1}{S}}$$

for which

$$\text{Indicated power } P_i = \frac{Sn}{(n-1)} \dot{m}R(T_2 - T_1)$$

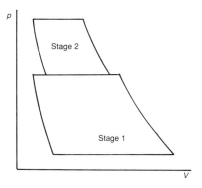

The efficiency is increased by using more than one stage if intercooling is used between the stages to reduce ideally the temperature of the air to that at the first stage inlet. The cylinders become progressively smaller as the pressure increases and volume decreases.

3.11.3 *Roots blower*

This has two rotors with 2, 3 or 4 lobes which rotate in opposite directions so that the lobes mesh. Compression takes place at approximately constant volume.

Work input per revolution $W = p_1 V_s (r - 1)$

where: $r = \dfrac{p_2}{p_1}$.

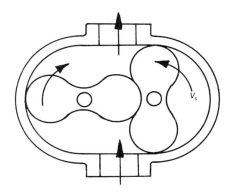

Isentropic work $W_i = p_1 V_s \left(\dfrac{\gamma}{\gamma - 1}\right)(r^{\left(\frac{\gamma - 1}{\gamma}\right)} - 1)$

Efficiency $\eta = \dfrac{W_i}{W} = \dfrac{\gamma}{\gamma - 1}\dfrac{(r^{\left(\frac{\gamma - 1}{\gamma}\right)} - 1)}{(r - 1)}$

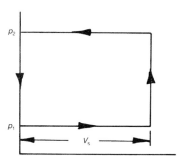

Typical efficiencies

r	1.2	1.6	2.0
η	0.95	0.84	0.77

Pressure ratio $\leqslant 2.0$ for one stage
$\leqslant 3.0$ for two stages

Size: 0.14–$1400\, \text{m}^3\, \text{min}^{-1}$

3.11.4 *Vane compressor*

The simplest type consists of a rotor mounted eccentrically in a cylindrical casing. The rotor has a number of radial slots in which are mounted sliding vanes, often of non-metallic material, between which the air is trapped. Reduction in the volume between vanes as the rotor rotates produces compression. Higher pressures may be attained by using more than one stage. The work is done partly isentropically and partly at constant volume. Assuming ideal conditions:

Isentropic work done $W_i = \dfrac{\gamma}{(\gamma - 1)} p_1 V_s (r^{\left(\frac{\gamma - 1}{\gamma}\right)} - 1)$

where: $r = \dfrac{p_i}{p_1}$

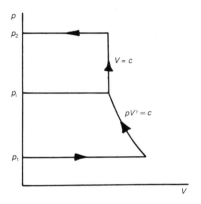

Constant-volume work done $W_v = \dfrac{(p_2 - p_i) V_s}{r_1^{\frac{1}{\gamma}}}$

where $r_1 = \dfrac{p_2}{p_i}$

Total work done per revolution $W_t = W_i + W_v$

Pressure ratio: $\leqslant 8.5$ normally
20 in special cases.

Size: $\leqslant 150 \, \text{m}^3 \, \text{min}^{-1}$

A two-stage vane compressor is shown in the figure.

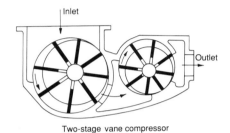

Two-stage vane compressor

3.12   Reciprocating air motor

Reciprocating air motors are used extensively for tools such as breakers, picks, riveters, vibrators and drillers. They are useful where there is fire danger such as in coal mines. The operating cycle is the reverse of that for the reciprocating compressor.

3.12.1   *Power and flow rate*

Referring to the p–V diagram:

Power $P = N\left[p_1(V_1 - V_6) + \dfrac{(p_1 V_1 - p_2 V_2)}{n - 1} \right.$

$\left. \qquad - p_3(V_3 - V_4) - \dfrac{(p_5 V_5 - p_4 V_4)}{n - 1} \right]$

where $n =$ index of expansion and compression.

Mass flow rate of air $\dot{m} = N\left(\dfrac{p_1 V_1}{R T_1} - \dfrac{p_4 V_4}{R T_4} \right)$

where: $\dfrac{p_5}{p_4} = \left(\dfrac{V_4}{V_5} \right)^n$ and $\dfrac{p_1}{p_2} = \left(\dfrac{V_2}{V_1} \right)^n$.

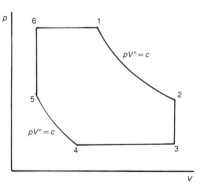

Cut-off ratio $= \dfrac{V_1 - V_6}{V_3 - V_6}$

Clearance ratio $= \dfrac{V_5}{V_3 - V_5}$

3.13 Refrigerators

Two basic types are considered, the 'vapour compression refrigerator' and the 'gas refrigerator'. The former consists of a compressor followed by a condenser where the refrigerant is liquified at high pressure. It is then expanded in a 'throttle valve' to a lower pressure and temperature and finally evaporated in an 'evaporator' before re-entry into the compressor. The cycle is similar to the Rankine cycle in reverse.

The gas cycle is the reverse of a closed gas-turbine cycle, i.e. the constant pressure or Joule cycle.

3.13.1 *Vapour compression cycle*

The process can be shown on the temperature entropy (T–s) chart for the appropriate refrigerant, e.g. ammonia or Freon.

(1) Compression
  Work $W = h_2 - h_1$
  where: $h_1 = h_g$ at p_1, $h_2 =$ enthalpy at p_2, $s_2 = s_1$
(since isentropic compression).

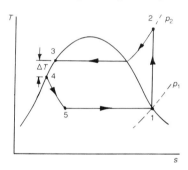

(2) Condensation at constant pressure p_2.
(3) Under-cooling from $T_3 (= T_s$ at p_2) to T_4.
  Degree of undercooling $\Delta T = T_3 - T_4$
(4) Throttling from 4 to 5. Therefore $h_5 = h_4$ and $h_4 = h_f$ at T_4.
(5) Evaporation at pressure p_1.

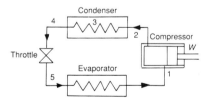

Refrigeration effect $RE = h_1 - h_5$

Coefficient of performance $COP = \dfrac{RE}{W}$

Heat removed $Q = \dot{m}RE$
where: $\dot{m} =$ mass flow rate of refrigerant.

3.13.2 *Pressure–enthalpy chart*

The pressure–enthalpy chart is a more convenient way of showing refrigeration cycles. Work in and refrigeration effect can be measured directly as the length of a line.

If p_1, p_2 and the under cooling temperature T_4 are known, the diagram can be easily drawn and RE and W scaled off as shown.

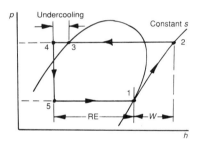

3.13.3 *Gas refrigeration cycle*

Referring to the T–s diagram:

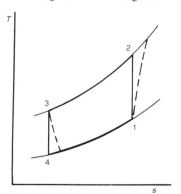

Refrigeration effect $RE = c_p(T_1 - T_3) + c_p\eta_t(T_3 - T_4)$

Work in $W = c_p \dfrac{(T_2 - T_1)}{\eta_c} - c_p\eta_t(T_3 - T_4)$

Coefficient of performance $COP = \dfrac{RE}{W}$

where: $\dfrac{T_1}{T_2} = \dfrac{T_4}{T_3} = \left(\dfrac{p_1}{p_2}\right)^{\left(\frac{\gamma-1}{\gamma}\right)}$, $\eta_t =$ turbine isentropic efficiency, $\eta_c =$ compressor isentropic efficiency.

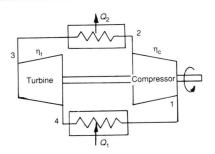

3.14 Heat transfer

Heat may be transmitted by conduction, convection or radiation.

3.14.1 *Conduction*

Heat transfer by conduction is the transfer of heat from one part of a substance to another without appreciable displacement of the molecules of the substance, e.g. heat flow along a bar heated at one end. This section deals with conduction of heat through a flat wall, a composite wall, a cylindrical wall and a composite cylindrical wall. A table of thermal-conductivity coefficients is given.

3.14.2 *Conduction through wall*

Let:
$k =$ conductivity of wall, $Wm^{-1}K^{-1}$
$A =$ area of wall, m^2
$x =$ thickness of wall, m
$t =$ temperature (°C)
$q =$ heat flow rate, W
$h =$ heat transfer coefficient, $Wm^{-2}K^{-1}$
$U =$ overall heat transfer coefficient, $Wm^{-2}K^{-1}$
$R =$ thermal resistance KW^{-1}

Heat flow $q = \dfrac{kA}{x}(t_1 - t_2)$

Overall heat transfer coefficient $U = \dfrac{k}{x}$

Therefore, $q = UA(t_1 - t_2)$

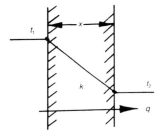

Thermal resistance $R = \dfrac{x}{kA} = \dfrac{1}{UA}$

Conduction from fluid to fluid through wall

In this case the surface coefficients are taken into account.

$$q = Ah_a(t_a - t_1) = \dfrac{kA}{x}(t_1 - t_2) = Ah_b(t_2 - t_b)$$

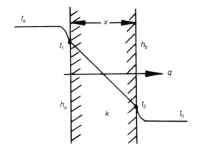

$$U = \frac{1}{\frac{1}{h_a} + \frac{x}{k} + \frac{1}{h_b}}$$

$$q = U A (t_a - t_b)$$

$$R = \frac{1}{h_a A} + \frac{1}{h_b A} + \frac{x}{k A} = R_a + R_b + R$$

Conduction through composite wall

$$q = U A (t_a - t_b)$$

$$U = \frac{1}{(R_a + R_1 + R_2 + \ldots R_b) A}$$

$$R = R_a + R_1 + R_2 + \ldots R_b$$

where: $R_1 = \dfrac{x_1}{k_1 A_1}$, $R_2 = \dfrac{x_2}{k_2 A_2}$, etc.

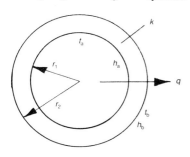

cylinder wall

$$q = \frac{2\pi k (t_1 - t_2) L}{\ln \frac{r_2}{r_1}} = \frac{k A_m}{x}(t_1 - t_2)$$

$$A_1 = 2\pi r_1 L; \ A_2 = 2\pi r_2 L; \ A_m = \frac{A_2 - A_1}{\ln \frac{r_2}{r_1}}$$

$x = r_2 - r_1$, $L = $ Length of cylinder

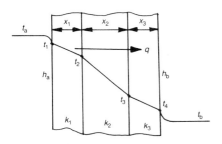

Conduction through composite cylinder fluid to fluid

A typical example is a lagged pipe.

$$q = \frac{(t_a - t_b)}{R}$$

$$R = \frac{1}{h_a A_a} + \frac{x_1}{k_1 A_{m1}} + \frac{x_2}{k_2 A_{m2}} + \ldots + \frac{1}{h_b A_b}$$

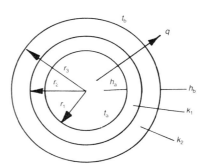

3.14.4 Heat transfer from fins

The heat flow depends on the rate of conduction along the fin and on the surface heat-transfer coefficient. The theory involves the use of hyperbolic functions.

Fin of constant cross-section with insulated tip

Let:
 $L = $ fin length
 $A = $ fin cross-sectional area
 $P = $ perimeter of fin
 $k = $ conductivity
 $h = $ surface heat-transfer coefficient
 $t_a = $ air temperature
 $t_r = $ fin root temperature

Heat flow from fin, $q = k A (t_r - t_a) m \tanh m L$

where: $m = \sqrt{\dfrac{hP}{kA}}$.

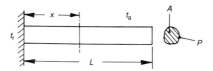

Fin efficiency $\eta = \dfrac{\text{Heat flow from fin}}{\text{Heat flow if fin all at } t_r}$

$\quad\quad = \dfrac{q}{hPL(t_r - t_a)}$

If fin has constant cross-section and is insulated at the end:

Efficiency $\eta = \dfrac{\tanh mL}{mL}$

Temperature profile along fin:

Temperature at distance x from root

$t_x = t_a + (t_r - t_a)\dfrac{\cosh m(L-x)}{\cosh mL}$

Fins on a circular pipe

Constant thickness:

Efficiency $\eta = \dfrac{q}{hA_s(t_r - t_a)}$

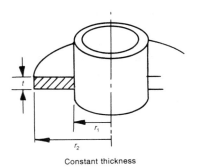

Constant thickness

where: A_s = surface area = $\pi(r_2^2 - r_1^2) + 2\pi r_2 t$.

Efficiency is plotted against the function $\sqrt{\dfrac{hL^3}{kA}}$ in the

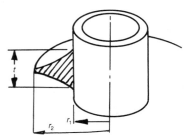

Hyperbolic section

figure where L = fin length = $(r_2 - r_1)$ and A = cross-sectional area = tL.

Hyperbolic section circular fins: curves are given for hyperbolic fins using the appropriate values of A_s and A.

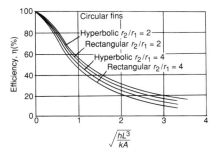

Straight fins

Similar efficiency curves are given in the figures for straight fins of various shapes.

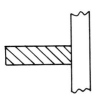

Constant thickness

Triangular

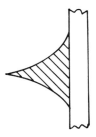

Parabolic (concave)

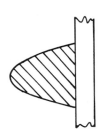

Parabolic (convex)

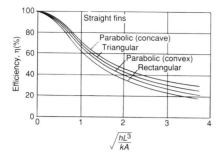

3.14.5 *Thermal conductivity coefficient*

The following table gives values of conductivity for solids, liquids and gases.

Thermal conductivity coefficients ($W\,m^{-1}\,K^{-1}$) at 20°C and 1 bar

Metals		Liquids		Plastics	
Aluminium	239	Benzene	0.16	Acrylic (Perspex)	0.20
Antimony	18	Carbon tetrachloride	0.11	Epoxy	0.17
Brass (60/40)	96	Ethanol	0.18	Epoxy glass fibre	0.23
Cadmium	92	(ethyl alcohol)		Nylon 6	0.25
Chromium	67	Ether	0.14	Polyethylene:	
Cobalt	69	Glycerine	0.29	low density	0.33
Constantan	22	Kerosene	0.15	high density	0.50
Copper	386	Mercury	8.80	PTFE	0.25
Gold	310	Methanol	0.21	PVC	0.19
Inconel	15	(methyl alcohol)			
Iron, cast	55	Oil: machine	0.15	*Refrigerants at critical*	
Iron, pure	80	transformer	0.13	*temperature*	
Lead	35	Water	0.58		
Magnesium	151			Ammonia (132.4°C)	0.049
Molybdenum	143			Ethyl chloride (187.2°C)	0.095
Monel	26	*Gases*		Freon 12 (112°C)	0.076
Nickel	92	Air	0.024	Freon 22 (97°C)	0.10
Platinum	67	Ammonia	0.022	Sulphur dioxide (157.2°)	0.0087
Silver	419	Argon	0.016		
Steel: mild	50	Carbon dioxide	0.015	*Insulating materials*	
stainless	25	Carbon monoxide	0.023		
Tin	67	Helium	0.142	Asbestos cloth	0.13
Tungsten	172	Hydrogen	0.168	Balsa wood (average)	0.048
Uranium	28	Methane	0.030	Calcium silicate	0.05
Zinc	113	Nitrogen	0.024	Compressed straw slab	0.09
		Oxygen	0.024	Corkboard	0.04
		Water vapour	0.016	Cotton wool	0.029
				Diatomaceous earth	0.06
				Diatomite	0.12
				Expanded polystyrene	0.03/0.04

Thermal conductivity coefficients (W m^{-1} K^{-1}) at 20°C and 1 bar (*continued*)

Miscellaneous materials		Insulating materials, cont.	
Asphalt	1.26	Felt	0.04
Bitumen	0.17	Glass fibre quilt	0.043
Breeze block	0.10–0.20	Glass wool quilt	0.040
Brickwork: common	0.6–1.0	Hardboard	0.13
dense	1.6	Kapok	0.034
Carbon	1.7	Magnesia	0.07
Concrete: lightweight	0.1–0.3	Mineral wool quilt	0.04
medium	0.4–0.7	Plywood	0.13
dense	1.0–1.8	Polyurethane foam	0.03
Firebrick (600°C)	1.09	Rock wool	0.045
Glass: crown	1.05	Rubber, natural	0.130
flint	0.84	Sawdust	0.06
Pyrex	1.30	Slag wool	0.042
Ice	2.18	Urea formaldehyde	0.040
Limestone	1.10	Wood	0.13–0.17
Mica	0.75	Wood wool slab	0.10–0.15
Cement	1.01		
Paraffin wax	0.25		
Porcelain	1.05		
Sand	0.06		
Sandstone	3.00		
Slate	2.01		

3.14.6 *Convection*

Convection is the transfer of heat in a fluid by the mixing of one part of the fluid with another. Motion of the fluid may be caused by differences in density due to temperature differences as in 'natural convection' (or 'free convection'), or by mechanical means, such as pumping, as in 'forced convection'.

3.14.7 *Dimensionless groups*

In the study of heat transfer by convection it is convenient to plot curves using dimensionless groups. Those commonly used are:

Reynold's number $\mathrm{Re} = \dfrac{\rho C L}{\mu}$

Nusselt number $\mathrm{Nu} = \dfrac{hL}{k}$

Prandtl number $\mathrm{Pr} = \dfrac{c\mu}{k}$

Stanton number $\mathrm{St} = \dfrac{h}{\rho c C} = \dfrac{\mathrm{Nu}}{\mathrm{Re}\,\mathrm{Pr}}$

Grashof number $\mathrm{Gr} = \dfrac{\beta g \rho^2 L^3 \theta}{\mu^2}$

where:
ρ = fluid density
μ = fluid viscosity
k = fluid conductivity
c = fluid specific heat
β = fluid coefficient of cubical expansion
C = fluid velocity
g = acceleration due to gravity
L = characteristic dimension
h = heat transfer coefficient
θ = fluid temperature difference

3.14.8 *Natural convection*

Natural convection from horizontal pipe

Nusselt number $\mathrm{Nu} = \dfrac{hL}{k}$

$\text{Nu} = 0.47(\text{PrGr})^{0.25}$ for $\text{PrGr} = 10^5$ to 10^8

$\text{Nu} = 0.10\ (\text{PrGr})^{0.33}$ for $\text{PrGr} > 10^8$

Approximate heat transfer coefficient:

$$h = 1.32\left(\frac{\theta}{d}\right)^{0.25} \text{ for } \text{Gr} = 10^4 \text{ to } 10^9$$

$$h = 1.25\theta^{0.33} \text{ for } \text{Gr} = 10^9 \text{ to } 10^{12}$$

where:

θ = temperature difference between cylinder and fluid

d = diameter of cylinder

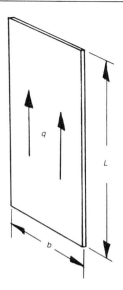

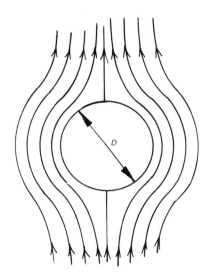

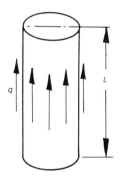

Natural convection from a vertical plate or cylinder

$\text{Nu} = 0.56(\text{GrPr})^{0.25}$ for $\text{PrGr} = 10^5$ to 10^9

$\text{Nu} = 0.12\ (\text{GrPr})^{0.33}$ for $\text{PrGr} > 10^9$

Approximately:

$$h = 1.42\left(\frac{\theta}{L}\right)^{0.25} \text{ for } \text{Gr} = 10^4 \text{ to } 10^9$$

$$h = 1.31\theta^{0.33} \text{ for } \text{Gr} = 10^9 \text{ to } 10^{12}$$

Horizontal plate facing upwards

Characteristic dimension $L = \dfrac{a+b}{2}$

$\text{Nu} = 0.54(\text{GrPr})^{0.25}$ for $\text{GrPr} = 10^5$ to 10^8

$\text{Nu} = 0.14(\text{GrPr})^{0.33}$ for $\text{GrPr} > 10^8$

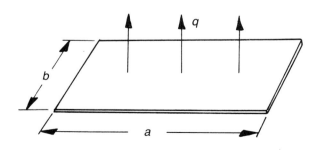

Horizontal plate facing downwards

$\mathrm{Nu} = 0.25(\mathrm{GrPr})^{0.25}$ for $\mathrm{GrPr} > 10^5$

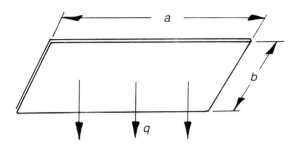

3.14.9   *Forced convection*

Laminar flow in pipe

$\mathrm{Nu} = 3.65$ and $h = 3.65\dfrac{k}{d}$

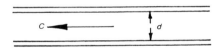

Turbulent flow over cylinder

Generally: $\mathrm{Nu} = 0.26\mathrm{Re}^{0.6}\mathrm{Pr}^{0.3}$
For gases: $\mathrm{Nu} = 0.24\mathrm{Re}^{0.6}$

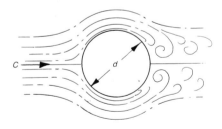

Turbulent flow over banks of pipes

Generally: $\mathrm{Nu} = 0.33C_h\mathrm{Re}^{0.6}\mathrm{Pr}^{0.3}$
For gases: $\mathrm{Nu} = 0.30C_h\mathrm{Re}^{0.6}$

In-line pipes: $C_h \simeq 1.0$
Staggered pipes: $C_h \simeq 1.1$

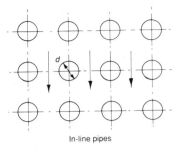

In-line pipes

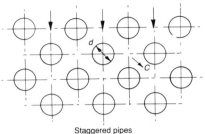

Staggered pipes

Turbulent flow over flat plate

Let:
L = the distance from the leading edge over which heat
      is transferred
C = fluid velocity

For a small temperature difference:

$\mathrm{Nu} = 0.332\mathrm{Re}^{0.5}\mathrm{Pr}^{0.33}$

For a large temperature difference:

$$\mathrm{Nu} = 0.332\mathrm{Re}^{0.5}\mathrm{Pr}^{0.33}\left(\frac{T_p}{T_f}\right)^{0.177}$$

where: T_p = plate temperature, T_f = mean fluid temperature.

$$\mathrm{Re} = \frac{\rho CL}{\mu};\ \mathrm{Nu} = \frac{hL}{k}$$

Turbulent flow in pipe

Heat transfer coefficient $h = \dfrac{k\mathrm{Nu}}{d}$

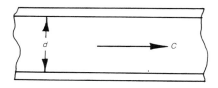

Reynold's number $\mathrm{Re} = \dfrac{\rho C d}{\mu}$

Nusselt number $\mathrm{Nu} = 0.0243\mathrm{Re}^{0.8}\mathrm{Pr}^{0.4}$
$\qquad\qquad\qquad = 0.02\mathrm{Re}^{0.8}$ for gases

For non-circular pipes use:

$$d = \frac{4 \times \text{Area of cross-section}}{\text{Inside perimeter}}$$

Heat transferred $q = hA\theta_{\mathrm{m}}$

where: $\theta_{\mathrm{m}} = \dfrac{\theta_1 - \theta_2}{\ln\dfrac{\theta_1}{\theta_2}}$

and θ_1 and θ_2 are the temperature differences at each end of a plate or tube between fluid and surface. θ_{m} is called the 'logarithmic mean temperature difference'.

3.14.10 Evaluation of Nu, Re and Pr

The fluid properties must be evaluated for a suitable mean temperature. If the temperature difference between the bulk of the fluid and the solid surface is small, use the 'mean bulk temperature' of the fluid, e.g. the mean of inlet and outlet temperatures for flow in a pipe. If the difference is large, use the 'mean film temperature' $t_{\mathrm{f}} = (\text{Mean bulk temperature} + \text{Surface temperature})/2$.

3.14.11 Radiation of heat

Radiated heat is electromagnetic radiation like light, radiowaves, etc., and does not require a medium for its propagation. The energy emitted from a hot body is proportional to the fourth power of its absolute temperature.

Symbols used:
q = radiated energy flow (watts)
T_1 = temperature of radiating body (K)
T_2 = temperature of surroundings (K)
A_1 = area of radiating body (m^2)

A_2 = area of receiving body (m^2)
e_1 = emissivity of radiating body ($= 1$ for black body)
e_2 = emissivity of surroundings
e = emissivity of intermediate wall
σ = Stefan–Boltzmann constant
$\qquad$ ($= 5.67 \times 10^{-8}$ W m^{-2} K^{-4})
f = interchange factor
F = geometric factor
h_r = heat transfer coefficient for radiation
$\qquad$ (W m^{-2} K^{-1})

Heat radiated from a body to surroundings

$q = \sigma e_1(T_1^4 - T_2^4)A_1$ (watts)

Taking into account emissivity of surroundings

$q = \sigma(e_1 T_1^4 - e_2 T_2^4)A_1$ (watts)

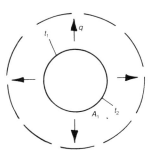

Interchange factor f

This takes into account the shape, size and relative positions of bodies.

(1) Large parallel planes: $f = \dfrac{e_1 e_2}{e_1 + e_2 - e_1 e_2}$

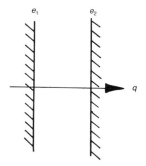

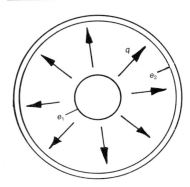

(2) Small body enclosed by another body: $f = e_1$

(3) Large body (1) enclosed by body (2):

$$f = \frac{e_1 e_2}{e_2 + \dfrac{A_1}{A_2}(e_1 - e_1 e_2)}$$

(4) Concentric spheres and concentric infinite cylinders: f as for (3)

(5) Parallel disks of different or same diameter: $f = e_1 e_2$

Geometric factor **F**

This takes into account the fact that not all radiation reaches the second body.

(a) For cases (1) to (4) above, $F = 1$.
(b) For case (5) with disks of radii r_1 and r_2 a distance x apart:

$$F = \frac{(x^2 + r_1^2 + r_2^2) - \sqrt{(x^2 + r_1^2 + r_2^2)^2 - 4r_1^2 r_2^2}}{2r_1^2}$$

Heat radiated including f *and* F

$$q = f F \sigma A_1 (T_1^4 - T_2^4)$$

Heat transfer coefficient

$$q = h_r A_1 (T_1 - T_2)$$
Therefore: $h_r = f F \sigma (T_1 + T_2)(T_1^2 + T_2^2)$

Parallel surfaces with intermediate wall

Let:
T = wall temperature
e = emissivity of wall

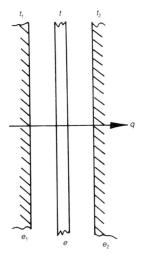

For side 1: $f_1 = \dfrac{e_1 e}{e_1 + e - e_1 e}$

For side 2: $f_2 = \dfrac{e_2 e}{e_2 + e - e_2 e}$

Intermediate temperature: $T^4 = \dfrac{f_1 T_1^4 + f_2 T_2^4}{f_1 + f_2}$

$$q = f_1 \sigma A(T_1^4 - T^4) = f_2 \sigma A(T^4 - T_2^4)$$

3.14.12 *Emissivity of surfaces*

Emissivity depends not only on the material but also to a large extent on the nature of the surface, being high for a matt surface (e.g. 0.96 for matt black paint) and low for a polished surface (e.g. 0.04 for polished aluminium).

Emissivity of surfaces (0–50°C except where stated)

Aluminium: oxidized	0.11, 0.12 (250°C)	Tile	0.97
polished	0.04, 0.05 (250°C)	Water	0.95
anodized	0.72, 0.79 (250°C)	Wood	0.90
Aluminium-coated paper,		Paint: white	0.95, 0.91 (250°C)
polished	0.20	black gloss	0.96, 0.94 (250°C)
Aluminium, dull	0.20	Paper	0.93
Aluminium foil	0.05 (average)	Plastics	0.91 (average)
Asbestos board	0.94	Rubber: natural, hard	0.91
Black body (matt black)	1.00	natural, soft	0.86
Brass: dull	0.22, 0.24 (250°C)	Steel: oxidized	0.79, 0.79 (250°C)
polished	0.03, 0.04 (250°C)	polished	0.07, 0.11 (250°C)
Brick, dark	0.90	Steel: stainless	
Concrete	0.85	weathered	0.85, 0.85 (250°C)
Copper: oxidized	0.87, 0.83 (250°C)	polished	0.15, 0.18 (250°C)
polished	0.04, 0.05 (250°C)	Steel: galvanized	
Glass	0.92	weathered	0.88, 0.90 (250°C)
Marble, polished	0.93	new	0.23, 0.42 (250°C)

3.15   Heat exchangers

In a heat exchanger, heat is transferred from one fluid to another either by direct contact or through an intervening wall. Heat exchangers are used extensively in engineering and include air coolers and heaters, oil coolers, boilers and condensers in steam plant, condensers and evaporators in refrigeration units, and many other industrial processes.

There are three main types of heat exchanger: the 'recuperator', in which the fluids exchange heat through a wall; the 'regenerative', in which the hot and cold fluids pass alternately through a space containing a porous solid acting as a heat sink; and 'evaporative', in which a liquid is cooled evaporatively and continuously, e.g. as in a cooling tower. The following deals with the recuperative type.

3.15.1   *Shell and tube heat exchangers*

One fluid flows through a series of pipes and the other through a shell surrounding them. Flow may be either 'parallel' (both fluids moving in the same direction) or 'counter flow' (fluids moving in opposite directions). Another possibility is the 'cross-flow' arrangement in which the flows are at right angles. Other types have more complex flows, e.g. the 'multi-pass' and 'mixed-flow' types. The following formulae give the heat transferred, the logarithmic mean temperature difference and the 'effectiveness'.

Symbols used:

U = overall heat transfer coefficient
A = surface area of tubes (mean)
h_a = heat transfer coefficient for hot side
h_b = heat transfer coefficient for cold side
θ = temperature difference (°C)
t = Temperature (°C)

$$\theta_1 = {}_1t_a - {}_1t_b;\ \theta_2 = {}_2t_a - {}_2t_b$$

Parallel flow

Logarithmic mean temperature difference $\theta_m = \dfrac{\theta_1 - \theta_2}{\ln \dfrac{\theta_1}{\theta_2}}$

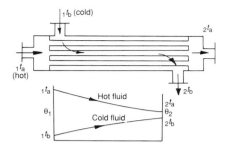

Heat transferred $q = U A \theta_m$

Overall coefficient $U = \dfrac{1}{\dfrac{1}{h_a} + \dfrac{1}{h_b}}$

Heat-exchanger effectiveness $E = \dfrac{_2t_b - _1t_b}{_1t_a - _1t_b}$

Note: if one of the fluids is a wet vapour or a boiling liquid, the temperature is constant and $_1t = _2t$.

Counter flow

The temperature range possible is greater than for the parallel-flow type. The same formulae apply.

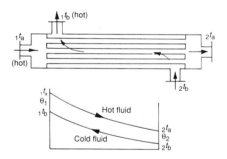

Cross-flow

Instead of using θ_m as above, $\theta_m K$ is used, where K is a factor obtained from tables.

$q = U A K \theta_m$

If one fluid is a wet vapour (constant temperature), θ_m is the same as for parallel-flow and counter-flow types.

If θ_1 and θ_2 are nearly the same, the arithmetic mean temperature difference is used:

$\theta_m = \dfrac{\theta_1 + \theta_2}{2}$

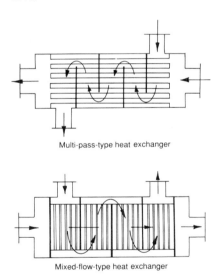

3.15.2   *Multi-pass and mixed-flow heat exchangers*

In some cases the values for θ_m for parallel- and counter-flow types may be used for these, with reasonable accuracy. Otherwise, correction factors must be used.

Multi-pass-type heat exchanger

Mixed-flow-type heat exchanger

3.15.3   *Steam condenser*

The steam condenser is a particular type of heat exchanger in which one fluid is usually cooling water and the other wet steam which condenses on the tubes carrying the cooling water. It is assumed that the steam temperature is constant throughout (i.e. at the saturation temperature). Formulae for cooling-water flow rate and the number and dimensions of the tubes are given.

Symbols used:

$\dot{m}_c$ = cooling water mass flow (kg s^{-1})
$\dot{m}_s$ = steam mass flow (kg s^{-1})
h_{fg} = latent heat of steam (kJ kg^{-1})
x = dryness fraction of steam
c = specific heat capacity of water
    (4.183 kJ kg^{-1} K^{-1} for fresh water)
h_o = overall heat transfer coefficient (kW m^{-2} K^{-1})
t_1 = water inlet temperature (°C)
t_2 = water outlet temperature (°C)
t_s = steam saturation temperature (°C)
C_t = velocity of water in tubes (m s^{-1})
A_t = area of tube bore (m^2)
D_t = outside diameter of tubes (m)
n_t = number of tubes per pass
n_p = number of tube passes
L = tube length (m)
A_s = surface area of tubes (m^2)
ρ = density of water (kg m^{-3})

Cooling water flow $\dot{m}_c = \dfrac{\dot{m}_s h_{fg}}{c(t_2 - t_1)}$

Overall heat transfer coefficient

$$h_o = 1.14 \left(\frac{C_t}{1.5}\right)^{0.5} \left(\frac{t + 18}{56}\right)^{0.25}$$

where: $t = (t_1 + t_2)/2$.

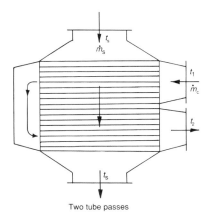

Two tube passes

Surface area of tubes $A_s = \dfrac{1.25 \dot{m}_s h_{fg} x}{h_o \theta_m}$

(assuming 25% allowance for fouling)

where: θ_m = logarithmic mean temperature difference

$$= \frac{(t_s - t_1) - (t_s - t_2)}{\ln\left(\dfrac{t_s - t_1}{t_s - t_2}\right)}$$ (assuming no undercooling of condensate)

Number of tubes per pass $n_t = \dot{m}_c / \rho A_t C_t$

Tube length $L = A_s / \pi D_t n_t n_p$

3.16 Combustion of fuels

3.16.1 *Air–fuel ratio and mixture strength*

The following deals with the combustion of solid, liquid and gaseous fuels with atmospheric air. The fuels are supposed to be composed only of carbon, hydrogen and sulphur, with perhaps oxygen and ash. The carbon, hydrogen and sulphur combine with the oxygen in the air; the nitrogen in the air remains unchanged.

The correct proportion of air for complete combustion is called the 'stoichiometric air/fuel ratio'. Usually the proportion of air is higher and the mixture is said to

be 'weak' or 'lean'. With less air the combustion is incomplete and the mixture is said to be 'rich' (see table).

Definitions:

Air/fuel ratio $R = \dfrac{\text{Amount of air}}{\text{Amount of fuel}}$

(by mass for solids and liquids and by volume for gases)
Stoichiometric air/fuel ratio R_s = ratio for complete combustion

Percentage excess air $E = \dfrac{(R - R_s)}{R_s} \times 100\%$

Mixture strength $M_s = \dfrac{R_s}{R} \times 100\%$

Weak mixture $M_s < 100\%$
Rich mixture $M_s > 100\%$

Therefore: $E = \dfrac{(100 - M_s)}{M_s} \times 100\%$

3.16.2 Combustion equations

The following are the basic equations normally used for combustion processes. A table of elements and compounds is given.

Carbon: $C + O_2 \rightarrow CO_2$; $2C + O_2 \rightarrow 2CO$
Hydrogen: $2H_2 + O_2 \rightarrow 2H_2O$
Sulphur: $S + O_2 \rightarrow SO_2$

Typical hydrocarbon fuels:
$C_4H_8 + 6O_2 \rightarrow 4CO_2 + 4H_2O$
$C_2H_6O + 3O_2 \rightarrow 2CO_2 + 3H_2O$

Carbon with air (assuming that air is composed of 79% nitrogen and 21% oxygen by volume):

$$C + O_2 + \frac{79}{21}N_2 \rightarrow CO_2 + \frac{79}{21}N_2 \text{ (by volume)}$$

$$12C + 32O_2 + \frac{28 \times 79}{21}N_2 \rightarrow 44CO_2 + \frac{28 \times 79}{21}N_2$$

(by mass)

since the molecular weights of C, O_2, CO_2 and N_2 are 12, 32, 44 and 28.

3.16.3 Molecular weights of elements and compounds

The molecular weights of elements and compounds used in combustion processes are listed in the table.

Element	Formula	Approximate molecular weight
Benzene	C_6H_6	78
Butane	C_4H_{10}	58
Carbon	C	12
Carbon monoxide	CO	28
Carbon dioxide	CO_2	44
Ethane	C_2H_6	30
Ethanol	C_2H_5OH	46
Ethene	C_2H_4	28
Hydrogen	H	2
Methane	CH_4	16
Methanol	CH_3OH	32
Nitrogen	N_2	28
Octane	C_8H_{18}	114
Oxygen	O_2	32
Pentane	C_5H_{12}	72
Propane	C_3H_8	44
Propene	C_3H_6	42
Sulphur	S	32
Sulphur monoxide	SO	48
Sulphur dioxide	SO_2	64
Water (steam)	H_2O	18

Engine exhaust and flue gas analysis

If the analysis includes the H_2O (as steam) produced by the combustion of hydrogen, it is known as a 'wet analysis'. Usually the steam condenses out and a 'dry analysis' is made.

3.16.4 Solid and liquid fuels

Let: $c = \%C$, $h = \%H_2$, $o = \%O_2$, $n = \%N_2$, $s = \%S$, all by mass.

Stoichiometric air/fuel ratio $R_s = \dfrac{(2.67c + 8h + s - o)}{23.3}$

If: $x = 0.84c + 0.3135s + 0.357n + 0.0728ER_s + 27.4R$
$y = x + 5h$ (using $E = 0$ for a stoichiometric air/fuel ratio)

Combustion products (% volume)

	CO_2	H_2O	SO_2	O_2	N_2
Wet analysis	$84\dfrac{c}{y}$	$500\dfrac{h}{y}$	$31.3\dfrac{s}{y}$	$\dfrac{7.28}{y}ER_s$	$\dfrac{35.7n+2740R}{y}$
Dry analysis	$84\dfrac{c}{x}$	0	$31.3\dfrac{s}{x}$	$\dfrac{7.28ER_s}{x}$	$\dfrac{35.7n+2740R}{x}$

3.16.5 Hydrocarbon fuels, solid and liquid

Weak mixture

Let: $c = \%C$, $h = \%H_2$, both by mass.

Then: $R_s = \dfrac{(2.67c + 8h)}{23.3}$

$x = 0.84c + 0.0728ER_s + 27.4R$
$y = x + 5h$

Combustion products (% volume)

	CO_2	H_2O	O_2	N_2
Wet analysis	$84\dfrac{c}{y}$	$500\dfrac{h}{y}$	$\dfrac{7.28}{y}ER_s$	$2740\dfrac{R}{y}$
Dry analysis	$84\dfrac{c}{x}$	0	$\dfrac{7.28}{x}ER_s$	$2740\dfrac{R}{x}$

Rich mixture $(M_s > 100\%)$

$n = \dfrac{31.3(c + 3h)}{M_s}$

$a = 0.532n - \dfrac{(c + 6h)}{12}$

$b = \dfrac{c}{12} - a$

$x = a + b + n$

$y = x + \dfrac{h}{2}$

Combustion products (% volume)

	CO_2	CO	H_2O	N_2
Wet analysis	$100\dfrac{a}{y}$	$100\dfrac{b}{y}$	$50\dfrac{h}{y}$	$100\dfrac{n}{y}$
Dry analysis	$100\dfrac{a}{x}$	$100\dfrac{b}{x}$	0	$100\dfrac{n}{x}$

Air/fuel ratio from the CO_2 in the exhaust for fuel consisting of C and H_2 by weight

$$R = 2.4\frac{\%C}{\%CO_2} + 0.072\%H_2$$

Ratio of carbon to hydrogen by mass from the dry exhaust analysis

$$r = \frac{\%C}{\%H_2} = \frac{(\%CO_2 + \%CO + \%CH_4)}{(8.858 - 0.422\%CO_2 - 0.255\%CO + 0.245\%CH_4 + 0.078\%H_2 - 0.422\%O_2)}$$

$$\%C = \frac{r}{(1+r)}100\%; \quad \%H_2 = \frac{100\%}{(1+r)}$$

3.16.6 *Liquid fuels of the type $C_pH_qO_r$*

Weak mixture

$$R_s = 4.292\frac{(32p + 8q - 16r)}{(12p + q + 16r)} \qquad\qquad x = p + 376\frac{n}{M_s} + \frac{En}{100}$$

$$n = p + \frac{q}{4} - \frac{r}{2} \qquad\qquad\qquad\qquad y = x + \frac{q}{2}$$

Combustion products (% volume)

	CO_2	H_2O	O_2	N_2
Wet analysis	$100\dfrac{p}{y}$	$50\dfrac{q}{y}$	$\dfrac{En}{y}$	$\dfrac{37\,600n}{M_s y}$
Dry analysis	$100\dfrac{p}{x}$	0	$\dfrac{En}{x}$	$\dfrac{37\,600n}{M_s x}$

Rich mixture

$$R_s = 4.292 \frac{(32p + 8q - 16r)}{(12p + q + 16r)} \qquad b = p - a$$

$$n = p + \frac{q}{4} - \frac{r}{2} \qquad x = a + b + \frac{376n}{M_s}$$

$$a = r - p - \frac{q}{2} + \frac{200n}{M_s} \qquad y = x + \frac{q}{2}$$

Combustion products (% volume)

	CO_2	CO	H_2O	N_2
Wet analysis	$100\frac{a}{y}$	$100\frac{b}{y}$	$\frac{50q}{y}$	$\frac{37\,600n}{M_s y}$
Dry analysis	$100\frac{a}{x}$	$100\frac{b}{x}$	0	$\frac{37\,600n}{M_s x}$

3.16.7 *Gaseous fuels*

For a mixture of gases such as H_2, O_2, CO, CH_4, etc., let V_1, V_2, V_3, etc., be the percentage by volume of gases, 1, 2, 3, etc., containing C, H_2 and O_2. V_n and V_c are the percentage volumes of N_2 and CO_2.

Let:
c_1, c_2, c_3, etc. = the number of atoms of carbon in each gas
h_1, h_2, h_3, etc. = the number of atoms of hydrogen in each gas
o_1, o_2, o_3, etc. = the number of atoms of oxygen in each gas

And let:
$S_c = c_1 V_1 + c_2 V_2 + \ldots$
$S_h = h_1 V_1 + h_2 V_2 + \ldots$
$S_o = o_1 V_1 + o_2 V_2 + \ldots$

$$k = S_c + \frac{S_h}{4} + \frac{S_o}{2}$$

Then:

$$R_s = \frac{k}{21}; \; R = R_s \left(1 + \frac{E}{100}\right)$$

$$x = 100R + \frac{S_o}{2} - \frac{S_h}{4} + V_n$$

$$y = x + \frac{S_h}{2}$$

Combustion products (% volume)

	CO_2	H_2O	O_2	N_2
Wet analysis	$100\dfrac{S_c+V_c}{y}$	$50\dfrac{S_h}{y}$	$\dfrac{100(21R-k)}{y}$	$\dfrac{100(Vn+79R)}{y}$
Dry analysis	$100\dfrac{S_c+V_c}{x}$	0	$\dfrac{100(21R-k)}{x}$	$\dfrac{100(Vn+79R)}{x}$

3.16.8 *Calorific value of fuels*

The calorific value of a fuel is the quantity of heat obtained per kilogram (solid or liquid) or per cubic metre (gas) when burnt with an excess of oxygen in a calorimeter.

If H_2O is present in the products of combustion as a liquid then the 'higher calorific value' (HCV) is obtained. If the H_2O is present as a vapour then the 'lower calorific value' (LCV) is obtained.

$$LCV = HCV - 207.4\%H_2 \text{ (by mass)}$$

Calorific value of fuels

	Higher calorific value	Lower calorific value
Solid (kJ kg^{-1}; 15°C)		
Anthracite	34 600	33 900
Bituminous coal	33 500	32 450
Coke	30 750	30 500
Lignite	21 650	20 400
Peat	15 900	14 500

Liquid (kJ kg^{-1}; 15°C)

Petrol (gasoline)	47 000	43 900
	average	average
Benzole (crude benzene)	42 000	40 200
Kerosene (paraffin)	46 250	43 250
Diesel	46 000	43 250
Light fuel oil	44 800	42 100
Heavy fuel oil	44 000	41 300
Residual fuel oil	42 100	40 000

Gas (MJ m^{-3}; 15°C; 1 bar)

Coal gas	20.00	17.85
Producer gas	6.04	6.00
Natural gas	36.20	32.60
Blast-furnace gas	3.41	3.37
Carbon monoxide	11.79	11.79
Hydrogen	11.85	10.00

3.16.9 *Boiler efficiency*

This may be based on either the HCV or the LCV.

$$\text{Boiler efficiency } E_b = \frac{\dot{m}_s(h_b - h_w)}{\dot{m}_f(\text{HCV or LCV})}$$

where:
$\dot{m}_s$ = mass flow of steam
$\dot{m}_f$ = mass flow of fuel
h_b = enthalpy of steam
h_w = enthalpy of feed water

Analysis of solid fuels

Fuel	Moisture (%mass)	%mass					Volatile matter (%mass of dry fuel)
		C	H_2	O_2	N_2	ash	
Anthracite	1	90.27	3.00	2.32	1.44	2.97	4
Bituminous coal	2	81.93	4.87	5.98	2.32	4.90	25
Lignite	15	56.52	5.72	31.89	1.62	4.25	50
Peat	20	43.70	6.48	44.36	1.52	4.00	65

Analysis of liquid fuels

Fuel	%mass			
	C	H_2	S	Ash, etc.
Petrol (gasolene)				
s.g. 0.713	84.3	15.7	0.0	—
s.g. 0.739	84.9	14.76	0.08	—
Benzole	91.7	8.0	0.3	—
Kerosene (paraffin)	86.3	13.6	0.1	—
DERV (diesel engine road vehicle fuel)	86.3	13.4	0.3	—
Diesel oil	86.3	12.8	0.9	—
Light fuel oil	86.2	12.4	1.4	—
Heavy fuel oil	86.1	11.8	2.1	—
Residual fuel oil	88.3	9.5	1.2	1.0

Analysis of gaseous fuels

Fuel	%volume								
	H_2	CO	CH_4	C_2H_4	C_2H_6	C_4H_8	O_2	CO_2	N_2
Coal gas	53.6	9.0	25.0	0.0	0.0	3.0	0.4	3.0	6.0
Producer gas	12.0	29.0	2.6	0.4	0.0	0.0	0.0	4.0	52.0
Natural gas	0.0	1.0	93.0	0.0	3.0	0.0	0.0	0.0	3.0
Blast-furnace gas	2.0	27.0	0.0	0.0	0.0	0.0	0.0	11.0	60.0

4 Fluid mechanics

4.1 Hydrostatics

4.1.1 *Buoyancy*

The 'apparent weight' of a submerged body is less than its weight in air or, more strictly, a vacuum. It can be shown that it appears to weigh the same as an identical volume having a density equal to the difference in densities between the body and the liquid in which it is immersed. For a partially immersed body the weight of the displaced liquid is equal to the weight of the body.

4.1.2 *Archimedes principle*

Submerged body

Let:
W = weight of body
V = volume of body = W/ρ_B
ρ_B = density of body
ρ_L = density of liquid

Apparent weight $W' = W - \rho_L V$
Then: $W' = V(\rho_B - \rho_L)$

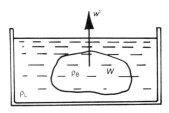

Floating body

Let:
V_B = volume of body
V_S = volume submerged

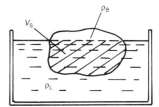

Weight of liquid displaced = Weight of body
$$\text{or } \rho_L V_S = \rho_B V_B$$

Therefore: $V_S = V_B \dfrac{\rho_B}{\rho_L}$ or $\dfrac{V_S}{V_B} = \dfrac{\rho_B}{\rho_L}$

4.1.3 *Pressure of liquids*

The pressure in a liquid under gravity increases uniformly with depth and is proportional to the depth and density of the liquid. The pressure in a cylinder is equal to the force on the piston divided by the area of the piston.

The larger piston of a hydraulic jack exerts a force greater than that applied to the small cylinder in the ratio of the areas. An additional increase in force is due to the handle/lever ratio.

4.1.4 *Pressure in liquids*

Gravity pressure $p = \rho g h$

where: ρ = fluid density, h = depth.

Units are: newtons per square metre (N m^{-2}) or pascals (Pa); $10^5 \, \text{N m}^{-2} = 10^5 \, \text{Pa} = 1 \, \text{bar} = 1000$ millibars (mbar).

Pressure in cylinder $p = \dfrac{F}{A}$

where: F = force on piston, A = piston area.

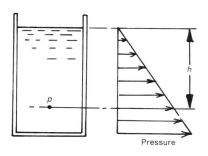

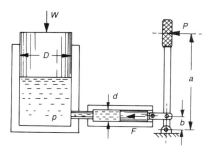

Piston area A

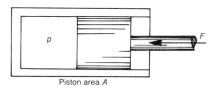

Symbols used:
ρ = density of liquid
A = plate area
x = depth of centroid
I = second moment of area of plate about a horizontal axis through the centroid
θ = angle of inclined plate to the horizontal

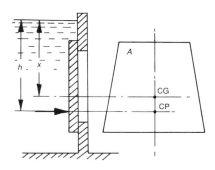

Hydraulic jack

A relatively small force F_h on the handle produces a pressure in a small-diameter cylinder which acts on a large-diameter cylinder to lift a large load W:

Pressure $p = \dfrac{4F}{\pi d^2} = \dfrac{4W}{\pi D^2}$, where $F = F_h \dfrac{a}{b}$

Load raised $W = F \dfrac{D^2}{d^2} = F_h \dfrac{a}{b} \dfrac{D^2}{d^2}$

Force on plate $F = \rho g x A$

Depth of centre of pressure $h = x + \dfrac{I}{Ax}$

$h = x + \dfrac{I \sin^2 \theta}{Ax}$ (for the inclined plate)

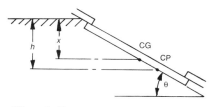

CG = centroid
CP = centre of pressure

4.1.5 *Pressure on a submerged plate*

The force on a submerged plate is equal to the pressure at the depth of its centroid multiplied by its area. The point at which the force acts is called the 'centre of pressure' and is at a greater depth than the centroid. A formula is also given for an angled plate.

4.2 Flow of liquids in pipes and ducts

The Bernoulli equation states that for a fluid flowing in a pipe or duct the total energy, relative to a height datum, is constant if there is no loss due to friction. The formula can be given in terms of energy, pressure or 'head'.

The 'continuity equation' is given as are expressions for the Reynold's number, a non-dimensional quantity expressing the fluid velocity in terms of the size of pipe, etc., and the fluid density and viscosity.

4.2.1 Bernoulli equation

Symbols used:
  p = pressure
  ρ = density
  h = height above datum
  V = velocity
  A = area

For an incompressible fluid ρ is constant, also the energy at 1 is the same as at 2, i.e.

$E_1 = E_2$

or $p_1/\rho + V_1^2/2 + gh_1 = p_2\rho + V_2^2/2 + gh_2 + \text{Energy loss}$
  (per kilogram)

In terms of pressure:
$p_1 + \rho V_1^2/2 + \rho gh_1 = p_2 + \rho V_2^2/2 + \rho gh_2 + \text{Pressure losses}$

In terms of 'head':
$p_1/\rho g + V_1^2/2g + h_1 = p_2/\rho g + V_2^2/2g + h_2 + \text{Head losses}$

Velocity pressure $p_v = \rho V^2/2$

Velocity head $h_v = V^2/2g$

Pressure head $h_p = p/\rho g$

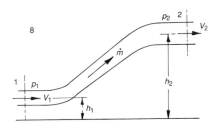

4.2.2 Continuity equation

If no fluid is gained or lost in a conduit:

Mass flow $\dot{m} = \rho_1 A_1 V_1 = \rho_2 A_2 V_2$

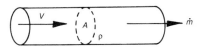

If $\rho_1 = \rho_2$ (incompressible fluid), then:
$A_1 V_1 = A_2 V_2$ or $Q_1 = Q_2$
where Q = volume flow rate

4.2.3 Reynold's number (non-dimensional velocity)

In the use of models, similarity is obtained, as far as fluid friction is concerned, when:

Reynold's number $\text{Re} = \rho \dfrac{VD}{\mu} = \dfrac{VD}{v}$

is the same for the model and the full scale version.

For a circular pipe:
D = diameter
μ = dynamic viscosity
v = kinematic viscosity

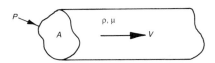

For a non-circular duct:

D = equivalent diameter $= \dfrac{4 \times \text{Area}}{\text{Perimeter}} = \dfrac{4A}{P}$

Types of flow

In a circular pipe the flow is 'laminar' below $\text{Re} \simeq 2000$ and 'turbulent' above about $\text{Re} = 2500$. Between these values the flow is termed 'transitional'.

4.2.4 Friction in pipes

The formula is given for the pressure loss in a pipe due to friction on the wall for turbulent flow. The friction factor f depends on both Reynold's number and the surface roughness k, values of which are given for different materials. In the laminar-flow region, the friction factor is given by $f = 16/\mathrm{Re}$, which is derived from the formula for laminar flow in a circular pipe. This is independant of the surface roughness.

For non-circular pipes and ducts an equivalent diameter (equal to 4 times the area divided by the perimeter) is used.

Let:

L = length (m)
D = diameter (m)
V = velocity (m s^{-1})
ρ = density (kg m^{-3})

Pressure loss in a pipe $p_{\mathrm{f}} = 4f \dfrac{L}{D} \rho \dfrac{V^2}{2}$ (N m^{-2})

Friction factor f   This depends on the Reynold's number

$$\mathrm{Re} = \frac{\rho V D}{\mu}$$

and the relative roughness k/D (for values of k, see table).

For non-circular pipes, use the equivalent diameter

$$D_{\mathrm{e}} = \frac{4 \times \mathrm{Area}}{\mathrm{Perimeter}} = \frac{4A}{P}$$

Example

For a water velocity of 0.5 m s^{-1} in a 50 mm bore pipe of roughness $k = 0.1$ mm, find the pressure loss per metre (viscosity $= 0.001$ N-s m^{-2} and $\rho = 1000$ kg m^{-3} for water).

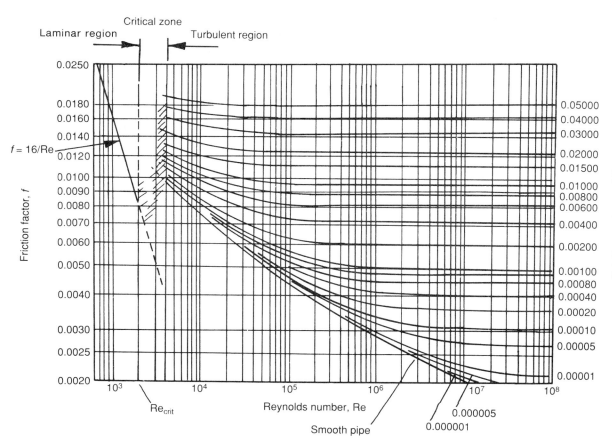

Reynold's number $\mathrm{Re} = \dfrac{1000 \times 0.5 \times 0.05}{0.001} = 2.5 \times 10^4$

Relative roughness $k/D = \dfrac{0.1}{50} = 0.002$

Friction factor (from chart) $f = 0.0073$

Pressure loss

$$p_f = 4 \times 0.0073 \times \frac{1}{0.05} \times \frac{1000 \times 0.5^2}{2} = 73 \,\mathrm{N\,m}^{-2}$$

Laminar (viscous) flow

For circular pipes only, the friction factor $f = 16/\mathrm{Re}$. This value is independant of roughness.

Typical roughness of pipes

Material of pipe (new)	Roughness, k (mm)
Glass, drawn brass, copper, lead, aluminium, etc.	'Smooth' $(k = 0)$
Wrought iron, steel	0.05
Asphalted cast iron	0.12
Galvanized iron, steel	0.15
Cast iron	0.25
Wood stave	0.2–1.0
Concrete	0.3–3.0
Riveted steel	1.0–10

4.2.5   Pipes in series and parallel

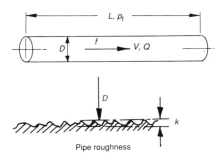

Pipe roughness

Pipes in series

The pressure loss is the sum of the individual losses:

Pressure loss $p_f = p_{f1} + p_{f2} + \dots$

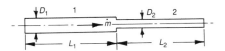

The mass flow rate is the same in all pipes, i.e.

$\dot{m} = \dot{m}_1 = \dot{m}_2 = \text{etc.}$
where: $\dot{m}_1 = \rho A_1 V_1$, etc. $\mathrm{kg\,s}^{-1}$

Pipes in parallel

The pressure loss is the same in all pipes:

Pressure loss $p_f = p_{f1} = p_{f2} = \text{etc.}$

The total flow is the sum of the flow in each pipe:
Total flow $\dot{m} = \dot{m}_1 + \dot{m}_2 + \dots$

where: $p_{f1} = 4f_1 \dfrac{L_1}{D_1} \rho \dfrac{V_1^2}{2}$, $p_{f2} = 4f_2 \dfrac{L_2}{D_2} \rho \dfrac{V_2^2}{2}$, etc.

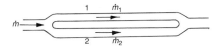

4.2.6   Pressure loss in pipe fittings and pipe section changes

In addition to pipe friction loss, there are losses due to changes in pipe cross-section and also due to fittings such as valves and filters. These losses are given in terms of velocity pressure $\rho(V^2/2)$ and a constant called the 'K factor'.

Sudden enlargement

Pressure loss $p_L = K\rho \dfrac{V_1^2}{2}$, where $K = \left(1 - \dfrac{A_1}{A_2}\right)^2$

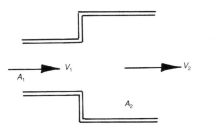

Sudden exit

Pressure loss $p_L = \rho \dfrac{V_1^2}{2}$, $(K = 1)$

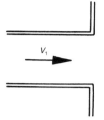

Sudden contraction

Pressure loss $p_L = K \rho \dfrac{V_2^2}{2}$

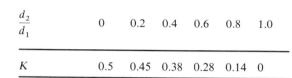

$\dfrac{d_2}{d_1}$	0	0.2	0.4	0.6	0.8	1.0
K	0.5	0.45	0.38	0.28	0.14	0

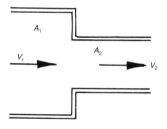

Sudden entry

Pressure loss $p_L = K \rho \dfrac{V_2^2}{2}$, where $K \simeq 0.5$

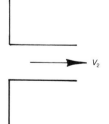

Losses in valves

Globe valve wide open $K = 10$
Gate valve wide open $K = 0.2$
Gate valve three-quarters open $K = 1.15$
Gate valve half open $K = 5.6$
Gate valve quarter open $K = 24$

Rounded entry

$K \simeq 0.05$

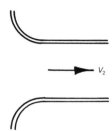

Re-entrant pipe

$K = 0.8$–1.0

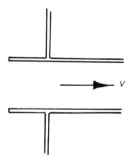

Bends

The factor K depends on R/D, the angle of bend θ, and the cross-sectional area and the Reynold's number. Data are given for a circular pipe with 90° bend. The loss factor takes into account the loss due to the pipe length.

R/D	0	1	2	4	6	8	10	16
K		1.0	0.4	0.2	0.18	0.2	0.27	0.33	0.4

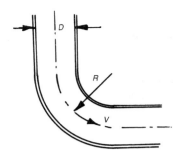

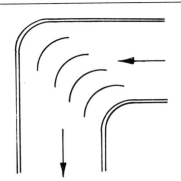

Cascaded bends

$K = 0.05$ aerofoil vanes, 0.2 circular arc plate vanes

Plate : $K = 0.2$

Aerofoil : $K = 0.05$

4.3   Flow of liquids through various devices

Formulae are given for the flow through orifices, weirs and channels. Orifices are used for the measurement of flow, weirs being for channel flow.

Flow in channels depends on the cross-section, the slope and the type of surface of the channel.

4.3.1   *Orifices*

Let:
C_d = coefficient of discharge
C_v = coefficient of velocity
C_c = coefficient of contraction
H = head
A = orifice area
A_j = jet area

$$Q = C_d A \sqrt{2gH}; \quad C_d = C_c C_v; \quad C_c = \frac{A_j}{A}; \quad C_v = \frac{V}{\sqrt{2gH}}$$

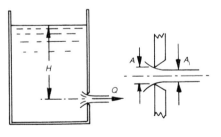

Values of C_d

Orifice type	C_d	Arrangement
a) Rounded entry	Nearly 1.0	
b) Sharp edged	0.61–0.64	
c) Borda re-entrant (running full)	About 0.72	
d) External mouthpiece	About 0.86	

4.3.2 *Weirs, vee notch and channels*

Unsuppressed weir

Flow $Q = 2.95 C_d (b - 0.2H) H^{1.5}$

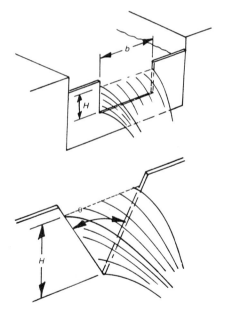

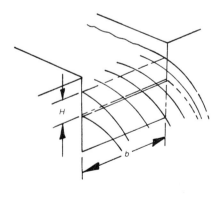

Suppressed weir

Flow $Q = 3.33 b H^{1.5}$

Vee notch

flow $Q = 2.36 C_d \tan \dfrac{\theta}{2} H^{2.5}$

where $C_d =$ discharge coefficient

Channels

Symbols used:
m = hydraulic mean radius = A/P
i = slope of channel
C = constant = $87/[1 + (K/\sqrt{m})]$
A = flow area
P = wetted perimeter

Mean velocity $V = C\sqrt{mi}$

Flow rate $Q = VA$

Values of K

Surface	K
Clean smooth wood, brick, stone	0.16
Dirty wood, brick, stone	0.28
Natural earth	1.30

Maximum discharge for given excavation

Channel	Condition	Arrangement
Rectangular Trapezoidal	$d = b/2$ Sides tangential to semicircle	

4.3.3 *Venturi, orifice and pipe nozzle*

These are used for measuring the flow of liquids and gases. In all three the restriction of flow creates a pressure difference which is measured to give an indication of the flow rate. The flow is always proportional to the square root of the pressure difference so that these two factors are non-linearly related. The venturi gives the least overall pressure loss (this is often important), but is much more expensive to make than the orifice which has a much greater loss. A good compromise is the pipe nozzle. The pressure difference may be measured by means of a manometer (as shown) or any other differential pressure device.

The formula for flow rate is the same for each type.

Let:
D = pipe diameter
d = throat diameter
ρ = fluid density
ρ_m = density of manometer fluid
p_1 = upstream pressure
p = throat pressure
C_d = coefficient of discharge
h = manometer reading

Flow rate $Q = C_d E \dfrac{\pi d^2}{4} \sqrt{\dfrac{2(p_1 - p_2)}{\rho}}$

Approach factor $E = \dfrac{1}{\sqrt{1 - \left(\dfrac{d}{D}\right)^4}}$

$(p_1 - p_2) = (\rho_m - \rho)gh$

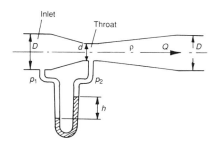

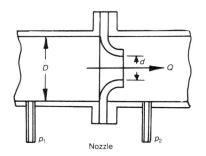

Nozzle

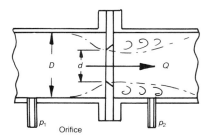

Orifice

Values of C_d

	C_d
Venturi	0.97–0.99
Orifice plate	0.60
Nozzle	0.92 to 0.98

4.4 Viscosity and laminar flow

4.4.1 *Viscosity*

In fluids there is cohesion and interaction between molecules which results in a shear force between adjacent layers moving at different velocities and between a moving fluid and a fixed wall. This results in friction and loss of energy.

The following theory applies to so-called 'laminar' or 'viscous' flow associated with low velocity and high viscosity, i.e. where the Reynold's number is low.

Definition of viscosity

In laminar flow the shear stress between adjacent layers parallel to the direction of flow is proportional to the velocity gradient.

Let:
V = velocity
y = distance normal to flow
μ = dynamic viscosity

Shear stress $\tau = \text{constant}\,\dfrac{\mathrm{d}V}{\mathrm{d}y} = \mu\dfrac{\mathrm{d}V}{\mathrm{d}y}$

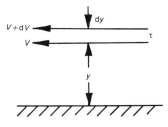

Flat plate moving over fixed plate of area A

Force to move plate $F = \tau A = \mu A\,\dfrac{V}{y}$

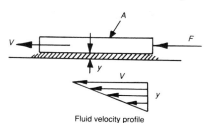

Fluid velocity profile

Kinematic viscosity

$$\text{Kinematic viscosity} = \frac{\text{Dynamic viscosity}}{\text{Density}}$$

$$\text{or } v = \frac{\mu}{\rho}$$

Dimensions of viscosity

Dynamic viscosity: $ML^{-1}T^{-1}$
Kinematic viscosity: $L^2 T^{-1}$

Units with conversions from Imperial and other units

Dynamic viscosity	Kinematic viscosity
SI unit: $N s m^{-2}$	SI unit: $m^2 s^{-1}$
$1 lbf\text{-}s ft^{-2} = 47.9 N s m^{-2}$	$1 ft^2 s^{-1} = 0.0929 m^2 s^{-1}$
$1 lbf\text{-}h ft^{-2} = 17.24 N s m^{-2}$	$1 ft^2 h^{-1} = 334 m^2 s^{-1}$
$1 \text{ poundal-s } ft^{-2} =$	
$1.49 N s m^{-2}$	
$1 lb ft^{-1} s^{-1} =$	
$1.49 kg ms^{-1}$	
$1 \text{ slug } ft^{-1} s^{-1} =$	
$47.9 kg ms^{-1}$	

Viscosity of water

Approximate values at room temperature:
$\mu = 10^{-3} N s m^{-2}$
$v = 10^{-6} m^2 s^{-1}$

Temperature (°C)	Dynamic viscosity ($\times 10^{-3} N s m^{-2}$)
0.01	1.755
20	1.002
40	0.651
60	0.462
80	0.350
100	0.278

4.4.2 *Laminar flow in circular pipes*

The flow is directly proportional to the pressure drop for any shape of pipe or duct. The velocity distribution in a circular pipe is parabolic, being a maximum at the pipe centre.

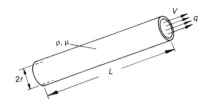

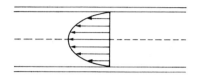

Velocity distribution

Flow $Q = \pi \dfrac{(p_1 - p_2)r^4}{8\mu L}$

Mean velocity $V = \dfrac{(p_1 - p_2)r^2}{8\mu L}$

Maximum velocity $V_m = 2V$

4.4.3 *Laminar flow between flat plates*

Flow $Q = \dfrac{(p_1 - p_2)Bt^3}{12\mu L}$

Mean velocity $V = \dfrac{(p_1 - p_2)t^2}{12\mu L}$

Maximum velocity $V_m = \frac{3}{2} V$

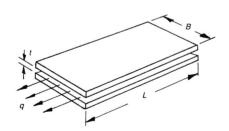

4.4.4 *Flow through annulus (small gap)*

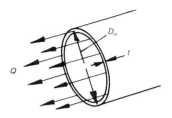

Mean velocity $V = \dfrac{Q}{\pi(R^2 - r^2)}$

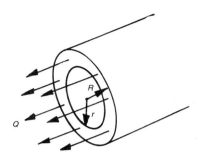

Use formula for flat plates but with $B = \pi D_m$, where D_m is the mean diameter.

Flow through annulus (exact formula)

Flow $Q = \dfrac{\pi}{8uL}(p_1 - p_2)(R^2 - r^2)\left[(R^2 + r^2) - \dfrac{(R^2 - r^2)}{\ln\dfrac{R}{r}}\right]$

4.5   Fluid jets

If the velocity or direction of a jet of fluid is changed, there is a force on the device causing the change which is proportional to the mass flow rate. Examples are of jets striking both fixed and moving plates.

Change of momentum of a fluid stream

Let:
$\dot{m}$ = mass flow rate = ρAV
V_1 = initial velocity
V_2 = final velocity
ρ = fluid density
A = flow area

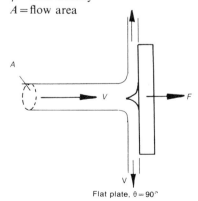

Flat plate, $\theta = 90°$

For flow in one direction, the force on a plate, etc., causing a velocity change is

$F = \dot{m}(V_1 - V_2)$

4.5.1   *Jet on stationary plates*

Jet on a flat plate

In this case $V_2 = 0$, and if $V_1 = V$

$F = \dot{m}V = \rho AV^2$ in direction of V_1

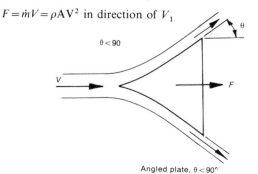

$\theta < 90$

Angled plate, $\theta < 90°$

Jet on angled plate

$F = \rho A V^2 (1 - \cos\theta)$ in direction of V_1

For $\theta = 90°$, $F = \rho A V^2$
For $\theta = 180°$, $F = 2\rho A V^2$.

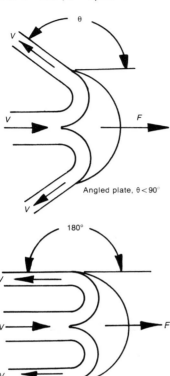

Angled plate, $\theta < 90°$

Angled plate, $\theta = 180°$

4.5.2   *Jet on moving plates*

Jet on a flat plate

$F = \rho A V (V - U)$
where: $U =$ plate velocity.

Power $P = FU = \rho A V U (V - U)$
$\qquad = \rho A V^3 r (1 - r)$

where: $r = \dfrac{U}{V}$.

Jet on angled plate

$F = \rho A V (V - U)(1 - \cos\theta)$ in direction of V
$P = \rho A V^3 r (1 - r)(1 - \cos\theta) \rho$

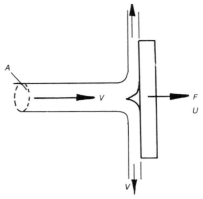

Moving flat plate

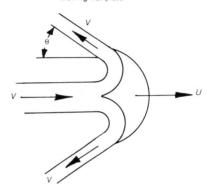

Moving angled plate

Example

If $r = \dfrac{U}{V} = 0.4$, $\theta = 170°$, $V = 10 \,\mathrm{m\,s^{-1}}$, $A = 4 \,\mathrm{cm^2}$ ($= 4 \times 10^{-4} \,\mathrm{m^3}$) and $\rho = 1000 \,\mathrm{kg\,m^{-3}}$. Then $P = 1000 \times 4 \times 10^{-4} \times 10^3 \times 0.4(1 - 0.4)(1 - \cos 170°) = 190.5$ watts

Jet on fixed curved vane

In the x direction: $F_x = \rho A V^2 (\cos\theta_1 + \cos\theta_2)$
In the y direction: $F_y = \rho A V^2 (\sin\theta_1 - \sin\theta_2)$

Jet on moving curved vane

$$F_x = \dot{m} V \left(\cos\alpha + \frac{\sin\alpha \, \cos\theta_2}{\sin\theta_1} - r \right)$$

where: $r = \dfrac{U}{V}$.

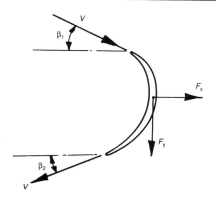

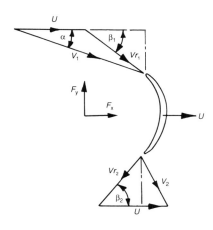

occurs when the boat speed is half the jet speed and maximum power is attained. When the water enters the front of the boat, maximum efficiency occurs when the boat speed equals the jet speed, that is, when the power is zero. A compromise must therefore be made between power and efficiency.

Let:
V = jet velocity relative to boat
U = boat velocity

$$r = \frac{U}{V}$$

$\dot{m}$ = mass flow rate of jet

Water enters side of boat

Thrust $F = \dot{m}V(1-r)$

Pump power $P = \dot{m}\dfrac{V^2}{2}$

Efficiency $\eta = 2r(1-r)$;
$\eta_{max} = 0.5$, at $r = 0.5$.

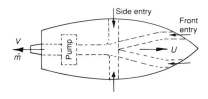

$$F_y = \dot{m}V\left(1 - \frac{\sin\theta_2}{\sin\theta_1}\right)\sin\alpha$$

$$P = \dot{m}V^2 r\left(\cos\alpha + \frac{\sin\alpha\,\cos\theta_2}{\sin\theta_1} - r\right)$$

Efficiency $\eta = 2r\left(\cos\alpha + \dfrac{\sin\alpha\,\cos\theta_2}{\sin\theta_1} - r\right)$

where: V = jet velocity, α = jet angle, θ_1 = vane inlet angle, θ_2 = vane outlet angle.

4.5.3 *Water jet boat*

This is an example of change in momentum of a fluid jet. The highest efficiency is obtained when the water enters the boat in the direction of motion. When the water enters the side of the boat, maximum efficiency

Water enters front of boat

Thrust $F = \dot{m}V(1-r)$

Pump power $P = \dot{m}\dfrac{(V^2-U^2)}{2} = \dot{m}\dfrac{V^2}{2}(1-r^2)$

Efficiency $\eta = \dfrac{2r}{(1+r)}$

$\eta = 0.667$, for $r = 0.5$.
$\eta = 1.0$, for $r = 1.0$.

Output power (both cases)

$P_o = \dot{m}V^2 r(1-r)$

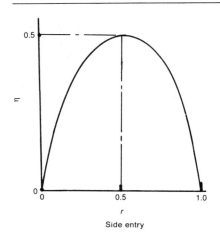

Side entry

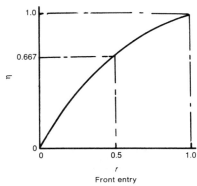

Front entry

$$P_{o\,max} = \dot{m}\frac{V^2}{4}, \text{ at } r = 0.5.$$

4.5.4 *Aircraft jet engine*

Let:
V = jet velocity relative to aircraft
U = aircraft velocity
$\dot{m}$ = mass flow rate of air
$\dot{m}_f$ = mass flow rate of fuel

Thrust $T = \dot{m}U - (\dot{m} + \dot{m}_f)V$
Output power $P = TU = \dot{m}U^2 - (\dot{m} + \dot{m}_f)UV$

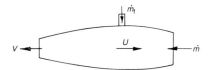

4.6 Flow of gases

Formulae are given for the compressible flow of a gas. They include isothermal flow with friction in a uniform pipe and flow through orifices. The velocity of sound in a gas is defined.

Symbols used:
p = pressure
L = pipe length
D = pipe diameter
T = temperature
C_d = discharge coefficient

V_1 = inlet velocity
R = gas constant
$\dot{m}$ = mass flow
f = friction coefficient
γ = ratio of specific heats
ρ = density

4.6.1 Isothermal flow in pipe

Pressure drop:

$$\Delta p = p_1 \left(1 - \sqrt{1 - \frac{8fLV_1^2}{2gDRT}} \right)$$

Mass flow $\dot{m} = \rho_1 V_1 \pi \dfrac{D^2}{4}$

where: $\rho_1 = \left(\dfrac{p_1}{RT} \right)$.

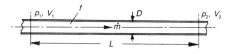

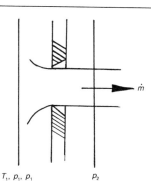

T_1, p_1, ρ_1 $\qquad\qquad$ p_2

4.6.2 Flow through orifice

Mass flow $\dot{m} = C_d A \sqrt{2g \left(\dfrac{\gamma}{\gamma - 1} \right) p_1 \rho_1 n^2 \left(1 - n^{\frac{\gamma - 1}{\gamma}} \right)}$

where: $n = p_2/p_1$; $\rho_1 = p_1/RT_1$

Maximum flow when $n = \left[\dfrac{2}{\gamma + 1} \right]^{\frac{\gamma}{\gamma - 1}} = 0.528$ for air.

4.6.3 Velocity of sound in a gas

$$V_s = \sqrt{\gamma p / \rho} = \sqrt{\gamma RT}$$

Mach number $M = \dfrac{V}{V_s}$

4.6.4 Drag coefficients for various bodies

The drag coefficient (non-dimensional drag) is equal to the drag force divided by the product of velocity pressure and frontal area. The velocity may be that of the object through the air (or any other gas) or the air velocity past a stationary object. Coefficients are given for a number of geometrical shapes and also for cars, airships and struts.

Drag coefficients for various bodies

Drag $D = C_d A \rho \dfrac{V^2}{2}$; ρ = fluid density; A = frontal area; V = fluid velocity.

Shape	$\dfrac{L}{d}$	C_d	$\dfrac{R_e}{10^4}$	A	Arrangement
Circular flat plate		1.12	100	$\dfrac{\pi d^2}{4}$	

Drag coefficients for various bodies (*continued*)

Shape	$\dfrac{L}{d}$	C_d	$\dfrac{R_e}{10^4}$	A	Arrangement
Rectangular flat plate	1 2 5 10 30 ∞	1.15 1.16 1.20 1.22 1.62 1.98	60	Ld	
Long semicircular concave surface		2.30	>0.1	Ld	
Long semicircular convex surface		1.20	>0.1	Ld	
Long circular cylinder		1.00 0.35	<20 >20	Ld	
Long square section flow on face		2.0	>0.1	Ld	
Long square section flow on edge		1.0	>0.1	$\sqrt{2}\,Ld$	

Drag coefficients for various bodies (*continued*)

Shape	$\dfrac{L}{d}$	C_d	$\dfrac{R_e}{10^4}$	A	Arrangement
(a) Cube flow on face		1.05	100	d^2	(a)
(b) Cube flow on edge		0.80	100	$\sqrt{2}\,d^2$	(b)
Sphere		0.45	< 20	$\dfrac{\pi d^2}{4}$	
		0.20	> 20		
Long elliptical section	8	0.24			
	4	0.32			
	2	0.46	10	Ld	

Drag coefficients for various bodies (*continued*)

Shape	$\dfrac{L}{d}$	C_d	$\dfrac{R_e}{10^4}$	A	Arrangement
Long symmetrical aerofoil	16 8 7 5 4	0.005 0.006 0.007 0.008 0.009	800	Ld	
Ellipsoid	5 2.5 1.25	0.06 0.07 0.13	100	$\dfrac{\pi d^2}{4}$	
Streamlined body of circular cross-section	3 4 5 6	0.049 0.051 0.060 0.072	500	$\dfrac{\pi d^2}{4}$	
Solid hemisphere flow on convex face		0.38	0.1	$\dfrac{\pi d^2}{4}$	
Solid hemisphere flow on on flat face		1.17	0.1	$\dfrac{\pi d^2}{4}$	

Drag coefficients for various bodies (*continued*)

Shape	$\dfrac{L}{d}$	C_{d}	$\dfrac{R_{\mathrm{e}}}{10^4}$	A	Arrangement
Hollow hemisphere flow on convex face	0.80	0.1	$\dfrac{\pi d^2}{4}$	
Hollow hemisphere flow on concave face	1.42	0.1	$\dfrac{\pi d^2}{4}$	
(a) High-drag car	>0.55	50	—	(a)
(b) Medium-drag car	0.45	50	—	(b)
(c) Low-drag car	<0.30	50	—	(c)

4.7 Fluid machines

4.7.1 *Centrifugal pump*

A centrifugal pump consists of an impeller with vanes rotating in a suitably shaped casing which has an inlet at the centre and usually a spiral 'volute' terminating in an outlet branch of circular cross-section to suit a pipe.

Fluid enters the impeller axially at its centre of rotation through its 'eye' and is discharged from its rim in a spiralling motion having received energy from the rotating impeller. This results in an increase in both pressure and velocity. The kinetic energy is mostly converted to pressure energy in the volute and a tapered section of the discharge branch.

Some pumps have a ring of fixed (diffuser) vanes into which the impeller discharges. These reduce the velocity and convert a proportion of the kinetic energy into pressure energy.

Symbols used:

D_1 = mean inlet diameter of impeller
D_2 = outlet diameter of impeller
b_1 = mean inlet width of impeller
b_2 = outlet width of impeller
t = vane thickness at outlet
β_1 = vane inlet angle
β_2 = vane outlet angle
N = impeller rotational speed
K = whirl coefficient
Q = flow
H = head
Z = number of vanes
ρ = fluid density
1 refers to impeller inlet
2 refers to impeller outlet
3 refers to diffuser outlet
P = power
V_t = tangential velocity
V_w = whirl velocity
V_f = flow velocity
V_r = velocity relative to vane
V = absolute velocity of fluid
η_h = hydraulic efficiency
η_v = volumetric efficiency
η_m = mechanical efficiency
η_o = overall efficiency
α = diffuser inlet angle
d_2 = diffuser inlet width
d_3 = diffuser outlet width
b = diffuser breadth (constant)
a_2 = diffuser inlet area = bd_2
a_3 = diffuser outlet area = bd_3
V_3 = diffuser outlet velocity
p = pressure rise in pump

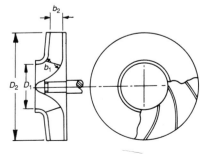

Head

Referring to velocity triangles

Theoretical head $H_{th} = \dfrac{(V_{w2}V_{t2} - V_{w1}V_{t1})}{g}$

It is usually assumed that V_{w1} is zero, i.e. there is no 'whirl' at inlet. The outlet whirl velocity V_{w2} is reduced by a whirl factor K to KV_{w2} ($K<1$). Then:

Actual head $H = \dfrac{KV_{w2}V_{t2}\eta_h}{g}$

where η_h = hydraulic efficiency. Or:

Pressure rise $p = \rho K V_{w2} V_{t2} \eta_h$

$$\begin{aligned} \text{Flow } Q &= V_{f1}A_1 = V_{f2}A_2 \\ &= \pi D_1 b_1 V_{f1} \eta_v \\ &= \pi b_2 \left(D_2 - \frac{Zt}{\sin \beta_2} \right) V_{f2} \eta_v \end{aligned}$$

where η_v = volumetric efficiency

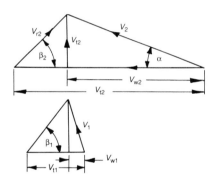

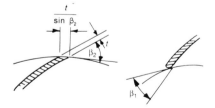

Velocity relationships

$V_{t1} = \pi D_1 N_1$; $V_{t2} = \pi D_2 N_2$
$V_{f1} = V_{t1} \tan \beta_1$; $V_2 = \sqrt{V_{w2}^2 + V_{f2}^2}$
$V_{w2} = V_{t2} - V_{f2} \cot \beta_2$

Power and efficiency

Overall efficiency $\eta_o = \eta_m \eta_v \eta_h$

Input power $P = \rho \dfrac{gHQ}{\eta_o}$

Inlet angles

Diffuser (fixed vanes):

Inlet angle $\alpha = \tan^{-1} \dfrac{V_{f2}}{V_{w2}}$

Outlet velocity $V_3 = V_2 \dfrac{a_2}{a_3}$

Vane:

Inlet angle $\beta_1 = \tan^{-1} \dfrac{V_{f1}}{V_{t1}}$ (assuming no whirl)

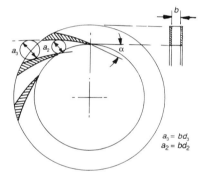

$a_3 = bd_3$
$a_2 = bd_2$

Pump volute

Velocity in volute $V_v = \dfrac{Q}{A_4}$

where: $A_4 = $ maximum area. Then:

$A_1 = \dfrac{A_4}{4}$; $A_2 = \dfrac{A_4}{2}$; $A_3 = \dfrac{3A_4}{4}$

Pump outlet velocity $V_0 = V_v \dfrac{A_4}{A_o}$

where: $A_o = $ outlet area.

Pressure head at outlet $H_o = H - \dfrac{V_o^2}{2g} - K_v \dfrac{V_v^2}{2g}$

where: $K_v = $ diffuser and volute discharge coefficient.

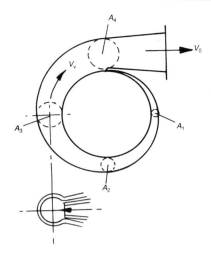

Static and total efficiencies

Static head $= H_o$ 　　　 Total head $H_t = H_o + \dfrac{V_o^2}{2g}$

Static pressure $= p_o = \rho g H_o$ 　　 Total pressure $= p_t = \rho g H_t$

Static efficiency $= \eta_s = \dfrac{p_o Q}{P}$ 　　 Total efficiency $= \eta_t = \dfrac{p_t Q}{P}$

4.7.2 *Pump characteristics*

Pump characteristics are plotted to a base of flow rate for a fixed pump speed. Head (or pressure), power and efficiency are plotted for different speeds to give a family of curves. For a given speed the point at which maximum efficiency is attained is called the 'best efficiency point' (B.E.P.). If the curves are plotted non-dimensionally a single curve is obtained which is also the same for all geometrically similar pumps.

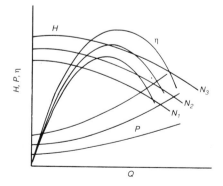

Head (H), power (P) and efficiency (η) are plotted against flow at various speeds (N) and the B.E.P. can be determined from these.

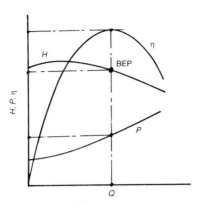

Non-dimensional characteristics

To give single curves for any speed the following non-dimensional quantities, (parameters) are plotted (see figure):

Head parameter $X_h = gH/N^2D^2$
Flow parameter $X_f = Q/ND^3$
Power parameter $X_p = P/\rho N^3 D^5$

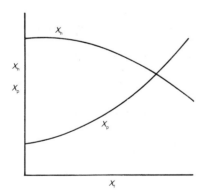

4.7.3 Cavitation

If the suction pressure of a pump falls to a very low value, the fluid may boil at a low pressure region (e.g. at the vane inlet). A formula is given for the minimum suction head, which depends on the fluid density and

vapour pressure at the operating temperature and also on the 'specific speed'.

Symbols used:
  ρ = fluid density
  p_a = atmospheric pressure
  p_v = vapour pressure of liquid at working temperature
  V_s = suction pipe velocity
  h_f = friction head loss in suction pipe plus any other losses
  H_a = pump head
  σ_c = cavitation constant which depends on vane design and specific speed

Minimum safe suction head
$H_{min} = p_a/\rho g - (\sigma_c H_a + V_s^2/2g + h_f + p_v/\rho g)$

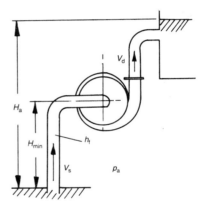

Range of σ_c:
Safe region $\sigma_c > 0.0005 N_s^{1.37}$, where N_s = specific speed.
Dangerous region $\sigma_c < 0.00022 N_s^{1.33}$
A 'doubtful zone' exists between the two values.

4.7.4 Centrifugal fans

The theory for centrifugal fans is basically the same as that for centrifugal pumps but there are differences in construction since fans are used for gases and pumps for liquids. They are usually constructed from sheet metal and efficiency is sacrificed for simplicity. The three types are: the radial blade fan (paddle wheel fan); the backward-curved vane fan, which is similar in design to the centrifugal pump; and the forward-curved vane fan which has a wide impeller and a large number of vanes. Typical proportions for impellers, maximum efficiencies and static pressures are given together with the outlet-velocity diagram for the impeller.

Centrifugal fan types

Type and application	Arrangement	b/D	Max. efficiency (%)	No. of vanes	Static pressure (cm H_2O)	Velocity triangle
Radial vanes: (paddle wheel), mill exhaust		0.35–0.45	60–70	6–8	40–46	$\beta_2 = 90°$ $v_{w2} = v_2$
Backward-curved vanes: air conditioning		0.25–0.45	75–90	8–12	12–15	$\beta_2 < 90°$ $v_{w2} < v_2$
Forward-curved vanes: ventilation		0.50–0.60	55–60	16–20	7–10	$\beta_2 > 90°$ $v_{w2} > v_2$

4.7.5 Impulse (Pelton) water turbine

This is a water turbine in which the pressure energy of the water is converted wholly to kinetic energy in one or more jets which impinge on buckets disposed around the periphery of a wheel. The jet is almost completely reversed in direction by the buckets and a high efficiency is attained. Formulae are given for the optimum pipe size to give maximum power, and for the jet size for maximum power (one jet).

Symbols used:
θ = bucket angle
H = available head
H_{tot} = total head
H_f = friction head
D = mean diameter of bucket wheel
D_p = pipe diameter
d = jet diameter
ρ = water density
f = pipe friction factor
L = length of pipe
N = wheel speed
C_v = jet velocity coefficient
V = jet velocity
V_p = pipe velocity
η_o = overall efficiency

Available head $H = (H_{tot} - H_f)$
Shaft power $P = \rho g H \eta_o$
Jet velocity $V = C_v \sqrt{2gH}$
Mean bucket speed $U = \pi D N$

Flow through jet $Q = \dfrac{\pi d^2 V}{4}$

Hydraulic efficiency $\eta_h = 2r(1-r)(1 + k\cos\theta)$

where: $r = \dfrac{U}{V}$, θ = bucket angle (4–7°),

k = friction coefficient (about 0.9).

Maximum efficiency (at $r = 0.5$): $\eta_h(max) = \dfrac{(1 + k\cos\theta)}{2}$

Overall efficiency $\eta_o = \eta_h \eta_m$

Maximum power when $H_f = \dfrac{H_{tot}}{3} = \dfrac{4fLV_p^2}{2gD_p}$. Hence:

Optimum size of supply pipe $D_p = \sqrt[5]{\dfrac{fLQ^2}{H_{tot}}}$

(approximately)

Jet size for maximum power $d = \left(\dfrac{D_p^5}{8fL}\right)^{\frac{1}{4}}$

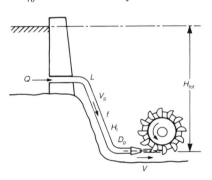

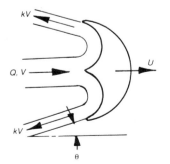

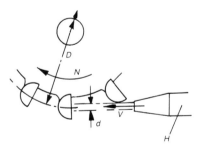

4.7.6 Reaction (Francis) water turbine

The head of water is partially converted to kinetic energy in stationary guide vanes and the rest is converted into mechanical energy in the 'runner'. The water first enters a spiral casing or volute and then into the guide vanes and a set of adjustable vanes which are used to control the flow and hence the power. The water then enters the runner and finally leaves via the 'draft tube' at low velocity. The draft tube tapers to reduce the final velocity to a minimum.

Velocity triangles

Radial velocities: $V_{f1} = Q/\pi b_1 D_1$ (inlet)
$V_{f2} = Q/\pi b_2 D_2$ (outlet)

Tangential velocities: $V_{t1} = \pi D_1 N$ (inlet)
$V_{t2} = \pi D_2 N$ (outlet)

Whirl velocities: $V_{w1} = gH\eta_h/V_{t1}$ (inlet, usually)
$V_{w2} = 0$ (outlet, usually)

Guide vane velocity: $V_1 = \sqrt{2gH}$

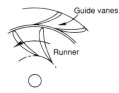

Vane and blade angles

Guide vanes: $\alpha = \tan^{-1} V_{f1}/V_{w1}$
Blade inlet: $\beta_1 = \tan^{-1} V_{f1}/(V_{t1} - V_{w1})$
Blade outlet: $\beta_2 = \tan^{-1} V_{f2}/V_{t2}$

Overall efficiency $\eta_{ov} = \eta_m \eta_h$
Shaft power $= \rho g H Q \eta_o$
Available head $H = H_{tot} - H_f - V_o^2/2g$
where: $V_o =$ draft tube outlet velocity.

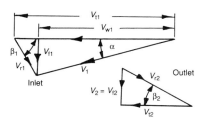

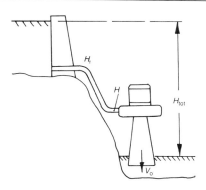

Specific speed of pumps and turbines

It is useful to compare design parameters and characteristics of fluid machines for different sizes. This is done by introducing the concept of 'specific speed', which is a constant for geometrically similar machines.

4.7.7 *Specific speed of pumps and turbines*

Symbols used:
$N =$ speed of rotation
$Q =$ flow
$H =$ head
$P =$ power

Specific speed of pump $N_s = \dfrac{N\sqrt{Q}}{H^{\frac{3}{4}}}$

Specific speed of turbine $N_s = \dfrac{N\sqrt{P}}{H^{\frac{5}{4}}}$

5

Manufacturing technology

5.1 Metal processes

Metals can be processed in a variety of ways. These can be classified roughly into casting, forming and machining.

The following table gives characteristics of different processes for metals, although some may also apply to non-metallic materials such as plastics and composites.

General characteristics of metal processes

Process	Economic quantity	Materials (typical)	Optimum size	Minimum section (mm)	Holes possible	Inserts possible
Sand casting	Small/large	No limit	1–100 kg	3	Yes	Yes
Die casting, gravity	Large	Al, Cu, Mg, Zn alloys	1–50 kg	3	Yes	Yes
Die casting, pressure	Large	Al, Cu, Mg, Zn alloys	50 g to 5 kg	1	Yes	Yes
Centrifugal casting	Large	No limit	30 mm to 1 m diameter	3	Yes	Yes
Investment casting	Small/large	No limit	50 g to 50 kg	1	Yes	No
Closed die forging	Large	No limit	3000 cm^3	3	Yes	No
Hot extrusion	Large	No limit	500 mm diameter	1	—	No
Hot rolling	Large	No limit	—	—	No	No
Cold rolling	Large	No limit	—	—	No	No
Drawing	Small/large	Al, Cu, Zn, mild steel	3 mm/6 m diameter	0.1	No	Yes
Spinning	One-off, large	Al, Cu, Zn, mild steel	6 mm/4.5 m diameter	0.1	No	Yes
Impact extrusion	Large	Al, Pb, Zn, Mg, Sn	6–100 mm diameter	0.1	—	No
Sintering	Large	Fe, W, bronze	80 g to 4 kg	0.5	Yes	Yes
Machining	One-off, large	No limit	—	—	Yes	Yes

5.2 Turning

5.2.1 *Single point metal cutting*

In metal cutting, a wedge-shaped tool is used to remove material from the workpiece in the form of a 'chip'. Two motions are required: the 'primary motion', e.g. the rotation of the workpiece in a lathe; and the 'secondary motion', e.g. the feed of a lathe tool. Single-point tools are used for turning, shaping, planing, etc., and multi-point tools are used for milling, etc. It is necessary to understand the forces acting on the tool and their effects on power requirement, tool life and production cost.

In the following tables of tool forces and formulae specific power consumption, metal removal rate, tool life, etc., are given. A graph shows the tool life plotted against cutting speed for high-speed steel, carbide and ceramic tools.

5.2.2 *Cutting tool forces*

Tool forces vary with cutting speed, feed rate, depth of cut and rake angle. Force may be measured experimentally by using a 'cutting tool dynamometer' in which the tool is mounted on a flexible steel diaphragm and its deflections in three planes measured by three electrical transducers. Three meters indicate the force, typically of 25 N up to, say, 2000 N. Graphs show typical characteristics.

Symbols used:
F_c = cutting force (in newtons)
F_r = radial force (in newtons)
F_f = feed force (in newtons)

Resultant force on tool in horizontal plane

$= \sqrt{F_f^2 + F_r^2}$ newtons

5.2.3 *Cutting power, P*

Let:
D = work diameter (mm)
d = depth of cut (mm)
N = number of revolutions per minute

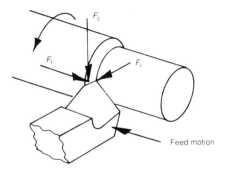

Feed motion

$$P = F_c \frac{v}{60} \text{ (watts)}$$

where: $v = \dfrac{\pi(D-d)N}{1000}$ (m min^{-1})

Metal removal rate $Q = \dfrac{\pi(D-d)dfN}{1000}$ (cm^3 min^{-1})

where: f = feed rate (mm rev^{-1}).

Specific power consumption $P_s = \dfrac{P}{Q}$ (watts cm^{-3} min^{-1})

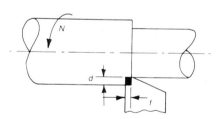

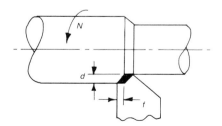

Typical values of P_s

Material	Specific power consumption, P_s
Plain carbon steel	34
Alloy steel	71
Cast iron	24
Aluminium alloy	12
Brass	25

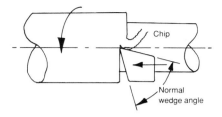

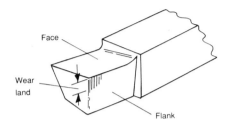

5.2.4 *Tool life, T*

$$T = \left(\frac{C}{v}\right)^{\frac{1}{n}} \text{(min)}$$

Values of C and n

Tool material	C	n	Wear land width (mm)	
			Roughing	Finishing
High-speed steel	60–100	0.08–0.15	1.5	0.25–0.38
Cemented carbide	200–330	0.16–0.5	0.75	0.25–0.38
Ceramic	330–600	0.40–0.6	0.25–0.38	0.25–0.38

5.2.5 *Tool characteristics*

Force versus cutting speed

F_c is constant over normal range of cutting speed.
F_f increases slowly with cutting speed.

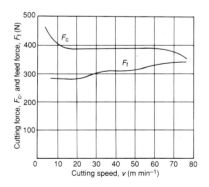

Force versus depth of cut

F_c increases with depth of cut.
F_f increases at decreasing rate with depth of cut.

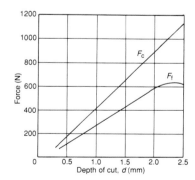

Force versus rake angle

F_c and F_f fall slowly with rake angle.

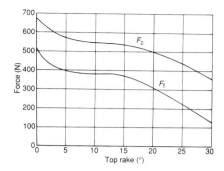

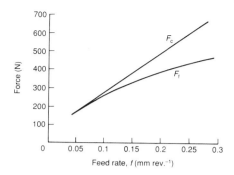

Force versus feed rate

F_c increases linearly with feed rate.
F_f increases in a curve with decreasing rate.

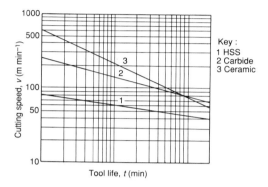

5.2.6 Cutting speeds

Turning cutting speeds (m min^{-1})

Material	Tool material			
	High-speed steel	Super-high-speed steel	Stellite	Tungsten carbide
Aluminium alloys	70–100	90–120	>200	>350
Brass, free cutting	70–100	90–120	170–250	350–500
Bronze	40–70	50–80	70–150	150–250
Grey cast iron	35–50	45–60	60–90	90–120
Copper	35–70	50–90	70–150	100–300
Magnesium alloys	85–135	110–150	85–135	85–135
Monel metal	15–20	18–25	25–45	50–80
Mild steel	35–50	45–60	70–120	—
High tensile steel	5–10	7–12	20–35	—
Stainless steel	10–15	12–18	30–50	—
Thermosetting plastic	35–50	45–60	70–120	100–200

5.2.7   Turning of plastics

Turning of plastics – depth of cut, feed, and cutting speed

Material	Condition	Depth of cut (mm)	Feed (mm rev^{-1})	Cutting speed (m min^{-1})		
				HSS	Brazed carbide	Throw-away carbide tip
Thermoplastics, polyethylene, polypropylene, TFE fluorocarbon	Extruded, moulded or cast	4	0.25	50	145	160
High-impact styrene, modified acrylic	Extruded, moulded or cast	4	0.25	53	160	175
Nylon, acetals and polycarbonate	—	4	0.25	50	160	175
Polystyrene	Moulded or extruded	4	0.25	18	50	65
Soft grades of thermosetting plastic	Cast, moulded or filled	4	0.25	50	160	175
Hard grades of thermosetting plastic	Cast, moulded or filled	4	0.25	48	145	160

HSS, high-speed steels.

5.2.8   Typical standard times for capstan and turret lathe operations

Operation	Time (s)	Operation	Time (s)
Change speed	3	Engage feed	1.5
Change feed	3	Feed to bar stop	3.5
Index tool post	3.5	Chuck in, 3-jaw chuck	4.5

5.2.9   Lathe-tool nomenclature and setting

There are many types of lathe tool, the principal ones being: bar turning; turning and facing; parting-off; facing; boring; and screw cutting. Some are made from a bar of tool steel, others with high-speed steel tips welded to carbon steel shanks and some with tungsten carbide tips brazed to a steel shank. A tool holder with interchangeable tips can also be used.

Tool features

For cutting to take place the tool must have a 'front clearance angle' which must not be so large that the tool is weakened. There must also be a 'top rake angle' to increase the effectiveness of cutting. The value of this angle depends on the material being cut. Typical values are given in the following table.

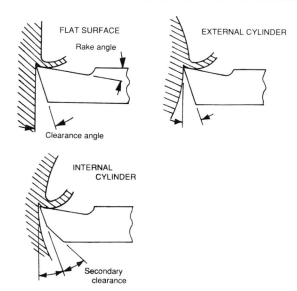

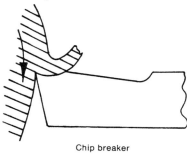

Chip breaker

Rake angle for different workpiece materials

Workpiece material	Tensile strength $(N\,mm^{-2})$	Tool rake angle $(°)$
High tensile steel	1550	-8
Nickel–chrome steel	1000–1150	-5
Steel	750	-3
Steel forging	450–600	-2
Brass and bronze	—	0
Cast iron	—	2
Mild steel	—	7
Free-cutting mild steel	—	10
Light alloys	—	12

Another feature is the 'chip breaker' which breaks long, dangerous and inconvenient streamers of 'swarf' into chips.

Plan approach angle

To reduce the load on the tool for a given depth of cut the cutting edge can be angled to increase its length. Note the direction of chip flow – if the angle is too large there is a danger of chatter.

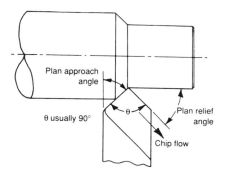

Other features

In addition to front clearance and top rake, there are side clearance and side rake. A small nose radius improves cutting and reduces wear.

Symbols used:
ϕ = top rake angle
α = front clearance angle
β = wedge angle
δ = plan relief or trail angle
ε = plan approach angle
θ = true rake angle
γ = true wedge angle
λ = side clearance angle
ψ = side rake angle.

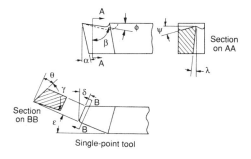

Single-point tool

Tool setting

The tool must not be set too high or too low, or inclined at an angle. The effects are shown in the figure.

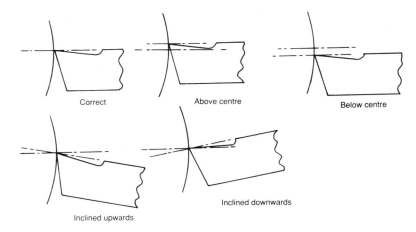

TOOL SETTING

Above centre: tool tends to rub.
Below centre: work tends to climb over tool.

Inclined upwards: tool rubs.
Inclined downwards: work tends to drag tool in.

5.2.10 *Parting-off tool*

This is used for 'parting-off' the workpiece from bar stock held in a chuck. Note that there is 'body clearance' on both sides as well as 'side clearance'. The tool is weak and must be used with care. It must be set on or slightly above centre. If set even slightly below centre the work will climb onto the tool before parting-off.

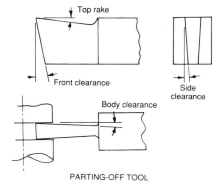

PARTING-OFF TOOL

5.3 Drilling and reaming

A twist drill is a manually or machine rotated tool with cutting edges to produce circular holes in metals, plastics, wood, etc. It consists of a hardened steel bar with usually two helical grooves or 'flutes' ending in two angled cutting edges. The flutes permit many regrinds and assist in removal of cuttings.

Drills vary in size from a fraction of a millimetre to over 10 cm. As with a lathe turning tool, the cutting edges must have top rake and clearance. Grinding is best done on a special drill grinding machine.

5.3.1 *Helix and point angles*

The helix angle is usually a standard size but 'quick' and 'slow' helix angles are used for particular materials. It is sometimes necessary, e.g. for brass and thin material, to grind a short length of straight flute, as shown. It is also sometimes necessary to thin down the web or core.

The point angle was traditionally about 120° (included angle), but other angles are now used to suit the material. The lip clearance also varies (see table).

5.3.2 *Core drills*

Core drills have three or four flutes and are used for opening out existing holes, e.g. core holes in castings.

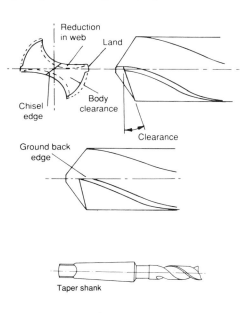

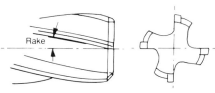

FOUR-FLUTE CORE DRILL

5.3.3 *Reamers*

A reamer is used to finish a hole accurately with a good surface finish. It is a periphery cutting tool, unlike the drill which is end cutting. Rake and clearance are required as shown; note that a reamer must be ground on the clearance face otherwise the size will be lost. Flutes may be straight or helical (usually left handed). A hand reamer requires a long slow taper, but machine reamers have a short 45° lead. The hole is drilled only slightly smaller than the reamer diameter, the allowance is about 0.015 mm per millimetre, but depends on the material. Taper reamers are used for finishing holes for taper pins.

Reamer

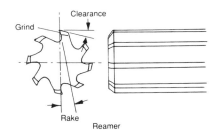

Reamer

Taper reamer

5.3.4 *Drilling parameters*

The tables below give drilling feeds and speeds including information on drilling plastics. Cutting lubricants for drilling, reaming and tapping are also given and tapping drill sizes for metric coarse threads. A table of suggested angles for drills is given.

Drilling feeds

Drill diameter (mm)	Feed (mm rev.$^{-1}$)	
	Hard materials*	Soft materials†
1.5	0.05	0.05
3.0	0.05	0.07
6.0	0.07	0.10
9.0	0.10	0.15
12.0	0.12	0.20
19.0	0.18	0.30
25.0	0.22	0.35

*Steels above 0.3 %C and alloy steels.
†Grey cast iron, steels below 0.3 %C, brass, bronze, aluminium alloys, etc.

High-speed-drill speeds

Material	Speed* (m s^{-1})
Cast iron	0.4–0.6
Mild steel	0.3–0.5
60/40 brass	0.8–1.0
Medium carbon steel	0.2–0.3

*Speed $= \pi DN/60\,000$ m s^{-1}, where $D =$ diameter (mm), $N =$ number of revolutions per minute.

Drilling plastics, cutting speeds and feeds

Material	Condition	Cutting speed (m min^{-1})	Feed (mm rev.$^{-1}$) for nominal hole diameter (mm) of:							
			1.5	3.0	6.0	12.0	20.0	25.0	30.0	50.0
Polyethylene, polypropylene, TFE-fluorocarbon	Extruded, moulded or cast	33	0.12	0.25	0.30	0.38	0.46	0.50	0.64	0.76
High-impact styrene, modified acrylic	Extruded, moulded or cast	33	0.05	0.10	0	0.15	0.15	0.20	0.20	0.25
Nylon, acetals, polycarbonate	Moulded	33	0.05	0.12	0.1	0.20	0.25	0.30	0.38	0.38
Polystyrene	Moulded or extruded	66	0.03	0.05	0	0.10	0.13	0.15	0.18	0.20
Soft grades of thermosetting plastic	Cast, moulded or filled	50	0.08	0.13	0	0.20	0.25	0.30	0.38	0.38
Hard grades of thermosetting plastic	Cast, moulded or filled	33	0.05	0.13	0.15	0.20	0.25	0.30	0.38	0.38

Cutting lubricants for drilling, reaming and tapping

Material	Drilling	Reaming	Tapping
Mild steel (hot and cold rolled)	Soluble oil, mineral oil, lard oil	Mineral lard oil	Soluble oil, lard oil
Tool steel (carbon and high speed)	Soluble oil, lard oil with sulphur	Lard oil	Sulphur base oil, mineral lard oil
Alloy steel	Soluble oil, mineral oil	Lard oil	Sulphur base oil, mineral lard oil
Brass and bronze	Dry, lard oil, paraffin mixture	Soluble oil	Soluble oil, lard oil
Copper	Soluble oil	Soluble oil	Soluble oil, lard oil
Aluminium	Paraffin, lard oil	Mineral lard oil	Soluble oil, mineral lard oil
Monel metal	Lard oil, sulphur base oil	Mineral lard oil, sulphur base oil	Mineral lard oil, sulphur base oil
Malleable iron	Soluble oil	Soluble oil	Soluble oil
Cast iron	Dry	Dry	Dry, lard oil for nickel cast iron

Tapping drill sizes for metric coarse threads

Nominal diameter (mm)	Thread pitch (mm)	Tap drill size (mm)	Nominal diameter (mm)	Thread pitch (mm)	Tap drill size (mm)
1.6	0.35	1.20	20.0	2.50	17.5
2.0	0.40	1.60	24.0	3.0	21.0
2.5	0.45	2.05	30.0	3.50	26.5
3.0	0.50	2.50	36.0	4.00	32.0
3.5	0.60	2.90	42.0	4.50	37.5
4.0	0.70	3.30	48.0	5.00	43.0
5.0	0.80	4.20	56.0	5.50	50.5
6.0	1.0	5.30	64.0	6.00	58.0
8.0	1.25	6.80	72.0	6.00	66.0
10.0	1.50	8.50	80.0	6.00	74.0
12.0	1.75	10.20	90.0	6.00	84.0
14.0	2.00	12.00	100.0	6.00	94.0
16.0	2.00	14.00			

Drill angles

Material	Helix angle	Point angle (°)	Lip clearance (°)
Aluminium alloy	Quick	140	12–15
Magnesium alloy	Standard	100	12–15
Brass	Slow	130	10–12
Copper	Quick	125	12–15
Bakelite	Slow	30	12–15
Manganese steel	Slow	130	7–10

5.4 Milling

5.4.1 *Milling process*

Milling machines produce mainly flat surfaces by means of a rotating cutter with multiple cutting edges. The two main types of machine are the horizontal and the vertical spindle. Milling cutters usually have teeth cut on the periphery and/or on the end of a disk or cylinder. Alternatively, 'inserted-tooth' cutters with replaceable teeth may be used. In horizontal milling 'up-cutting' is the usual practice, but 'down-cutting' may be used. The types of cutter are listed in the following table.

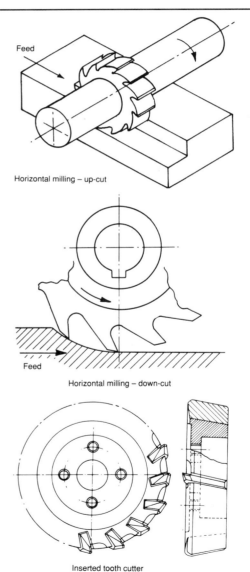

Horizontal milling – up-cut

Horizontal milling – down-cut

Inserted tooth cutter

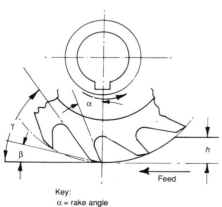

Key:
α = rake angle
β = primary clearance angle
γ = secondary clearance angle
h = depth of cut

Types of milling cutter

Type	Arrangement of teeth	Application	Size	Appearance
Cylindrical (slab or rolling)	Helical teeth on periphery	Flat surfaces parallel to cutter axis	Up to 160 × 160 mm	
Side and face	On periphery and both sides	Steps and slots	Up to 200 mm diameter, 32 mm wide	
Straddle ganged	On periphery and both sides	Cutting two steps	Up to 200 mm diameter, 32 mm wide	
Side and face staggered tooth	Teeth on periphery. Face teeth on alternate sides	Deep slots	Up to 200 mm diameter, 32 mm wide	
Single angle	Teeth on conical surface and flat face	Angled surfaces and chamfers	60–85° in 5° steps	

Types of milling cutter (*continued*)

Type	Arrangement of teeth	Application	Size	Appearance
Double angle	Teeth on two conical faces	Vee slots	45°, 60°, 90°	
Rounding	Concave quarter circle and flat face	Corner radius on edge	1.5–20 mm radius	
Involute gear cutter	Teeth on two involute curves	Involute gears	Large range	
End mill	Helical teeth at one end and circumferential	Light work, slots, profiling, facing narrow surfaces	⩽ 50 mm	

Types of milling cutter (*continued*)

Type	Arrangement of teeth	Application	Size	Appearance
Tee slot	Circumferential and both sides	Tee slots in machine table	For bolts up to 24 mm diameter	
Dovetail	On conical surface and one end face	Dovetail machine slides	38 mm diameter, 45° and 60°	
Shell end mill	Circumferential and one end	Larger work than end mill	40–160 mm diameter	
Slitting saw (slot)	Circumferential teeth	Cutting off or slitting. Screw slotting	60–400 mm diameter	
Concave–convex	Curved teeth on periphery	Radiusing	1.5–20 mm radius	

5.4.2 *Milling parameters*

Power for peripheral milling

Symbols used:
P = power (watts)
v = cutting speed (m s^{-1})
z = number of teeth
b = chip width (mm)
C = constant
f = feed per tooth (mm)
d = depth of cut (mm)
r = radius of cutter (mm)
x, y = indices
k = constant

$$P = kvzbCf^x\left(\frac{d}{r}\right)^y$$

Values of x, y, k and C are given in the tables.

Material	x	y	k
Steels	0.85	0.925	0.164
Cast iron	0.70	0.85	0.169

Material	C*	
Free machining carbon steel	980 (120 BHN)	1190 (180 BHN)
Carbon steels	1620 (125 BHN)	2240 (225 BHN)
Nickel–chrome steels	1460 (125 BHN)	2200 (270 BHN)
Nickel–molybdenum and chrome–molybdenum steels	1600 (150 BHN)	1960 (280 BHN)
Chrome–Vanadium steels	1820 (170 BHN)	2380 (190 BHN)
Flake graphite cast iron	635 (100 BHN)	1330 (263 BHN)
Nodular cast irons	1110 (annealed)	1240 (as cast)

*BHN numbers are hardness grades.

Milling cutting speeds

Let:
D = cutter diameter (mm),
N = number of revolutions per minute.

Cutting speed $v = \pi DN/1000$ (m min^{-1})

Milling cutting speeds at a feed rate of 0.2 mm per tooth

Metal being cut	Cutting speed ($m\ min^{-1}$)							
	Brazed cutters ISO carbide grade				Indexable inserts ISO carbide grade			
	P10	P30	P40	K20	P10	P30	P40	K20
Mild steel	150	130	100	20	200	170	130	—
Carbon steel 0.7%	120	90	75	—	150	90	75	—
Steel castings	60	45	50	—	80	75	50	—
Stainless steel	100	100	100	—	125	125	115	—
Grey cast iron	150	130	110	—	150	130	110	—
Aluminium alloy	—	—	—	600	—	—	—	600

Milling cutting speeds for high-speed steel cutters

Material being cut	Cutting speed ($m\ min^{-1}$)
Alloy steel	10
Cast iron	20
Low-carbon steel	28
Bronze	35
Hard brass	45
Copper	60
Aluminium alloy	100

Table feed rate

For the values given in the table below

Feed rate $f_t = fzN$ ($mm\ min^{-1}$)

where: f = feed/tooth (mm), z = number of teeth, N = number of revolutions per minute of cutter.

Typical values of feed per tooth (mm)

Material being cut	Face mills		Side and face mills		End mills		Saws	
	HSS	Carbide	HSS	Carbide	HSS	Carbide	HSS	Carbide
Aluminium alloy, brass, bronze	0.55	0.50	0.33 0.18	0.30 0.18	0.28	0.25	0.13	0.13
Copper	0.30	0.30	0.18	0.18	0.15	0.15	0.07	0.07
Cast iron	0.40	0.50	0.22	0.30	0.20	0.25	0.10	0.13
Low carbon steel	0.25	0.35	0.15	0.20	0.13	0.18	0.07	0.10
Alloy steel	0.20	0.35	0.13	0.20	0.10	0.18	0.05	0.10

HSS, high-speed steels.
These values should be lowered for finishing and increased for rough milling.

Metal removal rate in milling

Material being cut	Metal removal rate $(\mathrm{mm}^3\,\mathrm{kW}^{-1}\,\mathrm{min}^{-1})$
Mild steel	18 900
Alloy steel	10 500
Cast steel	12 600
Grey cast iron	12 600
Stainless steel	8 400
Copper	18 900
Aluminium	42 000
Magnesium	42 000
Titanium	10 500

5.5 Grinding

5.5.1 *Grinding machines*

Grinding machines produce flat, cylindrical and other surfaces by means of high-speed rotating abrasive wheels. Grinding is a means of giving a more accurate finish to a part already machined, but is also a machining process in its own right. The main types of machine are: the 'surface grinding machine' for flat surfaces; and the 'cylindrical grinding machine' for cylindrical surfaces. More complex shapes are produced by shaped wheels called 'contour grinding wheels'. 'Bench' and 'pedestal' grinders are used for tool sharpening, etc.

5.5.2 *Grinding wheels*

Typical materials for wheels are bonded abrasive powders such as aluminium oxide (Al_2O_3), silicon carbide (SiC) and diamond dust.

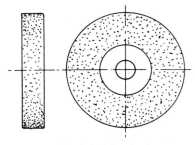

Standard grinding wheel

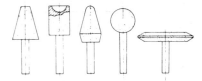

Contour grinding wheels

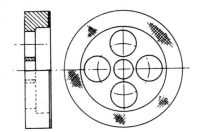

Steel wheel coated with abrasive

5.5.3 Grinding process calculations (cylindrical grinding)

Symbols used:
t = chip thickness (mm)
f = feed or depth of cut (mm)
p = pitch of grains (mm)
b = width of cut (mm)
P = power (watts)
v = wheel peripheral velocity (mm s^{-1})
u = work peripheral velocity (mm s^{-1})
d = wheel diameter (mm)
D = work diameter (mm)
F = tangential force on wheel (newtons)

Chip thickness $t = \dfrac{2pu}{v}\sqrt{\dfrac{(D \pm d)f}{Dd}}$

minus sign for internal grinding

Power $P = \dfrac{Fv}{1000}$

Energy per unit volume removed $E = \dfrac{P}{bfv}$ (J mm^{-3})

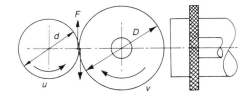

5.6 Cutting-tool materials

5.6.1 Carbon steels

Their use is restricted to the cutting of soft metals and wood. Performance is poor above 250 °C.

5.6.2 High-speed steels

These are used extensively, particularly for multi-point tools. They have been replaced to a large extent by carbides for single-point tools. Their main application is for form tools and complex shapes, e.g. for gear-cutting and broaching. They are also used for twist drills, reamers, etc.

5.6.3 Carbides

These consist of powdered carbides of tungsten, titanium, tantalum, niobium, etc., with powdered cobalt as binder. They are produced by pressing the powder in dies and sintering at high temperature. They are then ground to the final shape. They are generally used as tips and can operate up to 1000 °C.

5.6.4 Laminated carbide

These consist of a hard thin layer of titanium carbide bonded to a tungsten carbide body. The surface has very high strength at high temperature, whilst the body has high thermal conductivity and thus efficient removal of heat.

5.6.5 Diamonds

These are the hardest of all cutting materials with low thermal expansion and good conductivity. They are twice as good as carbides under compression. A good finish can be obtained with non-ferrous metals and final polishing can be eliminated. Diamonds are particularly good for cutting aluminium and magnesium alloys, copper, brass and zinc. They have a long life.

5.6.6 *Characteristics of steel tools*

Carbon steels (for softer metals and wood; poor performance above 250°C)

Composition	Characteristics	Applications
Plain carbon steel, 0.2%Mn, water hardening 0.7–0.8%C	High toughness, low hardness	Shear blades, chisels, turning mandrels
0.9%C	General purpose. Best combination of toughness and hardness	Large taps and reamers
1.0–1.4%C	High hardness, keen edge, low shock resistance	Taps, screw dies, twist drills, mills for soft metals, files
Carbon steel + vanadium 0.8–1.0%C, 0.2%Mn, 0.2%Va	Water hardening, takes keen edge, more shock resistant than plain carbon steel	Screw taps and dies, twist drills, reamers, broaches
Chrome steel 0.9%C, 0.2%Mn, 0.5%Cr	Water hardening, good abrasion resistance, takes high compression	Drawing dies, wood planes, chisels
High manganese steel 0.55–0.8%C, 0.6–0.8%Mn	Oil hardening, tougher but less hard, high shock resistance	Bending form dies, hammers, tool shanks

High-speed steels

Type	Composition (%)						Characteristics
	C	W	Cr	Va	Co	Mo	
Super	0.8	18/22	4.5	1.5	10/12	—	Highest temperature of HSS. Very hard but not so tough. Most expensive. For materials with tensile strength > 1225 MPa
General purpose	0.75	18	4.15	1.2	—	—	Tougher than super and cheaper, for materials over 1225 MPa tensile strength
General purpose tungsten/molybdenum	1.25	7	4.3	2.8	6	5.5	Better impact resistance and cheaper than general purpose HSS. High wear resistance
High vanadium	1.55	12.5	4.75	5.0	5	—	Best abrasion resistance. Used for highly abrasive materials

HSS, high-speed steels.

5.6.7 *Carbide and ceramic tools*

Carbides are graded according to series (see table) and
by a number from 01 (hardest) to 50 (toughest), e.g.
P01 and K40.

Carbides

Series	Material machined	Carbides
P	Steel, steel castings	W, Ta, Tt, Ni with Co binder
M	Cast iron, non-ferrous, plastic	W with Co binder
K	Heat resistant steels, stainless steels	W with Co binder

Sintered carbide tools – cutting conditions and positive rake

Material being cut	Cutting speed (m min^{-1})		Top rake (°)	
	Rough	Fine	Clean metal	Rough scaly metal
Steel:				
Low–medium carbon	75	120–210	8	0–4
Medium–high carbon	60	90–180	4	0–3
Nickel chrome	30	75–120	0.4	0
Cast iron:				
200 Brinell hardness	45–60	90–120		3.5–8
White heart	3–6	6–4		0
Copper	150–240	240–360		13–16
Brass	120–240	240–360		0–3.5
Bronze and gun metal	120–180	240–300		0–3.5
Aluminium alloy	90–150	180–225		13–16
Plastics	120–180	240–300		0
Glass	9–15	15–21		3.5

Ceramic tools (sintered aluminium oxide with grain refiners and binder)

Material being cut	Cutting speed (m min^{-1})
Cast iron	60–610
Steel	90–450
Aluminium	>610

5.7 General information on metal cutting

5.7.1 Cutting speeds and feed rates

	Cutting speed ($m s^{-1}$)								Feed rate ($mm rev.^{-1}$)	
	High-speed steel				Carbide, turning		Stellite, turning			
	Turning									
Material of workpiece	R	F	R & T	D	R	F	R	F	Rough	Fine
Mild steel	40	66	7.5–15	30	90	180	50	75	0.625–2.0	0.125–0.75
Cast steel	15	24	3.5	12	45	100	24	33	0.5–1.25	0.125–0.50
Stainless steel	15	18	3.0	12	27	45	22	25	0.5–1.00	0.075–0.175
Grey cast iron	18	27	3.5	13	60	100	23	45	0.4–2.5	0.20–1.00
Aluminium	90	50	15	72	240	360	120	180	0.1–0.5	0.075–0.25
Brass	75	100	18	60	180	270	90	150	0.375–2.0	0.20–1.25
Phosphor bronze	18	36	4.5	13	120	180	30	50	0.375–0.75	0.125–0.50

R, rough; F, fine; R & T, reaming and threading; D, drilling.

5.7.2 Power used and volume removed in metal cutting

Symbols used:
P = power (kW)
d = depth of cut (mm)
f = feed ($mm rev.^{-1}$)
v = cutting speed ($m min^{-1}$)
T = torque (N-m)
D = drill diameter (mm)
N = rotational speed ($rev. min^{-1}$)
w = width of cut (mm)
f_m = milling machine table feed ($mm min^{-1}$)
V = volume of metal removed ($cm^3 min^{-1}$)

Material	k_L	k_D	k_M	Material	k_L	k_D	k_M
Aluminium	700	0.11	0.9	Mild steel	1200	0.36	2.7
Brass	1250	0.084	1.6	Tool steel	3000	0.40	7.0
Cast iron	900	0.07	1.9				

Power

Turning: $P = \dfrac{k_{\text{L}} dfv}{60\,000}$

Drilling: $T = k_{\text{D}} f^{0.75} D^{1.8}$

$P = \dfrac{2\pi N T}{60\,000}$

Milling: $P = \dfrac{k_{\text{M}} dwf_{\text{m}}}{60}$

Volume of metal removed

Turning: $V = dfv$

Drilling: $V = \dfrac{\pi D^2 fN}{4000}$

Milling: $V = \dfrac{wdf_{\text{M}}}{1000}$

5.7.3 *Surface finish*

Different processes produce different degrees of finish on machined surfaces. These are graded from N1 with an average height of roughness of 0.025 μm, up to N12 roughness 50 μm. The manner in which a machined surface is indicated is shown.

Average height of roughness, $h_{\text{a}} = \dfrac{a + b + c + \ldots}{L}$

–where a, b, c, etc. = area on graph, and L = length of surface.

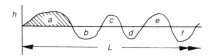

Roughness grade	N1	N2	N3	N4	N5	N6	N7	N8	N9	N10	N11	N12
$h_{\text{a}}(\mu\text{m})$	0.025	0.05	0.1	0.2	0.4	0.8	1.6	3.2	6.3	12.3	25	50

Finishing processes		Surface indication
Mill	Ream	
Bore	Broach	
Turn	Lap	
Grind	Hone, etc.	

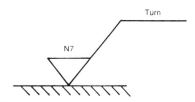

5.7.4 *Merchants circle for tool forces*

'Merchant's circle' is a well-known construction for the analysis of cutting forces for a single-point tool. If the cutting and feed forces, the initial and final chip thickness and the tool rake angle are known, then the other forces, friction and shear angles can be found.

Known:
F_{c} = cutting force
F_{f} = feed force
t_1 = initial chip thickness
t_2 = final chip thickness
α = tool rake angle

The diagram can be drawn to give:
F_{s} = shear force
F_{r} = resultant force
F = friction force on tool face
F_{ns} = force normal to shear force
F_{n} = force normal to F
μ = coefficient of friction = F/F_{n}
θ = friction angle = $\tan^{-1} \mu$
ϕ = shear angle

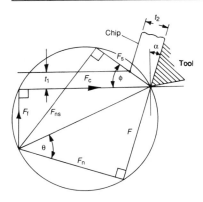

5.7.5 Machining properties of thermoplastics

Material	Rake angle (°)	Clearance angle (°)	Turning		Drilling		Milling	
			Cutting speed (m s^{-1})	Feed (mm rev.$^{-1}$)	Cutting speed (m s^{-1})	Feed (mm rev.$^{-1}$)	Cutting speed (m s^{-1})	Feed (mm s^{-1})
Nylon	0/−10	20/30	2.5–5.0	0.1–0.25	2.0–5.0	0.1–0.38	5	⩽4
PTFE	0/−5	20/30	1.0–2.5	0.05–0.25	1.25–5.0	0.1–0.38	5	⩽4
Polystyrene	0/−5	20/30	1.5–5.0	0.05–0.25	0.5–10	0.1–0.38	5	⩽4
Rigid PVC	0/−10	20/30	1.5–5.0	0.25–0.75	2.5–30	0.05–0.13	5	⩽4

5.7.6 Negative rake cutting

Material being cut	Roughing speed (m min^{-1})	Finishing speed (m min^{-1})	Feeds (mm/tooth)
Steel 0.15%C	230	300	Milling: 0.2–0.4
0.4%C	160	210	
0.8%C	120	135	
Steel castings	90	105	Turning: 0.25–0.5
Phosphor bronze and gun metal	300	420	
Copper	450	540	
Brass	600	900	
Aluminium and alloys	900	1200	

5.7.7 Calculation of machining cost

The 'total-time cost per workpiece' is made up of 'machine-time cost', 'non-productive-time cost' and 'tool cost'. 'Machining-time cost' is for actual machining and includes overheads and wages. 'Non-productive-time cost' covers 'setting-up' and 'loading- and unloading-time cost'. 'Tool cost' combines 'tool-change-time cost' and actual 'tool cost'. The former is the cost of changing the cutting edge, the latter is the cost of the cutting plus resharpening. When 'total cost' is plotted against 'cutting speed' an optimum speed for minimum cost is found.

Let:

C_m = machining-time cost per workpiece
C_n = non-productive-time cost per workpiece
C_c = tool-change-time cost per workpiece
C_t = tool cost per workpiece

Total cost of machining $C_{tot} = C_m + C_n + C_c + C_t$

(£/workpiece)

Total tool cost per workpiece $C_{tt} = C_c + C_t$

Let:

t_m = machining time per workpiece (min)
t_L = loading and unloading time per workpiece (min)
t_s = setting time per batch (min)
t_t = tool life (min)
t_c = tool change time (min)
t_{sh} = tool sharpening time (min)
R = cost rate per hour (£)
n_b = number per batch
n_s = number of resharpenings

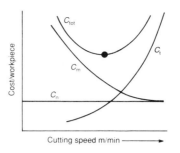

$$C_m = \frac{t_m R}{60}$$

$$C_n = \left(t_L + \frac{t_s}{n_b}\right)\frac{R}{60}$$

$$C_c = \frac{t_m t_c R}{60 t_t}$$

$$C_t = \frac{C_{tt}}{1 + n_s} + \frac{t_{sh} t_m R}{60 t_t}$$

5.7.8 Cutting fluids

It is necessary when machining to use some form of fluid which acts as a coolant and lubricant, resulting in a better finish and longer tool life. The fluid also acts as a rust preventative and assists in swarf removal. The following table lists various fluids and their advantages.

Cutting fluid applications

Group	Description	Advantages
Soluble oil	Oil, emulsifier and 2–10% water	Good coolant. Poor lubricant
Clear soluble oil	As above, with more emulsifier	Good coolant. Poor lubricant
Water based fluids	Solution of sodium nitride and triethanolamine	Good coolant. Poor lubricant
EP soluble oils	Soluble oils with EP additives, e.g. sulphur and/or chlorine	Fairly good lubricant

Cutting fluid applications (*continued*)

Group	Description	Advantages
Straight oils	Mineral or fatty oils (lard, sperm, olive, neat's foot, rape, etc.) alone or compounded	Good lubricant. Often unstable.
Sulphurized EP oils	Straight oils with sulphur, zinc oxide or other additives (0.2–0.8%S)	Average coolant. Good lubricant. Pressure resistant. Prevents welding of chip on tool
Sulphochlorinated EP oils	Mineral and fatty oil blends with sulphur and chlorine additives	More efficient than sulphurized oils. For most arduous conditions. Highly resistant to welding of chip on tool
Chlorinated materials	Carbon tetrachloride and trichlorethylene alone or blended with oils	Very good EP fluid. Highly dangerous to use
Gases and vapours	Air, oil mist, CO_2	Limited cooling power. Chip dispersed

EP, extreme pressure.

5.8 Casting

Casting is the forming of metal or plastic parts by introducing the liquid material to a suitably shaped cavity (mould), allowing it to solidify, and then removing it from the mould. Further processing is usually required.

5.8.1 *Sand casting*

In sand casting the mould is made in a 'moulding box' using a special sand and a wooden 'pattern'. Holes are produced by inserting previously made 'cores' of baked sand. Molten metal is poured into runners until it appears in risers. The casting is cleaned by chipping, grinding and sandblasting. Practically any metal can be cast.

SAND CASTING

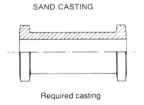

Required casting

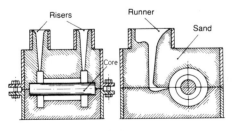

Moulding box

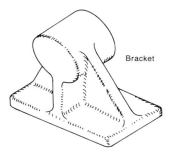

Bracket

INVESTMENT CASTING

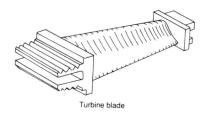

Turbine blade

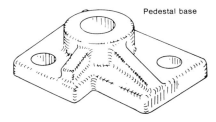

Pedestal base

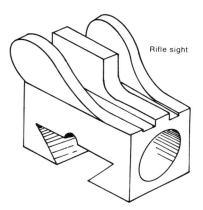

Rifle sight

5.8.2 Shell moulding

This is a form of sand casting done using a very fine sand mixed with synthetic resin. The pattern is made of machined and polished iron. The sand mixture is blown into a box containing the pattern which is heated to produce a hard, thin (6–10 mm) mould which is split and removed from the pattern and then glued together. It is a high-speed process, producing highly accurate castings.

5.8.3 Investment casting (lost wax casting)

Wax patterns are made from a permanent metal mould. The wax patterns are coated with ceramic slurry which is hardened and baked so that the wax is melted out. The cavity is filled with molten metal to give a precision casting. Any metal can be cast using this process.

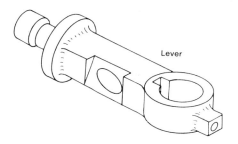

Lever

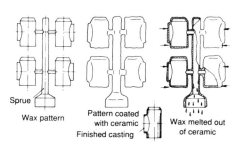

Sprue

Wax pattern

Pattern coated with ceramic

Finished casting

Wax melted out of ceramic

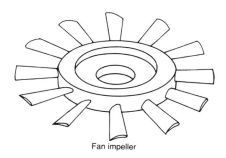

Fan impeller

5.8.4 *Die casting*

The mould is of steel in several parts dowelled together. Molten metal is fed by gravity or pressure and, when solid, is ejected by pins. Aluminium, copper, manganese and zinc alloy are suitable for casting by this method.

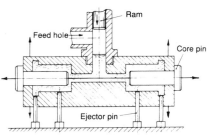

DIE CASTING

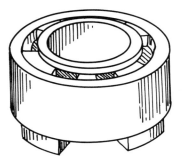

Shaft coupling part

5.8.5 *Centrifugal casting*

Cylindrical or circular components such as piston rings, cylinder liners, pipes, etc., may be cast in a rotating mould. Centrifugal pressure gives a fine grain casting. Any metal may be cast using this process.

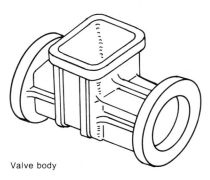

Valve body

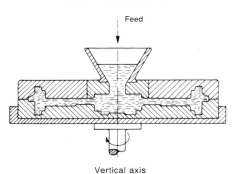

CENTRIFUGAL CASTING

Vertical axis

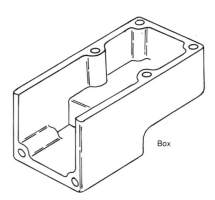

Box

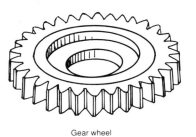

Gear wheel

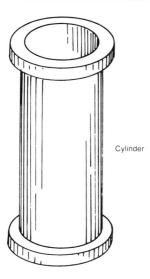

Cylinder

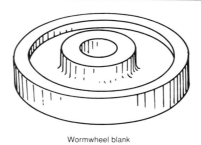

Wormwheel blank

5.9   Metal forming processes

5.9.1   *Hand forging and drop forging*

'Forging' is the forming of metal parts by hammering, pressing, or bending to the required shape, usually at red heat. 'Hand forging' involves the use of an anvil and special hammers, chisels and swages. A 'drop forging machine' uses pneumatic or hydraulic pressure to compress hot metal blanks between hard steel dies.

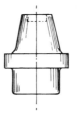

Forging with flash removed

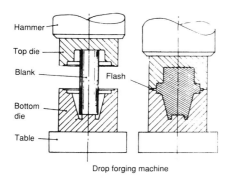

Drop forging machine

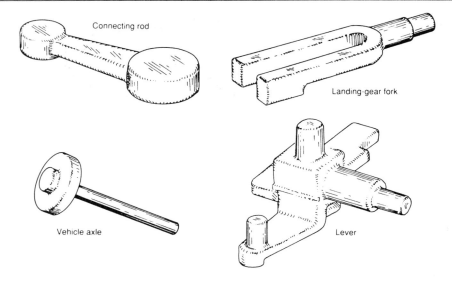

FORGINGS

5.9.2 *Drawing process*

This is the forming of flat metal blanks into box and cup-like shapes by pressing them with a shaped punch into a die. The process is used for cartridge cases, boxes, electrical fittings, etc.

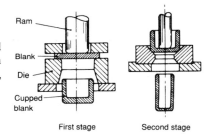

Deep drawing

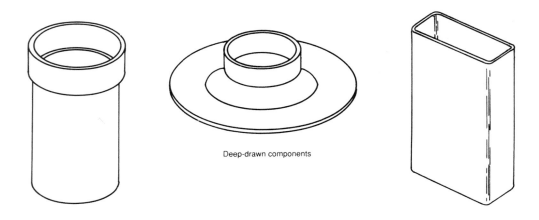

Deep-drawn components

5.9.3 *Extrusion*

Hot extrusion

A piece of red-hot bar or billet is placed in a cylinder and forced through a specially shaped die by a piston to produce long lengths of bar. Hollow sections can be made by placing a mandrel in the die orifice.

Cold extrusion

Soft metals such as aluminium and copper can be extruded cold. Practically all metals may be extruded.

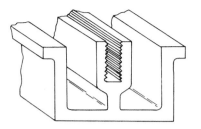

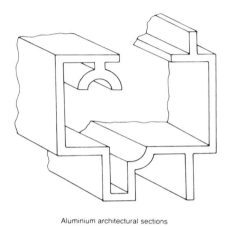

Aluminium architectural sections

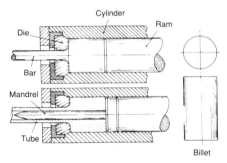

Hot extrusion

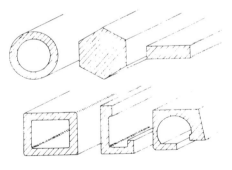

Hot-extruded sections

5.9.4 *Impact extrusion*

A metal which is plastic when cold may be extruded by the impact of a high-velocity punch. The metal of the blank flows up the sides of the punch to produce a cylinder. The process is used for manufacturing toothpaste tubes, ignition coil cans, etc.

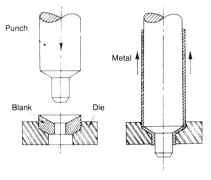

Impact extrusion

5.9.5 *Press work*

A press is used for a wide range of processes such as punching, piercing, blanking, notching, bending, drawing, and folding. It may be operated by means of a crank connected to a heavy flywheel or by hydraulic power. Formulae are given for various processes.

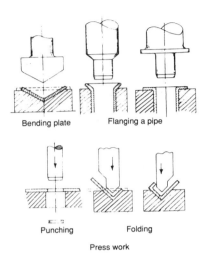

Bending plate   Flanging a pipe

Punching   Folding

Press work

Sheet metal work

In sheet metal work allowance must be made for bends depending on the thickness of the material, the radius of the bend and bend angle.

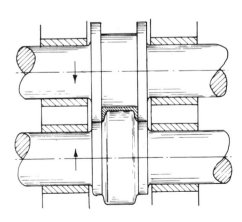

Rolling mill (rolling channel)

Rolling

In a rolling mill, red-hot ingots of steel or other metals are passed through successive pairs of specially shaped rollers to produce flat bar, sheet, I, T, channel, angle or other section bar. Final cold rolling may be carried out to give a better finish.

Universal Beams, Universal Columns, Joists, Angles, and Channels are made to British Standards BS 4: Part 1 and BS 4848: Part 4.

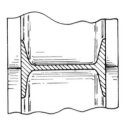

Rolls for I section

5.9.6 *Press tool theory*

Punching process

Symbols used:
F_{max} = maximum shear force
τ_u = ultimate shear stress
t = material thickness
x = penetration
p = perimeter of profile

Maximum shear force $F_{max} = \tau_u t p$.
Work done $W = F_{max} x$

Penetration ratio $c = \dfrac{x}{t}$

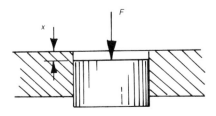

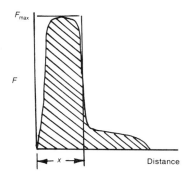

Shearing process

Shearing force $F = \dfrac{F_{max}}{\left(1 + \dfrac{h}{x}\right)}$

where: h = the 'shear'.

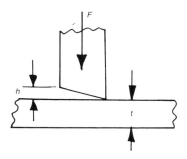

Bending process

Bending force $F_b = \tau_u L t$
Planishing force $F_f = \sigma_y L b$
where: σ_y = yield stress.

Initial length of strip $L_i = h - t - 2r + b + \dfrac{\pi}{2}\left(r + \dfrac{t}{2}\right)$

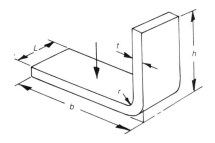

Drawing process

Blank diameter $D = \sqrt{d^2 + 4dh}$
Required force $F = \pi d t \sigma_u$
where: σ_u = ultimate tensile stress.

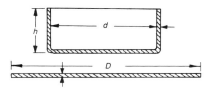

5.9.7 *Sheet metal work*

Allowance for right angle bend

Lengths a and b are reduced by an 'allowance' c, and

$$c = r + t - \frac{\pi}{4}\left(r + \frac{t}{2}\right)$$

When $r = 2t$ (as is often the case), $c = 1.037t$.

Allowance for bend with outside angle θ

$$c = (r + t)\tan\frac{\theta}{2} - \frac{\pi\theta}{360}\left(r + \frac{t}{2}\right), \ (\theta \text{ in degrees})$$

When $r = 2t$, $c = \left(3\tan\dfrac{\theta}{2} - 0.0218\theta\right)t$.

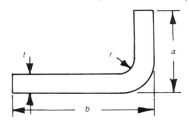

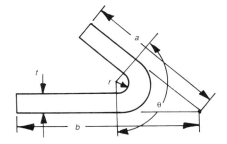

5.9.8 *Rolled sections*

Rolled sections are made to British Standards BS 4: Part 1 and BS 4848: Part 4.

Universal beams

$D \times B = 127$ mm $\times$ 76 mm to 914 mm $\times$ 419 mm.
t and T are of several sizes in each case.

Universal columns

$D \times B = 152$ mm $\times$ 152 mm to 356 mm $\times$ 406 mm.
t and T are in several sizes in each case.

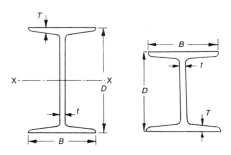

Beams, columns and joists

Joists

From 76 mm $\times$ 76 mm to 254 mm $\times$ 203 mm

Equal angles

$D \times B$ from 25 mm $\times$ 25 mm to 200 mm $\times$ 200 mm.
Several values of t in each case.

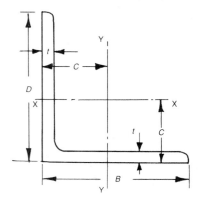

Unequal angles

$D \times B$ from 40 mm $\times$ 25 mm to 200 mm $\times$ 150 mm.

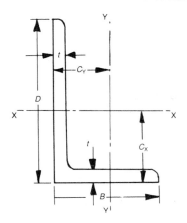

Channels

$D \times B$ from 76 mm $\times$ 38 mm to 432 mm $\times$ 102 mm.
One value of t in each case.

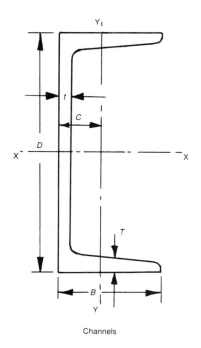

Channels

Miscellaneous rolled sections

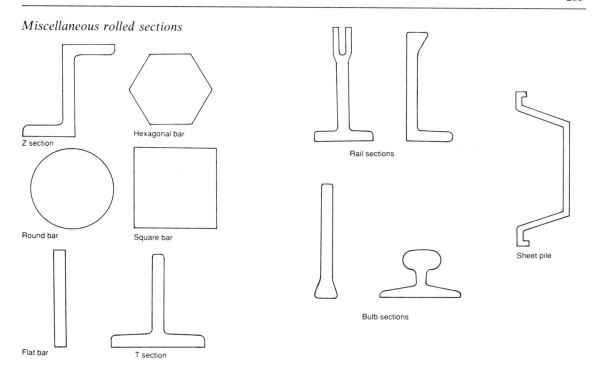

5.10   Soldering and brazing

In soldering and brazing, bonding takes place at a temperature below the melting points of the metals being joined. The bond consists of a thin film of low-melting-point alloy known as 'solder' or 'filler'.

5.10.1   *Solders and soldering*

For small parts, a 'soldering iron', which is heated by gas or an internal electric element, is used. For large joints a gas flame is used.

Soft solder

This is a mixture of lead, tin and sometimes antimony. Typical solders are 50% tin and 50% lead (melting range 182–215 °C), 60% tin and 40% lead (melting range 182–188 °C) and 95% tin and 5% antimony (melting range 238–243 °C). Solder is available in the form of bar or wire with cores of resin flux. Flux is used to prevent oxidation by forming a gas which excludes air from the joint. A solution of zinc chloride (killed spirits) or resin are commonly used as fluxes.

Silver solder

This is an alloy of silver, copper and zinc with a melting point of about 700 °C used mainly for joining brass and copper. It is in strip form and is used with a flux powder.

5.10.2 *Soldered joints*

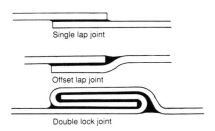

Single lap joint

Offset lap joint

Double lock joint

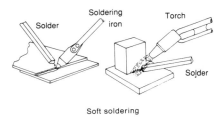

Soft soldering

5.10.3 *Brazing*

Above about 800 °C the process is called 'brazing' (or hard soldering). Brazing rod (50% copper and 50% zinc) is used for general work, with a flux consisting of borax mixed to a paste with water. A torch supplied with mains gas and compressed air is used. Taps control the flow and mixture. For large-scale production work, induction and furnace heating are used.

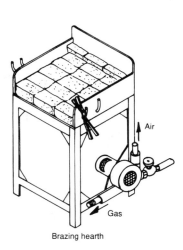

Brazing hearth

Gas–air brazing torch

5.10.4 *Brazed joints*

In the figure, several types of brazed joint are shown; the arrows indicate the direction of the load.

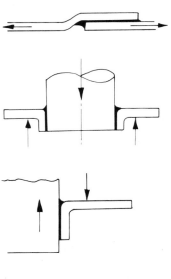

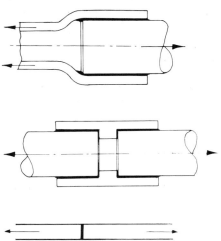

5.11 Gas welding

In gas welding the heat to melt the metal parts being welded is produced by the combination of oxygen and an inflammable gas such as acetylene, propane, butane, etc. Acetylene is the most commonly used gas; propane and butane are cheaper but less efficient.

5.11.1 *Oxyacetylene welding*

A flame temperature of about 3250 °C melts the metals which fuse together to form a strong joint. Extra metal may be supplied from a filler rod and a flux may be used to prevent oxidation. The gas is supplied from high pressure bottles fitted with special regulators which reduce the pressure to 0.13–0.5 bar. Gauges indicate the pressures before and after the regulators.

A torch mixes the gases which issue from a copper nozzle designed to suit the weld size. The process produces harmful radiation and goggles must be worn. The process is suitable for steel plate up to 25 mm thick, but is mostly used for plate about 2 mm thick.

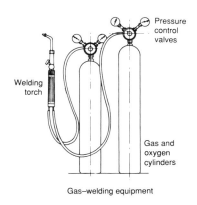

Gas–welding equipment

Gas welding – edge preparation, speed, and metal thickness

Welding rod diameter (mm)	Edge preparation	Method	Speed (mm min^{-1})	Metal thickness (mm)
1.5	(a)	Leftward	127–152 100–127	0.8 1.5
1.5–3	(b) 0.8–3 mm	Leftward	100–127 90–100	2.5 3.0
3–4	(c) 80° 1.6–3 mm	Leftward Rightward	75–90 60–75	4.0 4.8
3–4	(d) 3–4 mm	Rightward	50–60 35–40	6.0 8.0

Gas welding – edge preparation, speed, and metal thickness (*continued*)

Welding rod diameter (mm)	Edge preparation	Method	Speed (mm min^{-1})	Metal thickness (mm)
3–6.5	(e)	Rightward	30–35 22–25	9.5 12.5
6.5	(f)	Rightward	19–22 15–16 10–12	15.0 19.0 25.0

5.11.2 *Type of flame*

It is essential to have the correct type of flame which depends on the proportions of the gases.

Neutral flame

This is the type most used since it least affects the metal being welded. The almost transparent flame has a well defined blue core with a rounded end. Roughly equal amounts of gas are used.

Carburizing flame

This flame contains excess acetylene and hence carbon. Carbides are formed which produce brittleness. The flame is used when 'hard facing'. The blue core is surrounded by a white plume.

Oxidizing flame

This flame contains an excess of oxygen which pro-

duces brittle low-strength oxides. Use of this flame should be avoided when welding brass and bronze.

5.11.3 *Method of gas welding*

Two methods of gas welding are used: leftward and rightward.

Leftward welding

This is used for plate up to 4.5 mm thick and for non-ferrous metals. The torch is moved towards the filler rod and given a slight side-to-side motion.

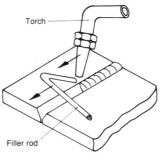

Torch

Filler rod

Leftward welding

Rightward welding

This is used for plate thicker than 4.5 mm. For larger plate the edges are chamfered to give an included angle of about 80°.

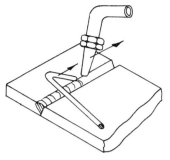

Rightward welding

5.11.4 *Fillers and fluxes*

The table below gives recommended filler rod materials and fluxes for gas welding.

Metal welded	Filler	Flux
Low carbon steels	Low carbon steel rod sometimes copper coated. 1.6–5 mm diameter	No flux required
Stainless steel	Special steel rod for each type. 1.6–3.2 mm diameter	Grey powder in paste with water (m.p. 910 °C). Weld cleaned with 5% caustic soda solution, then with hot water
Cast iron	High silicon cast iron rod. 5 or 6 mm square	Grey powder in paste with water (m.p. 850 °C). Excess removed by chipping and wire brushing
Brass or bronze	Silicon bronze sometimes flux coated. 1.6–6 mm diameter	Pale blue powder (m.p. 875 °C) in paste with alcohol. Cleaning is with boiling water and by brushing
Aluminium and alloys	Pure aluminium or alloy. 1.6–5 mm diameter	White powder in paste with water (m.p. 570 °C). Cleaning by dipping in 5% nitric acid solution and hot water wash
Copper	Copper–silver low melting point rods. 3.2 mm diameter	White powder in paste with water. Cleaning is with boiling water and by wire brushing

5.11.5 *Flame cutting*

Steel plate over 300 mm thick can be cut by this method, either manually or by automatic machine using templates for complicated shapes. Thin plates may be stacked so that many may be cut at one time.

The plate is first heated by a mixture of oxygen and acetylene until red hot and then a stream of oxygen alone is used to burn with the metal with intense heat. Propane and butane may be used with plain carbon steel, but are not as effective as oxygen. Cutting speeds of up to 280 mm min^{-1} are possible with 25-mm plate. Typical speeds are given in the table.

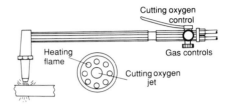

Oxyacetylene cutting torch

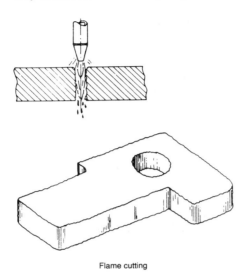

Flame cutting

Oxyacetylene cutting

Plate thickness (mm)	Nozzle diameter (mm)	Acetylene pressure (bar)	Oxygen pressure (bar)	Cutting speed (mm min^{-1})
6	0.8	0.14	1.8	430
13	1.2	0.21	2.1	360
25	1.6	0.14	2.8	280
50	1.6	0.14	3.2	200
75	1.6	0.14	3.5	200
100	2.0	0.14	3.2	150

5.12 Arc welding

5.12.1 *Description of arc welding*

The heat of fusion is generated by an electric arc struck between two electrodes, one of which is the workpiece and the other a 'welding rod'. The welding rod is made of a metal similar to the workpiece and is coated with a solid flux which melts and prevents oxidation of the weld. The rod is used to fill the welded joint. Power is obtained from an a.c. or d.c. 'welding set' providing a regulated low-voltage high-current supply to an 'electrode holder' and 'earthing clamp'. The work is done on a steel 'welding table' to which the work is clamped and to which the earthing clamp is attached to complete the circuit.

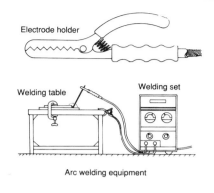

Electrode holder

Welding table

Welding set

Arc welding equipment

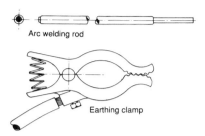

Arc welding rod

Earthing clamp

5.12.2 *Arc welding processes*

Joint condition – fusion
Manual metal arc
Carbon arc
Submerged arc
Tungsten inert gas (TIG)
Metal inert gas (MIG)
Open arc, automatic
Atomic hydrogen
Arc stud welding
Spot welding
Roller spot welding
Projection welding
Electroslag
Thermit

Laser welding
Plasma welding
Electron beam welding

Joint condition – solid phase
Butt welding
Flash butt welding
Friction welding
Ultrasonic welding
Sintering

Joint condition – solid/liquid
Brazing

5.12.3 *Types of weld*

The *fillet weld*, the most used, is formed in the corner of overlapping plates, etc. In the interests of economy, and to reduce distortion, *intermittent welds* are often used for long runs, with correct sequencing to minimize distortion. *Tack welds* are used for temporary holding before final welding.

Plug welds and *slot welds* are examples of fillet welds used for joining plates. For joining plates end to end, *butt welds* are used. The plates must have been suitably prepared, e.g. single or double V or U, or single and double bevel or J. To avoid distortion, especially with thick plates, an unequal V weld may be used. the smaller weld being made first.

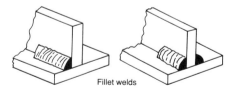

Fillet welds

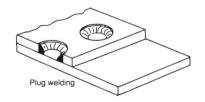

Plug welding

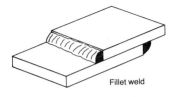

Fillet weld

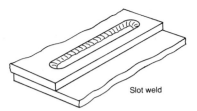

Slot weld

Resistance welding is used to produce *spot welds* and *stud welding* by passing an electric current through the two metal parts via electrodes. In *seam welding* the electrodes are wheels.

BUTT WELDS

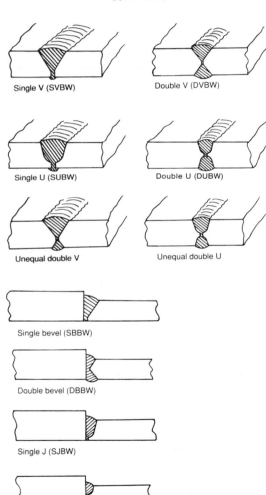

Single V (SVBW)

Double V (DVBW)

Single U (SUBW)

Double U (DUBW)

Unequal double V

Unequal double U

Single bevel (SBBW)

Double bevel (DBBW)

Single J (SJBW)

Double J (DJBW)

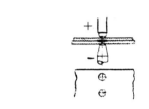

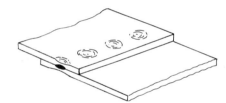

Resistance spot welding

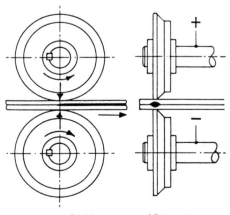

Resistance seam welding

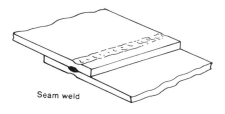

Seam weld

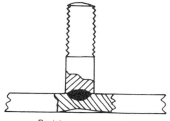

Resistance stud welding

5.12.4 *Weld symbols*

Weld symbols (BS 499)

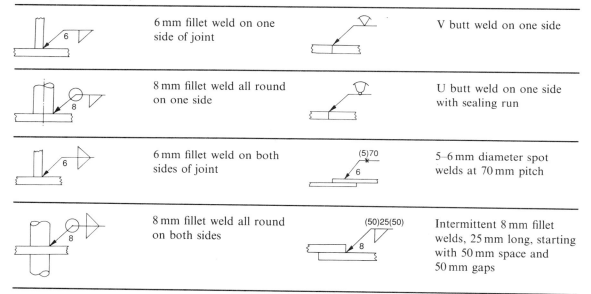

6 mm fillet weld on one side of joint	V butt weld on one side
8 mm fillet weld all round on one side	U butt weld on one side with sealing run
6 mm fillet weld on both sides of joint	5–6 mm diameter spot welds at 70 mm pitch
8 mm fillet weld all round on both sides	Intermittent 8 mm fillet welds, 25 mm long, starting with 50 mm space and 50 mm gaps

5.12.5 *Gas-shielded metal arc welding*

In this process an inert gas such as argon is used as a flux; the electrode is a continuously fed consumable wire. Two processes are used: 'metal inert gas' (MIG) and 'tungsten inert gas' (TIG).

Welding processes

A table is given of all the welding processes, together with recommendations for the use of a number of these.

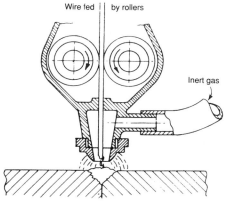

Wire fed | by rollers

Inert gas

Gas metal arc welding

Recommended welding processes

Process	Low carbon steel	Medium carbon steel	Low alloy steel	Austenitic stainless steel	Ferritic and stainless steel	High temperature and high strength alloy steel	Cast iron	Aluminium and alloys	Magnesium and alloys	Copper and alloys	Nickel and alloys	Titanium and alloys
Manual metal arc	R	R	R	R	R	R	S	S	U	N	R	U
Submerged arc	R	R	R	R	S	S	N	N	U	N	S	U
TIG	S	S	S	R	S	S	S	R	R	R	R	R
MIG	S	S	S	R	S	S	N	R	S	R	R	S
Flash welding	R	R	R	R	S	S	N	S	N	S	S	S
Spot welding	R	R	R	R	S	S	U	R	S	S	R	S
Oxyacetylene welding	R	R	S	S	S	S	R	S	N	S	S	U
Furnace brazing	R	R	S	S	S	N	N	R	N	S	R	S
Torch brazing	S	S	N	S	S	N	R	R	N	R	R	S

R = recommended; S = satisfactory; N = not recommended; U = unsuitable.

Edge preparation

Plates below 8 mm thick may be butt welded without preparation; with thicker plate the edges must be chamfered to obtain good penetration. The groove is then filled by depositing a number of runs of weld. The double V uses less material for thick plates and also reduces thermal distortion. A U preparation approaches a uniform weld width.

Arc welding – edge preparation

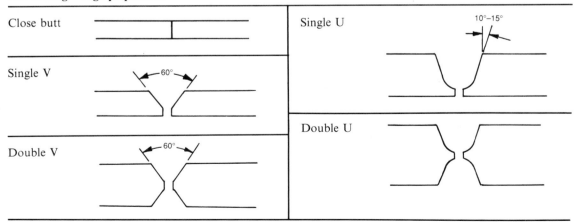

Close butt

Single V

Double V

Single U

Double U

Positions of welding

In addition to 'flat' welding, which is the ideal position, three other positions are used: horizontal, vertical and overhead. If one member is vertical and one horizontal the position is called horizontal–vertical. In the last case a number of passes must be made to overcome the tendency for molten metal to run out. (See figure.)

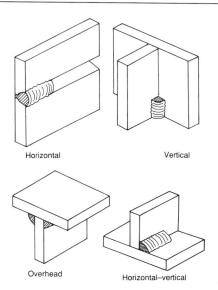

Horizontal   Vertical

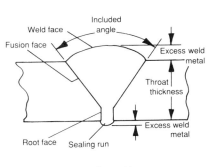

Flat

Overhead   Horizontal–vertical

5.12.9 *Welding terminology, throat size and allowable stress*

Welding practice

The relevant British Standards are BS 4360: Part 2, BS 639, BS 1719, BS 1856, BS 2642, BS 449 and BS 499.

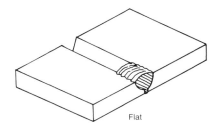

Butt weld

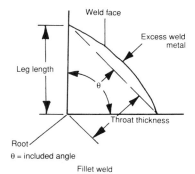

Fillet weld

Effective throat size (t = throat thickness, L = leg length)

Fillet angle, θ (°)	60–90	91–100	101–106	107–113	114–120
t/L	0.7	0.65	0.6	0.55	0.5

Allowable stress for welded structural steels

Grade	43	50	55
Stress ($N\,mm^{-2}$)	115	160	195

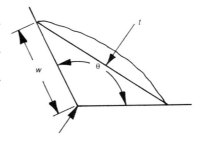

5.13   Limits and fits

It is impossible to make components the exact size and an allowance or 'tolerance' must be made which depends on the process and the application. The tolerance results in two extremes of size which must be maintained. The tolerances of two fitting parts, e.g. a shaft in a bearing, determines the type of 'fit' and makes interchangeability possible.

British Standard BS 4500: Part 1: 1969, 'ISO Limits and Fits', gives a comprehensive system relating to holes and shafts; it can, however, be used for other components, e.g. a key in a keyway.

5.13.1   *Terminology*

Taking the example of holes and shafts, there is a 'basic size' and then maximum and minimum sizes for each, their differences being the tolerances. Their differences from the basic size are called the 'maximum and minimum deviations'.

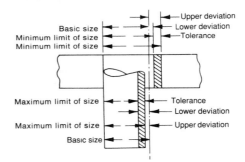

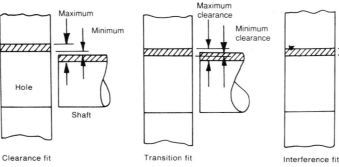

Clearance fit

Transition fit

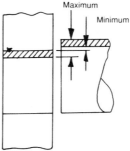

Interference fit

Types of fit

The fit describes the manner in which two parts go together. A 'clearance fit' means that the shaft will always be smaller than the hole. An 'interference fit' means that the shaft will always be larger than the hole and a fitting force will be necessary. A 'transition fit' means that there may be either clearance or interference.

Tolerance

BS 4500 gives 18 'tolerance grades' numbered IT01, IT0, IT1, IT2, up to IT16. The actual tolerance depends on the size of the component (see table below).

5.13.2 *Selected Fits*

BS 4500 'Selected Fits' Gives a much smaller range of fits, the hole tolerance is denoted by the letter H and the shaft by a lower-case letter (see table). For conventionally manufactured parts, the five fits given are usually sufficient (see table).

Selected fits (BS 4500)

Hole	H7 H8 H9 H11
Shaft	c11 d10 e9 f7 g6 h6 k6 n6 p6 s6

Reduced range of fits for conventionally manufactured parts

Type of fit	Shaft tolerance	Hole tolerance	Description of fit
Clearance	f7	H8	Running
Clearance	g6	H7	Sliding
Transition	k6	H7	Keying
Interference	p6	H7	Press
Interference	s6	H7	Push or shrink

5.13.3 *Example of symbols and sizes on drawing*

Preliminary design drawing

It is convenient to use symbols, e.g. 45 mm shaft and 'transition' fit. Tolerance is given as: ϕ 45H7/k6.

Production drawing

For a 30 mm diameter shaft, fit H9/d10:

Hole maximum limit of size = 30.012 mm
Hole minimum limit of size = 30.00 mm
Therefore tolerance = 0.012 mm.
Shaft maximum limit of size = 30.015 mm
Shaft minimum limit of size = 30.002 mm

Therefore, tolerance = 0.013 mm

On the drawing these parameters would be given as (rounding off to nearest 0.01 mm):

Hole: 30.01   *Shaft*: 30.02
      30.00   30.00

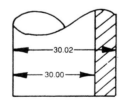

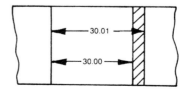

6 Engineering materials

6.1 Cast irons

6.1.1 Grey iron

Grey iron is so called because of the colour of the fracture face. It contains 1.5–4.3% carbon and 0.3–5% silicon plus manganese, sulphur and phosphorus. It is brittle with low tensile strength, but is easy to cast.

Properties of some grey irons (BS 1452)

Grade	Tensile strength $(N\,mm^{-2})$	Compressive strength $(N\,mm^{-2})$	Transverse strength $(N\,mm^{-2})$	Hardness, BHN*	Modulus of elasticity $(GN\,m^{-2})$
10	160	620	290–370	160–180	76–104
17	260	770	450–490	190–250	110–130
24	370	1240	620–700	240–300	124–145

*BHN = Brinell hardness number.

6.1.2 Spheroidal graphite (SG) iron

This is also called nodular iron because the graphite is in the form of small spheres or nodules.

These result in higher ductility which can be improved further by heat treatment. Mechanical properties approach those of steel combined with good castability.

Properties of some SG irons (BS 2789)

Grade	Tensile strength $(N\,mm^{-2})$	0.5% permanent set stress $(N\,mm^{-2})$	Hardness BHN*	Minimum elongation (%)
SNG24/17	370	230	140–170	17
SNG37/2	570	390	210–310	2
SNG47/2	730	460	280–450	2

*BH = Brinell hardness number.

6.1.3 Malleable irons

These have excellent machining qualities with strength similar to grey irons but better ductility as a result of closely controlled heat treatment. There are three types: *white heart* with superior casting properties; *black heart* with superior machining properties; and *pearlitic* which is superior to the other two but difficult to produce.

Properties of some malleable irons

Type	Grade	Minimum tensile strength ($N\,mm^{-2}$)	Yield point strength ($N\,mm^{-2}$)	Hardness, BHN*	Elongation (%)
White heart, BS 309	W22/24	310–340	180–200	248 (max.)	4
	W24/8	340–370	200–220	248 (max.)	6
Black heart, BS 310	B18/6	280	170	150 (max.)	6
	B20/10	310	190	150 (max.)	10
	B22/14	340	200	150 (max.)	14
Pearlitic, BS 3333	P28/6	430	—	143–187	6
	P33/4	460	—	170–229	4

*BHN = Brinell hardness number.

6.1.4 Alloy irons

The strength, hardness, wear resistance, temperature resistance, corrosion resistance, machinability and castability of irons may be improved by the addition of elements such as nickel, chromium, molybdenum, vanadium, copper and zirconium.

6.2 Carbon steels

6.2.1 Applications of plain carbon steels

These are alloys of iron and carbon, chemically combined, with other elements such as manganese, silicon, sulphur, phosphorus, nickel and chromium. Properties are governed by the amount of carbon and the heat treatment used. Plain carbon steels are broadly classified as: low carbon (0.05–0.3%C), with high ductility and ease of forming; medium carbon (0.3–0.6%C), in which heat treatment can double the strength and hardness but retain good ductility; and high carbon (>0.6%C), which has great hardness and high strength and is used for tools, dies, springs, etc.

Applications of plain carbon steels

% Carbon	Name	Applications
0.05	Dead mild	Sheet, strip, car bodies, tinplate, wire, rod, tubes
0.08–0.15	Mild	Sheet, strip, wire, rod, nails, screws, reinforcing bars
0.15	Mild	Case carburizing type
0.10–0.30	Mild	Steel plate, sections, structural steel
0.25–0.40	Medium carbon	Bright drawn bar
0.30–0.45	Medium carbon	High tensile tube, shafts
0.40–0.50	Medium carbon	Shafts, gears, forgings, castings, springs
0.55–0.65	High carbon	Forging dies, springs, railway rails
0.65–0.75	High carbon	Hammers, saws, cylinder liners
0.75–0.85	High carbon	Chisels, die blocks for forging
0.85–0.95	High carbon	Punches, shear blades, high tensile wire
0.95–1.10	High carbon	Knives, axes, screwing taps and dies, milling cutters

Properties of carbon steels (BS 970)

Type	Composition (%)			Mechanical properties			Applications, etc.
	C	Si	Mn	Tensile strength ($N\,mm^{-2}$)	Elongation (%)	Hardness, BHN*	
070 M20	0.2	—	0.7	400	21	150	Easily machinable steels suitable for light stressing. Weldable
070 M26	0.26	—	0.7	430	20	165	Stronger than En2. Good machinability. Weldable
080 M30	0.3	—	0.8	460	20	165	Increased carbon improves mechanical properties, but slightly less machinable
080 M36	0.36	—	0.8	490	18	180	Tough steel used for forgings, nuts and bolts, levers, spanners, etc.
080 M40	0.4	—	0.8	510	16	180	Medium carbon steel, readily machinable
080 M46	0.46	—	0.8	540	14	205	Used for motor shafts, axles, brackets and couplings
080 M50	0.5	—	0.8	570	14	205	Used where strength is more important than toughness, e.g. machine tool parts
216 M28	0.28	0.25	1.3	540	10	180	Increased manganese content gives enhanced strength and toughness

Properties of carbon steels (BS 970) (*continued*)

| Type | Composition (%) | | | Mechanical properties | | | Applications, etc. |
	C	Si	Mn	Tensile strength (N mm^{-2})	Elongation (%)	Hardness, BHN*	
080 M15	0.15	0.25	0.8	460	16	—	Case-hardening steel. Used where wear is important, e.g. gears and pawls
060A96†	0.99–1.0	0.1–0.7	0.5–0.7	1300	—	500	High carbon spring steel

*BHN = Brinell hardness number.
†To BS 950.

Tempering temperature and colour for carbon steels

Temperature (°C)	Colour	Application
220	Pale yellow	Hacksaw blades
230	Light yellow	Planing and slotting tools, hammers
240	Straw yellow	Milling cutters, drills, reamers
250	Dark yellow	Taps, dies, shear blades, punches
260	Brown-yellow	Wood drills, stone-cutting tools
270	Brown-purple	Axe blades, press tools
280	Purple	Cold chisels, wood chisels, plane blades
290	Dark purple	Screw drivers
300	Dark blue	Wood saws, springs
450–700	Up to dark red	Great toughness at expense of hardness

6.3 Alloy steels

6.3.1 *Classification*

Alloy steels differ from carbon steels in that they contain a high proportion of other alloying elements. The following are regarded as the minimum levels:

Element	%	Element	%	Element	%
Aluminium	0.3	Lead	0.1	Silicon	2.0
Chromium	0.5	Manganese and silica	2.0	Sulphur and phosphorus	0.2
Cobalt	0.3	Molybdenum	0.1	Tungsten	0.3
Copper	0.4	Nickel	0.5	Vanadium	0.1

Alloy steels are classified according to increasing proportion of alloying elements and also phase change during heating and cooling as follows:

low alloy steels
medium alloy steels
high alloy steels

and according to the number of alloying elements as follows:

ternary – one element
quarternary – two elements
complex – more than two elements

6.3.2   *General description*

Low alloy steels

These generally have less than 1.8% nickel, less than 6% chromium, and less than 0.65% molybdenum. The tensile strength range is from 450–620 N mm^{-2} up to 850–1000 N mm^{-2}.

Medium alloy steels

These have alloying elements ranging from 5–12%. They do not lend themselves to classification. They include: nickel steels used for structural work, axles, shafts, etc.; nickel–molybdenum steels capable of being case-hardened, which are used for cams, camshafts, rolling bearing races, etc.; and nickel–chrome–molybdenum steels of high strength which have good fatigue resistance.

High alloy steels

These have more than 12% alloying elements. A chromium content of 13–18% (stainless steel) gives good corrosion resistance; high wear resistance is obtained with austenitic steel containing over 11% manganese. Some types have good heat resistance and high strength.

6.3.3   *Effect of alloying elements*

Aluminium

This acts as a deoxidizer to increase resistance to oxidation and scaling. It aids nitriding, restricts grain growth, and may reduce strength unless in small quantities. The range used is 0–2%.

Chromium

A range of 0.3–4%, improves wear, oxidation, scaling resistance, strength and hardenability. It also increases high-temperature strength, but with some loss of ductility. Chromium combines with carbon to form a wear-resistant microstructure. Above 12% the steel is stainless, up to 30% it is used in martensitic and ferritic stainless steel with nickel.

Cobalt

Cobalt provides air hardening and resistance to scaling. It improves the cutting properties of tool steel with 8–10%. With chromium, cobalt gives certain high alloy steels high-temperature scaling resistance.

Copper

The typical range is 0.2–0.5%. It has limited application for improving corrosion resistance and yield strength of low alloy steels and promotes a tenacious oxide film.

Lead

Up to 0.25% is used. It increases machineability in plain carbon steels rather than in alloy steels.

Manganese

The range used is 0.3–2%. It reduces sulphur brittleness, is pearlitic up to 2%, and a hardening agent up to 1%. From 1–2% it improves strength and toughness and is non-magnetic above 5%.

Molybdenum

The range used is 0.3–5%. It is a carbide forming element which promotes grain refinement and increases high-temperature strength, creep resistance, and hardenability. Molybdenum reduces temper brittleness in nickel–chromium steels.

Nickel

The range used is 0.3–5%. It improves strength, toughness and hardenability, without affecting ductility. A high proportion of it improves corrosion resistance. For parts subject to fatigue 5% is used, and above 27% the steel is non-magnetic. Nickel promotes an austenitic structure.

Silicon

The usual range is 0.2–3%. It has little effect below 3%. At 3% it improves strength and hardenability but reduces ductility. Silicon acts as a deoxidizer.

Sulphur

Up to 0.5% sulphur forms sulphides which improve machineability but reduces ductility and weldability.

Titanium

This is a strong carbide forming element. In proportions of 0.2–0.75% it is used in maraging steels to make them age-hardening and to give high strength. It stabilizes austenitic stainless steel.

Tungsten

This forms hard stable carbides and promotes grain refining with great hardness and toughness at high temperatures. It is a main alloying element in high speed tool steels. It is also used for permanent-magnet steels.

Vanadium

This is a carbide forming element and deoxidizer used with nickel and/or chromium to increase strength. It improves hardenability and grain refinement and combines with carbon to form wear-resistant microconstituents. As a deoxidizer it is useful for casting steels, improving strength and hardness and eliminating blowholes, etc. Vanadium is used in high-speed and pearlitic chromium steels.

6.3.4 *Typical properties of alloy steels*

Typical properties of alloy steels

Content	Type	Specification	Tensile strength ($N\,mm^{-2}$)	Fatigue limit ($N\,mm^{-2}$)	Weldability	Corrosion resistance	Machine-ability	Formability
Low	1%Cr, Mo	709M40	1240	540	PH/FHTR	PR	F/HTR	F
	1.75%Ni, Cr, Mo	817M40	1550	700	PH/FHTR	PR	P/HTR	F
	4.25%Ni, Cr, Mo	835M30	1550	700	PH/FHTR	PR	P/HTR	F
	3%Cr, Mo, V	897M39	1310 (1780)	620	PH/FHTR	PR	P/HTR	F
	5%Cr, Mo, V	AISI H11	2010 (A2630)	850 (A1880)	PH/FHTR	PR	P/HTR	F
Medium	9%Ni, Co	HP9/4/45 Republic Steel	1390 1850	—	FHTR	PR	P/HTR	F
	12–14%Cr	410S21	1160	340	P/FHTR	F	F/HTR	F
	Cr, W, Mo, V	Vascojet MA Vanadium alloy steel	2320 (A3090)	960	PH/FHTR	PR	P/HTR	F
High	13%Cr, Ni, Mo	316S12	620	260	G	G	F	G
	19%Cr, Ni, Mo	317S16	650	260	G	G	F	G
	15%Cr, Ni, Mo, V	ESSHETE 1250 S. Fox	590	—	G/FHTR	G/HT	F	—
	17%Cr, Ni	AISI 301	740 (CR1240)	280	F	F	F	G
	17% Cr, Ni, Al	17/7 PH Armco	1480	—	F	F	F	G
	14%Cr, Ni, Cu, Mo, Nb	REX 627 Firth Vickers	1470	540	FHTR	F	F	F
	15%Cr, Ni, Mo, V	AM 355 Allegheny Ludlum	1480	740	FHTR	F	F	F
	18%Ni, Co, Mo	300 grade maraging INCO	1930	—	G/FHTR	PR	F	P
	18%Ni, Co, Mo	250 grade maraging	1700	660	G/FHTR	PR	F	P

A = ausformed, MA = martempered, CR = cold rolled, P = poor, F = fair, G = good, PH = preheat required, PR = protection required, HT = at high temperature, HTR = when heat treated, FHTR = final heat treatment required.

6.3.5 Cast high-alloy steels

BS specification	Type	Composition (%)							Tensile strength (N mm^{-2})	Yield stress (N mm^{-2})	Elongation (%)
		Cu	Si	Mn	Ni	Cr	Mo	C			
3100 BW 10	Austenitic manganese steel	—	1.0	11.0	—	—	—	1.0	—	—	—
	Possess great hardness hence used for earth moving equipment pinions, sprockets, etc., where wear resistance is important.										
3100 410 C 21	13% chromium steel	—	1.0	1.0	1.0	13.5	—	0.15	540	370	15
	Mildly corrosion resistant. Used in paper industry										
3100 302 C25	Austenitic chromium–nickel steel	—	1.5	2.0	8.0	21.0	—	0.08	480	210	26
	Cast stainless steel. Corrosion resistant and very ductile.										
3100 315 C16	Austenitic chromium–nickel–molybdenum steel	—	1.5	2.0	10.0	20.0	1.0	0.08	480	210	22
	Cast stainless steel with higher nickel content giving increased corrosion resistance. Molybdenum gives increased weldability										
3100 302 C35	Heat-resisting alloy steel	—	2.0	2.0	10.0	22.0	1.5	0.4	560	—	3
3100 334 C11		—	3.0	2.0	65.0	10.0	1.0	0.75	460	—	3
	Can withstand temperatures in excess of 650 °C. Temperature at which scaling occurs raised due to chromium										

6.3.6 Weldable structural steel for hollow sections

Mechanical properties of weldable structural steel for hollow sections (BS 4360: 1972)

Grade	Tensile strength (N mm^{-2})	Yield strength* (N mm^{-2})	Elongation (%)
43C	430/540	255	22
43D	430/540	255	22
43E	430/540	270	22
50B	490/620	355	20
50C	490/620	355	20
50D	490/620	355	20
55C	550/700	450	19
55E	550/700	450	19

*Up to 16 mm thickness.

6.4   Stainless steels

6.4.1   *Types of stainless steel*

Stainless steels comprise a wide range of iron alloys containing more than 10% chromium. They are classified as austenitic, ferritic and martensitic.

Austinitic stainless steels

A standard composition is 18%Cr, 8%Ni (18/8 steel). These steels have high resistance to corrosion, good weldability, high toughness, especially at low temperature, and excellent ductility. They may be hardened by cold working and are non-magnetic. Special properties are produced by the addition of molybdenum, cadmium, manganese, tungsten and columbium.

Ferritic stainless steels

The chromium content is normally 16–20% with corrosion resistance better than martensitic but inferior to austenitic steels. They are used for presswork because of their high ductility, but are subject to brittle failure at low temperature. They have moderate strength and limited weldability and are hardenable by heat treatment. The low carbon content makes them suitable for forming without cracking. They are magnetic and have low coefficients of thermal expansion.

Martensitic stainless steels

The chromium content is 12–18% and the nickel content is 1–3%. These steels are the least corrosion resistant of all. They are unsuitable for welding or cold forming. They have moderate machineability and are used where high resistance to tempering at high temperature is important, e.g. for turbine blades. They can be heat treated to improve properties and can be made with a wide range of properties. They are used for cutlery.

6.4.2   *Selection of stainless steels*

The applications of the different stainless steels are listed below.

Austenitic

Window and door frames. Roofing and guttering. Chemical plant and tanks. Domestic hot water piping. Spoons, forks, knife handles. Kitchen utensils. Washing machines. Hospital equipment. Car hub caps, rim embellishers and bumpers. Wheel spokes. Welding rods and electrodes. Wire ropes. Yacht fittings, masts and marine fittings. Nuts, bolts, screws, rivets, locking wire, split pins. Shafts. Coil and leaf springs.

Ferritic

Mouldings and trim for cars, furniture, television sets, gas and electric cookers, refrigerators, etc. Coinage. Spoons and forks. Domestic iron soles. Vehicle silencers. Driving mirror frames. Fasteners. Parts to resist atmospheric corrosion. Heat-resistant parts, e.g. oil-burner sleeves and parts working up to 800 °C.

Martensitic

Structural components. Tools. High temperature turbine parts. Flat and coil springs. Scales, rulers, knives, spatulas. Kitchen tools and appliances where high strength and hardness are required with moderate corrosion resistance. Surgical and dental instruments. Record player spindles. Fasteners.

6.4.3 *Properties of typical types*

BS code no.	Remarks	Condition	Yield stress ($N\,mm^{-2}$)	Tensile strength ($N\,mm^{-2}$)	Elongation (%)	Composition (%)		
						C	Ni	Cr
Stainless Iron 1 (416S21)	Martensitic steel, easy to manipulate	AD	430	510	12	0.9–0.15	1 (max.)	11.5/13.5
		S	280	400	35	0.9–0.15	1 (max.)	11.5/13.5
Stainless Iron 1 (416S29)	Similar to above, but harder	AD	465	540	10	0.14/0.2	1 (max.)	11.5/13.5
		S	280	430	—	0.14/0.2	1 (max.)	11.5/13.5
		AH	850	1080	—	0.14/0.2	1 (max.)	11.5/13.5
Stainless Iron W	Weldable martensitic steel	AD	465	540	10	—	—	—
		S	250	450	35	—	—	—
		AH	700	930	—	—	—	—
20/2 (431S29)	Martensitic steel harder to work than stainless iron but greater resistance, especially to sea water	AD	770	850	10	0.12/0.2	2/3	15/20
		S	590	740	25	0.12/0.2	2/3	15/20
		H	1005	1080	15	0.12/0.2	2/3	15/20
Stainless steel 17 (430S15)	Ferritic stainless steel more corrosion resistant than stainless iron	S	310	510	25	0.1 (max.)	0.5	16/18
Stainless Steel 20 (430S16)	Similar to above, but a little more corrosion resistant	S	340	540	25	0.1 (max.)	0.5	16/18
Stainless Steel 27	Ferritic steel with excellent resistance to scaling at high temperature	S	390	560	20	—	—	—
18/8 (302S25)	Austenitic steel, good for working and welding. Must be softened after welding	AD	620	700	25	0.12	8.11	17/19
		S	230	540	50	0.12	8.12	17/19

6.4.3 *Properties of typical types* (*continued*)

BS code no.	Remarks	Condition	Yield stress ($N\,mm^{-2}$)	Tensile strength ($N\,mm^{-2}$)	Elongation (%)	Composition (%)		
						C	Ni	Cr
18/8 low (304S15)	As above, but low carbon content. Need not be softened after welding	AD	540	620	25	0.06 (max.)	8/11	17/19
		S	230	540	30	0.06 (max.)	8/11	17/19
18/8/T 18/12/Ni (347S17)	Special welding qualities, need not be softened after welding. 18/12/Nb contains niobium	AD	700	770	20	0.08	9/12	17/19
		S	280	590	40	0.08	9/12	17/19
18/8/M (316S16)	For resistance to certain concentrations of acetic and sulphuric acids	AD	700	770	20	0.07	10/13	16.5/18.5
		S	330	660	40	0.07	10/13	16.5/18.5
18/8/MT	As above but need not be softened after welding	AD	700	770	20	—	—	—
		S	330	660	40	—	—	—
'316'	Similar to 18/8/M	AD	700	770	20	—	—	—
		S	310	620	40	—	—	—
25/20	Austenitic steel with good heat-resisting properties	AD	700	775	30	—	—	—
		S	340	620	45	—	—	—
23/16/T	Similar to 25/20, can be welded without subsequent softening	AD	700	770	30	—	—	—
		S	390	660	40	—	—	—
16/6/H	An austenitic/ martensitic steel suitable for hardening	S	310	850	30	—	—	—
		H	1080	1240	15	—	—	—

S = softened, H = hardened, AD = as drawn, AH = air hardened.

6.5   British Standard specification of steels

The relevant standard is BS 970 'Wrought Steels'. The standard is in six parts:

Part 1 Carbon and carbon manganese steels including free-cutting steels
Part 2 Direct hardening alloy steels
Part 3 Steels for Case Hardening
Part 4 Stainless, heat resisting and spring steels
Part 5 Carbon and alloy spring steels
Part 6 SI metric values (for use with Parts 1 to 5)

Each steel is designated by six symbols:

First three digits
000–199: Carbon and carbon–manganese steels. Digits represent 100 times the percentage of manganese.
200–240: Free cutting steels. Second and third digits represent 100 times the percentage of sulphur.
250: Silicon–manganese steel

300–449: Heat-resistant, stainless and valve steels
500–999: Alloy steels

Letter
The letters A, M, H and S indicate if the steel is supplied to – chemical analysis, mechanical properties, hardenability requirements, or is stainless, respectively.

Last two digits
These give 100 times the percentage of carbon, except for stainless steels.

Example

070M20: A plain carbon steel with 0.2% carbon and 0.7% manganese. The mechanical properties, i.e. tensile strength, yield strength, elongation and hardness, are given in the standard.

6.6   Non-ferrous metals

6.6.1   *Copper and copper alloys*

Electrolytically refined copper (99.95% pure) is used for components requiring high conductivity. Less pure copper is used for chemical plant, domestic plumbing,

etc. Copper is available in the form of wire, sheet, strip, plate, round bar and tube.

Copper is used in many alloys, including brasses, bronzes, aluminium bronze, cupronickel, nickel–silver and beryllium–copper.

Composition and mechanical properties of some copper alloys

	Composition (%)				Mechanical properties			
Type and uses	Cu	Zn	Others	Condition	0.1% proof stress ($N\,mm^{-2}$)	Tensile strength ($N\,mm^{-2}$)	Elongation (%)	Vickers hardness
Muntz metal: die stampings, and extrusions	60	40	—	Extruded	110	350	40	75
Free-cutting brass: high-speed machining	58	39	3 Pb	Extruded	140	440	30	100

Composition and mechanical properties of some copper alloys (*continued*)

| Type and uses | Composition (%) | | | Condition | Mechanical properties | | | |
	Cu	Zn	Others		0.1% proof stress (N mm^{-2})	Tensile strength (N mm^{-2})	Elongation (%)	Vickers hardness
Cartridge brass: severe cold working	70	30	—	Annealed Work hardened	75 500	270 600	70 5	65 180
Standard brass: presswork	65	35	—	Annealed Work hardened	90 500	320 690	65 4	65 185
Admiralty gunmetal: general-purpose castings	88	2	10 Sn	Sand casting	120	290	16	85
Phosphor bronze: castings and bushes for bearings	remainder		10 Sn, 0.03–0.25 P	Sand casting	120	280	15	90

Applications of copper and copper alloys

Type and composition	Condition	Tensile MN/m^2	Product	Use
Pure copper				
99.95%Cu	O H	220 350	Sheet, strip wire	High conductivity electrical applications
98.85%Cu	O H	220 360	All wrought forms	Chemical plant. Deep drawn, spun articles
99.25%Cu + 0.5%As	O H	220 360	All wrought forms	Retains strength at high temperatures. Heat exchangers, steam pipes
Brasses				
90%Cu, 10%Zn– gilding metal	O H	280 510	Sheet, strip and wire	Imitation jewellery, decorative work
70%Cu, 30%Zn– cartridge brass	O H	325 700	Sheet, strip	High ductility for deep drawing
65%Cu, 35%Zn– standard brass	O H	340 700	Sheet, strip and extrusions	General cold working alloy
60%Cu, 40%Zn– Muntz metal	M	375	Hot rolled plate and extrusions	Condenser and heat exchanger plates
59%Cu, 35%Zn, 2%Mn, 2%Al, 2%Fe	M	600	Cast and hot worked forms	Ships screws, rudders
58%Cu, 39%Zn, 3%Pb– free cutting	M	440	Extrusions	High speed machine parts

Applications of copper and copper alloys (*continued*)

Type and composition	Condition	Tensile MN/m^2	Product	Use
Bronzes				
95.5%Cu, 3%Sn, 1.5Zn	O	325	Strip	Coinage
	H	725		
5.5%Sn, 0.1%Zn, Cu	O	360	Sheet, strip and wire	Springs, steam turbine blades
	H	700		
10%Sn, 0.03–0.25P, Cu–phosphor bronze	M	280	Castings	Bushes, bearings and springs
10%Sn, 0.5%P, Cu	M	280	Castings	General-purpose castings and bearings
10%Sn, 2%Zn, Cu–Admiralty gunmetal	M	300	Castings	Pressure-tight castings, pump, valve bodies
Aluminium bronze				
95%Cu, 5%Al	O	400	Strip and tubing	Imitation jewellery, condenser tubes
	H	770		
10%Al, 2.5%Fe, 2–5%Ni, Cu	M	700	Hot worked and cast products	High-strength castings and forgings
Cupronickel				
75%Cu, 25%Ni	O	360	Strip	British 'silver' coinage
	H	600		
70%Cu, 30%Ni	O	375	Sheet and tubing	Condenser tubes, good corrosion resistance
	H	650		
29%Cu, 68%Ni, 1.25%Fe, 1.25%Mn	O	550	All forms	Chemical plant, good corrosion resistance
	H	725		
Nickel–silver				
55%Cu, 27%Zn, 18%Ni	O	375	Sheet and strip	Decorative use and cutlery
	H	650		
Beryllium–copper 1.75–2.5%Be, 0.5%Co, Cu	WP	1300	Sheet, strip, wire, forgings	Non-spark tools, springs

O = annealed, M = as manufactured, H = fully work hardened, WP = solution heat treated and precipitation hardened.

6.6.2 *Aluminium and aluminium alloys*

Pure aluminium is available in grades from 99% to 99.99% purity. It is soft and ductile but work hardens. Pure aluminium is difficult to cast.

Alloying elements improve properties as follows:

Copper: increases strength and hardness. Makes heat treatable.
Magnesium: increases hardness and corrosion resistance.

Manganese: increases strength.
Silicon: lowers melting point, increases castability.
Silicon and magnesium: gives a heat-treatable alloy.
Zinc: increases strength and hardness.
Zinc and magnesium: increases strength; makes heat treatable.
Bismuth: increases machinability.
Lead: increases machinability.
Boron: increases electrical conductivity.
Nickel: increases strength at high temperature.
Titanium: increases strength and ductility.
Chromium, vanadium and zirconium: also used.

Classification of aluminium alloys

Aluminium alloys may be classified as follows.

(1) Wrought alloys: (a) heat-treatable
  (b) non-heat-treatable

(2) Casting alloys: (a) heat-treatable
  (b) non-heat-treatable

Wrought aluminium alloys

Composition (%)	Condition	0.1% proof stress (N mm^{-2})	Tensile strength (N mm^{-2})	Elongation (%)	Machineability	Cold forming
Non-heat-treatable alloys						
Aluminium 99.99	Annealed	—	90 (max.)	30	Poor	Very good
	Half hard	—	100–120	8		
	Full hard	—	130	5		
Cu 0.15, Si 0.6, Fe 0.7, Mn 1.0, Zn 0.1, Ti 0.2, Al 97.2	Annealed	—	115 (max.)	30	Fair	Very good
	Quarter hard	—	115–145	12		
	Half hard	—	140–170	7		
	Three-quarters hard	—	160–190	5		
	Full hard		180	3		
Cu 0.1, Mg 7.0, Si 0.6, Fe 0.7, Mn 0.5, Zn 0.1, Cr 0.5, Ti 0.2, Al balance	Annealed	90	310–360	18	Good	Fair
Heat-treatable alloys						
Cu 3.5–4.8, Mg 0.6, Si 1.5, Fe 1.0, Mn 1.2, Ti 0.3, Al balance	Solution treated	—	380	—	Good	Good
	Fully heat treated	—	420	—	Very good	Poor
Cu 0.1, Mg 0.4–1.5, Si 0.6–1.3, Fe 0.6, Mn 0.6, Zn 0.1, Cr 0.5, Ti 0.2, Al balance	Solution treated	110	185	18	Good	Good
	Fully heat treated	230	280	10	Very good	Fair

Aluminium alloys for sheet, strip, extrusions and forgings

Specification no.	Composition (%)	Condition	Tensile strength ($N\,mm^{-2}$)	Type of product and use
1	99.99 Al	O	45	Sheet, strip. Linings for chemical and food plant
1A	99.80 Al	O	60	Sheet, strip. Linings for chemical and food plant
1C	99.0 Al	O $\frac{1}{2}$H H	90 120 150	Sheet, strip, wire, extruded sections. Hollow ware, kitchen ware, bus-bars, decorative panelling
N3	Al, 1.25% Mn	O $\frac{1}{2}$H H	110 160 210	Sheet, strip, extruded sections. Hollow ware, roofing, panelling, scaffolding, tubes
N4	Al, 2 Mg	O $\frac{1}{2}$H	210· 250	Sheet, plate, tubes, extrusions. Stronger deep-drawn articles, ship and boat construction, other marine applications
N5	Al, 3.5 Mg	O $\frac{1}{2}$H	230 280	
N6	Al, 5 Mg	O $\frac{1}{2}$H	280 320	
H10	Al, 0.7 Mg, 1.0 Si	W WP	270 325	Sheet, forgings, extrusions. Structural components for road and rail vehicles
H14	Al, 4.5 Cu, 0.75 Mg, 0.5 Mn	T	440	Sheet, forgings, extrusions. Highly stressed aircraft parts, general engineering parts
H15	Al, 4.5 Cu, 0.75 Mg, 0.5 Mn	WP	500	Tube. Highly stressed aircraft parts, general engineering parts
H16	Al, 1.75 Cu, 2 Mg, 7 Zn	WP	620	Sheet, extrusions. Aircraft construction

O = annealed, $\frac{1}{2}$H = half hard, H = fully work hardened, M = as manufactured, W = solution treated only, WP = solution treated and precipitation hardened, T = solution heat treated and naturally aged.

Aluminium casting alloys*

Composition (%)	Condition	0.2% proof stress ($N mm^2$)	Tensile strength ($N mm^{-2}$)	Elongation (%)	Hardness, BHN†	Machinability
As cast alloys						
Cu, 0.1, Mg 3–6, Si	Sand cast	60	160	5	50	Difficult
10–13, Fe 0.6, Mn	Chill cast	70	190	7	55	Difficult
0.5, Ni 0.1, Sn 0.05,	Die cast	120	280	2	55	—
Pb 0.1, Al balance						
Cu 0.7–2.5, Mg 0.3, Si	Chill cast	100	180	1.5	85	Fair
9–11.5, Fe 1.0, Mn	Die cast	150	320	1	85	—
0.5, Ni 1.0, Zn 1.2,						
Al balance						
Heat treatable						
Cu 4–5, Mg 0.1, Si	Chill cast	—	300	9	—	Good
0.25, Fe 0.25, Mn	Fully heat	—	300	9	—	Good
0.1, Ni 0.1, Zn 0.1,	treated					
Al balance						

*These alloys are used for food, chemical plant, marine castings and hydraulics.
†BHN = Brinell hardness number.

6.7 Miscellaneous metals

Antimony

A brittle lustrous white metal used mainly as an alloying element for casting and bearing alloys and in solders.

Beryllium

A white metal similar in appearance to aluminium. Brittle at room temperature. Has many applications in the nuclear field and for electronic tubes. With copper and nickel it produces alloys with high strength and electrical conductivity. Beryllium iron has good corrosion and heat resistance.

Cadmium

A fairly expensive soft white metal like tin. Used for plating and electrical storage batteries. It has good resistance to water and saline atmospheres and is useful as plating for electrical parts since it takes solder readily.

Chromium

A steel-grey soft but brittle metal. Small traces of carbide give it extreme hardness. It is used extensively in alloys and for electroplating and is also used for electrical resistance wire and in magnet alloys.

Lead

A heavy, soft, ductile metal of low strength but with good corrosion resistance. It is used for chemical equipment, roofing, cable sheathing and radiation shielding. It is also used in alloys for solder and bearings.

Lead–tin alloys

These are used as 'soft solders', often with a little antimony for strength.

Tinman's solder   Approximately 2 parts of tin to 1 part of lead. Used for electrical jointing and tinplate-can sealing.

Plumber's solder   Approximately 2 parts of lead to 1 part of tin. Used for wiping lead pipe joints.

Type metal   Contains about 25% tin, with lead and some antimony. Has negligible shrinkage.

Bearing metal   Lead based 'white metal' contains lead, tin, antimony and copper, etc.

Magnesium

A very light metal, only one-quarter the weight of steel and two-thirds that of aluminium, but not easily cold worked. Usually alloyed with up to 10% aluminium and often small amounts of manganese, zinc and zirconium. Used for aircraft and internal combustion engine parts, nuclear fuel cans and sand and die castings. Magnesium and its alloys corrode less in normal temperatures than does steel.

Manganese

A silvery white hard brittle metal present in most steels. It is used in manganese bronze and high nickel alloys and to improve corrosion resistance in magnesium alloys.

Nickel

Nickel has high corrosion resistance. It is used for chemical plant, coating steel plate and electroplating as a base for chromium. Nickel is used for many steel, iron and non-ferrous alloys.

Nickel–base alloys

Monel   Used for steam turbine blades and chemical plant. Composition: 68%Ni, 30%Cu, 2%Fe.

Inconel   Good at elevated temperatures, e.g. for cooker heater sheaths. Composition: 80%Ni, 14%Cr, 6%Fe.

Nimonic   A series of alloys based on 70–80%Ni, with small amounts of Ti, Co, Fe, Al and C. Has high resistance to creep and is used for gas turbine discs and blades, and combustion chambers. Strong up to 900 °C.

Platinum

A soft ductile white metal with exceptional resistance to corrosion and chemical attack. Platinum and its alloys are widely used for electrical contacts, electrodes and resistance wire.

Silver

A ductile malleable metal with exceptional thermal and electrical conductivity. It resists most chemicals but tarnishes in a sulphurous atmosphere. It is used for electrical contacts, plating, bearing linings and as an alloying element.

Tin

A low-melting-point metal with silvery appearance and high corrosion resistance. It is used for tinplate, bearing alloys and solder.

Titanium

An expensive metal with low density, high strength and excellent corrosion resistance. It is used in the aircraft industry, generally alloyed with up to 10% aluminium with some manganese, vanadium and tin. Titanium is very heat resistant.

Tungsten

A heavy refractory steel-grey metal which can only be produced in shapes by powder metallurgy (m.p. 3410 °C). It is used as an alloying element in tool and die steels and in tungsten carbide tool tips. It is also used in permanent magnets.

Zinc

Pure zinc has a melting point of only 400 °C so is good for die casting, usually with 1–2%Cu and 4%Al to increase strength. Used for carburettors, fuel pumps, door handles, toys, etc., and also for galvanizing sheet steel, nails and wire, and in bronze.

6.8 Spring materials

6.8.1 *Carbon steels*

Hard-drawn spring steel

Low cost; general purpose; low stress; low fatigue life. Temperatures below 120 °C. Tensile strength up to 1600 N mm^{-2}.

Piano (music) wire

Tougher than hard-drawn spring steel; high stress (tensile strength up to 2300 N mm^{-2}); long fatigue life; used for 'small springs'. Temperatures below 120 °C.

Oil-tempered spring steel

General purpose springs; stress not too high; unsuitable for shock or impact loads. Popular diameter range 3–15 mm.

6.8.2 *Alloy steels*

Chrome–vanadium steel

Best for shock and impact loads. Available in oil-tempered and annealed condition. Used for internal combustion engine valve springs. Temperatures up to 220 °C.

Silicon–manganese steel

High working stress; used for leaf springs; temperatures up to 220 °C.

Silicon–chromium steel

Better than silicon–manganese; temperatures up to 220 °C.

Stainless steels

Cold drawn; tensile strength up to 1200 N mm^{-2}. Temperatures from sub-zero to 290 °C, depending on type. Diameters up to 5 mm.

6.8.3 *Non-ferrous alloys*

Spring brass (70/30)

Low strength, but cheap and easily formed. Good electrical conductivity.

Phosphor bronze (5%Sn)

High strength, resilience, corrosion resistance and fatigue strength. Good electrical conductivity. Tensile strength 770 N mm^{-2}. Wire diameters 0.15–7 mm. Used for leaf and coil switch springs.

Beryllium–copper ($2\frac{1}{4}$%)

Formed in soft condition and hardened. High tensile strength. Used for current-carrying brush springs and contacts. Tensile strength 1300 N mm^{-2}.

Inconel

Nickel based alloy useful up to 370 °C. Exceedingly good corrosion resistance. Diameters up to 7 mm. Tensile strength up to 1300 N mm^{-2}.

6.8.4 *Moduli of spring materials*

Material	Modulus of rigidity, G, GN m^{-2}	Modulus of elasticity, E, GN m^{-2}
Carbon steel	80	207
Chrome–vanadium steel	80	207
18/8 Stainless steel	63	193
70/30 Brass	38	103
Phosphor bronze	36	97
Beryllium–copper	40–48	110–128
Inconel	76	214
Monel	66	179
Nickel–silver	38	110

6.9 Powdered metals

Powdered metal technology is used widely to produce components which are homogeneous, have controlled density, are inclusion free and of uniform strength.

They can be subject to secondary treatment such as forging, repressing, resintering, and heat treatment.

6.9.1 Process

(1) Production of metal powder, mixing for alloys and additives if required.
(2) Compacting in a shaped die with pressure of 400–800 N mm^{-2} to give required density.
(3) Sintering at high temperatures to bond particles, e.g. 1100 °C for iron and 1600 °C for tungsten.
(4) Sizing and finishing.

POWDERED-METAL COMPONENTS

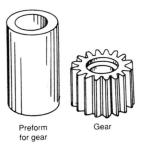

Preform   Gear
for gear

6.9.2 Metals used

Iron and copper The most used metals.
High-melting-point metals For example, platinum and tungsten.
Aluminium Special atmosphere and lubricant required because of the formation of the oxide.
Tin bronze Used for 'self-lubricating' bearings.
Stainless steel Used for filters.

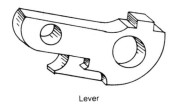

Lever

6.9.3 Advantages

(1) For use in alloys where metals are insoluble.
(2) For high-melting-point metals, e.g. tungsten.
(3) Virtually no waste.
(4) Little or no finishing required.
(5) Controlled density and strength.
(6) Relatively inexpensive production method.

Rotor

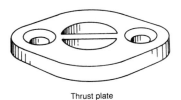

Thrust plate

6.10 Low-melting-point alloys

Name	Composition (%)				Melting point (°C)
	Sn	Pb	Bi	Cd	
—	37.5	50.0	12.5	0	178
—	50.0	40.0	10.0	0	162
—	25.0	50.0	25.0	0	149

Name	Composition (%)				Melting point (°C)
	Sn	Pb	Bi	Cd	
—	40.0	40.0	20.0	0	145
—	33.33	33.33	33.33	0	123
—	20.0	40.0	40.0	0	113
Rose's alloy	22.0	28.0	50.0	0	100
Newton's alloy	18.75	31.25	50.0	0	95
Darcet's alloy	25.0	25.0	50.0	0	93
—	50.0	25.0	0	25.0	86
—	9.25	34.5	50.0	6.25	77
Wood's metal	12.5	25.0	50.0	12.5	73
Lipowitz' alloy	13.33	26.67	50.0	10.0	70

6.11 Miscellaneous information on metals

Physical properties of common engineering materials

Material	Application	PS/YS	Tensile strength $(\mathrm{N\,mm^{-2}})$	E $(\mathrm{GN\,m^{-2}})$	G $(\mathrm{GN\,m^{-2}})$	v	α $(\times 10^6\,\mathrm{K^{-1}})$	ρ $(\mathrm{kg\,m^{-3}})$
Steel 070M20	Structures, lightly stressed parts, bolts, brackets, levers	240	430	207	80	0.3	11	7850
Steel 080M40	Shafts and machine details requiring strength and wear resistance	250–400	510–650	207	80	0.3	11	7850
Steel 070M55	Gears, machine tools and hard parts	310–570	620–980	207	80	0.3	11	7850
Steel 060A96	Springs	—	1300	207	80	0.3	11	7850
Steel 331S40	Internal combustion engine valves	—	1100–1700	207	80	0.3	11	7850
Aluminium alloy NS4	Plate, sheet and strip	60	170	70	27	0.32	23	2700
Aluminium alloy NF8M	Forgings	130	280	70	27	0.32	23	2700
Aluminium alloy HE15TB	Rolled sections	230	370	70	27	0.32	23	2700

Physical properties of common engineering materials (*continued*)

Material	Application	PS/YS	Tensile strength ($N\,mm^{-2}$)	E ($GN\,m^{-2}$)	G ($GN\,m^{-2}$)	v	α ($\times 10^6\,K^{-1}$)	ρ ($kg\,m^{-3}$)
Grey cast iron	Brittle. Castings not subject to heavy impact	—	150/400 (tension) 600/1200 (compression)	130	48	—	12	7200
Malleable cast iron blackheart	Foot pedals, small cast parts, bends before fracture	180	260 (tension) 780 (compression)	170	68	0.26	11	7350
Spheroidal graphite iron	Similar to malleable cast iron	240–420	380–740	170	68	0.26	11	7350
Brass, cold drawn	Bearings	—	168	100	34	0.32	20	8400
Phosphor bronze, rolled	Castings in contact with water. Non-magnetic springs	—	410	116	43	0.33	17	8800
Timber	Frames	—	3–5 (along grain) 35–60 (across grain)	8–16	—	—	—	350–800
Fibre glass	Cowls, motor bodies	—	100 (tension) 150 (compression)	10*	—	—	20	1500
Acetal resin	Mouldings	—	70	4.7 (compression) 3.6* (tension)	—	0.35	13.5	1420
Nylon	Bearings	—	80	1.6*	—	—	100	1100
Polystyrene	Moulded components	—	45 (tension) 110 (compression)	3*	—	—	70	1070

PS/YS = proof stress ($N\,mm^{-2}$)/yield stress ($N\,mm^{-2}$), E = Young's modulus, G = shear modulus, v = Poisson's ratio, α = coefficient of linear expansion, ρ = density.
*Do not obey Hooke's law.

Chemical symbols for metals and alloying elements

Aluminium	Al	Gold	Au	Selenium	Se
Antimony	Sb	Mercury	Hg	Silicon	Si
Arsenic	As	Indium	In	Sulphur	S
Barium	Ba	Magnesium	Mg	Tantalum	Ta
Beryllium	Be	Manganese	Mn	Tellurium	Te
Bismuth	Bi	Molybdenum	Mo	Tin	Sn
Carbon	C	Nickel	Ni	Titanium	Ti
Cadmium	Cd	Phosphorus	P	Tungsten	W
Cobalt	Co	Lead	Pb	Uranium	U
Chromium	Cr	Platinum	Pt	Vanadium	V
Copper	Cu	Plutonium	Pu	Zinc	Zn
Iron	Fe	Radium	Ra	Zirconium	Zr
Gallium	Ga	Rhodium	Rh		
Germanium	Ge	Silver	Ag		

Typical Brinell hardness numbers (BHN) for metals

Material	BHN
Soft brass	60
Mild steel	130
Annealed chisel steel	235
White cast iron	415
Nitrided surface	750

Comparison of hardness numbers

Rockwell C scale	Vicker's pyramid	Brinell hardness number	Rockwell C scale	Vicker's pyramid	Brinell hardness number	Rockwell C scale	Vicker's pyramid	Brinell hardness number
68	1030	—	49	515	468	30	299	286
67	975	—	48	500	458	29	291	279
66	935	—	47	485	447	28	284	272
65	895	—	46	470	436	27	277	266
64	860	—	45	456	426	26	271	260
63	830	—	44	442	416	25	265	255
62	800	—	43	430	406	24	260	250
61	770	—	42	418	396	23	255	245
60	740	—	41	406	386	22	250	240
59	715	609	40	395	376	21	245	235
58	690	594	39	385	366	20	240	230
57	670	579	38	375	356	—	220	210
56	650	564	37	365	346	—	200	190
55	630	549	36	355	337	—	180	171
54	610	534	35	345	328	—	160	152
53	590	519	34	335	319	—	140	133
52	570	504	33	325	310	—	120	114
51	550	492	32	315	302	—	100	95
50	532	480	31	307	294	—	—	—

Properties of pure metals

Metal	m.p. (°C)	ρ (kg m^{-3})	E (GN m^{-2})	G (GN m^{-2})	RSHC	α ($\times 10^6$ °C^{-1})	ρ_o ($\mu\Omega$-m)	α_r (mΩ °C^{-1})	ECE (mg °C^{-1})
Aluminium	659	2 700	70	27	0.21	23	245	450	0.093
Copper	1083	8 900	96	38	0.09	17	156	430	0.329
Gold	1063	19 300	79	27	0.03	14	204	400	0.681
Iron	1475	7 850	200	82	0.11	12	890	650	0.193
Lead	327	11 370	16	—	0.03	29	1900	420	1.074
Mercury	—	13 580	—	—	0.03	60	9410	100	1.039
Nickel	1452	8 800	198	—	0.11	13	614	680	0.304
Platinum	1775	21 040	164	51	0.03	9	981	390	0.506
Silver	961	10 530	78	29	0.06	19	151	410	1.118
Tungsten	3400	19 300	410	—	0.03	4.5	490	480	0.318
Zinc	419	6 860	86	38	0.09	30	550	420	0.339

m.p. = melting point, ρ = density, E = Young's modulus, G = shear modulus, RSHC = relative specific heat capacity, α = coefficient of linear expansion, ρ_o = resistivity at 0 °C, α_r = resistance temperature coefficient at 0 °C, ECE = electrochemical equivalent.

6.12 Corrosion of metals

6.12.1 Corrosion prevention

Corrosion may be prevented by considering the following points.

Material selection

Metals and alloys which resist corrosion in a particular environment can be used. Proximity of metals with large potential difference, e.g. a copper pipe on a steel tank, should be avoided. Galvanic protection can be used, e.g. by use of a 'sacrificial anode' of zinc close to buried steel pipe or a ship's hull.

Appropriate design

Crevices which hold water, e.g. bad joints and incomplete welds, should be avoided as should high tensile stresses in material subject to stress corrosion. Locked-in internal stress due to forming should be avoided.

Modified environment

Metals can be enclosured against a corrosive atmosphere, water, etc. Drying agents, e.g. silica gel, and corrosion inhibitors, e.g. in central-heating radiators can be used.

Protective coating

Metals can be coated to make them impervious to the atmosphere, water, etc., by use of a coating of grease, plasticizer, bitumen, resins, polymers, rubber latex, corrosion-resistant paints or metal coating.

6.12.2 Corrosion resistance of metals

Ferrous metals

Stainless steels Generally the best of all metals. All types have good resistance to atmospheric corrosion except gases such as chlorine and sulphur. Some types are suitable up to 1100 °C. Some resist sulphuric acid and some nitric acids, but not hydrochloric or hydrofluoric acids. All resist uncontaminated organic solvents and foods and also alkalis at room temperature, but not bleaches. They resist neutral water, but stress corrosion cracking may occur above 66 °C.

Alloy steels Chrome steel has good resistance which is improved by the addition of nickel; it can be used in sea water. Iron–nickel steel has good resistance with over 20% nickel plus 2–3% carbon; it is used in a marine environment.

Iron and carbon steel These readily corrode in air and especially sea water. They are subject to stress corrosion cracking and internal stress corrosion, and require protection by painting, plating, tinning, galvanizing, etc.

Copper and copper based alloys

Copper An oxide coating prevents corrosion from water and atmosphere, e.g. water pipes.

Brass 'Yellow brass' (>15%Zn) is subject to 'dezincification' in hot water. 'Red brass' (85%Cu minimum) is much better. Resistance is improved by the addition of arsenic or antimony.

Bronzes Over 5% tin gives better resistance than brass, especially to sea water and stress corrosion cracking. Aluminium bronze is good at elevated temperatures. Silicon bronze is as good but also has weldability; it is used for tanks.

Cupronickel This has the best resistance of all copper alloys and is used for heat-exchanger tubes.

Other metals and alloys

Nickel alloys These are generally extremely resistant to caustics up to high temperature, and to neutral water and sea water. They resist some acids. Alloys such as Inconel have good resistance up to 1170 °C which increases with chromium content. Nickel alloys have high resistance to stress corrosion cracking. Different alloys have resistance to different acids. Nickel alloys are used for tanks, heat exchangers, furnace parts, and chemical plant.

Magnesium and magnesium alloys These have better resistance than steel in the atmosphere, but are inferior to aluminium. They corrode in salty air. They are fairly resistant to caustics, many solvents and fuels, but not to acids.

Titanium and titanium alloys These have excellent resistance to e.g. seawater and aqueous chloride solutions over a wide temperature range. Most alloys resist nitric acid. When alloyed with noble metals such as palladium they will resist reducing acids. These materials are high in the galvanic series and so should not be used with other metals.

Zinc An oxide film gives reasonable resistance to water and normal atmosphere.

Aluminium An oxide coating gives good resistance to water and atmosphere, but stress corrosion cracking occurs.

6.12.3 *Stress corrosion cracking*

Under tensile stress and in a corrosive environment some metals develop surface cracks called 'stress corrosion cracking' which is time dependent and may take months to develop. It is avoided by minimizing stress and/or improving the environment.

Environments causing stress corrosion cracking

Material	Environment
Steels	Caustic solutions
Stainless steels 50–60 °C	Chloride solutions
Aluminium and alloys	Chloride solutions
Copper alloys	Ammonia atmosphere, sometimes neutral water
Acrylics	Chlorinated solvents

6.12.4 *Galvanic corrosion*

For a pair of metals, that highest up the 'galvanic table' is the 'negative electrode' or 'cathode'; that lower down is the 'positive electrode' or 'anode'. The anode loses metal, i.e. corrodes, whilst the cathode remains unchanged. The greater the potential, the greater the rate of corrosion. Hydrogen is assumed to have zero potential.

Galvanic table for pure metals (relative to hydrogen)

Metal		Potential difference (v)	
Gold	NOBLE	+1.70	CATHODIC
Platinum		+0.86	
Silver		+0.80	
Copper		+0.34	
Hydrogen		0	
Lead		−0.13	
Tin		−0.14	
Nickel		−0.25	
Cadmium		−0.40	
Iron		−0.44	
Chromium		−0.74	
Zinc		−0.76	
Aluminium		−1.67	
Magnesium		−2.34	
Sodium		−2.71	
Calcium	BASE	−2.87	ANODIC

6.13 Plastics

The term 'plastic' is used for materials based on polymers to which other materials are added to give the desired properties. 'Fillers' increase strength, 'plasticizers' reduce rigidity, and 'stabilizers' protect against ultraviolet radiation.

'Thermoplastic' polymers soften when heated and can be reshaped, the new shape being retained on cooling. The process can be repeated continuously.

Thermosetting polymers (or thermosets) cannot be softened and reshaped by heating. They are plastic at some stage of processing but finally set and cannot be resoftened. Thermosets are generally stronger and stiffer than thermoplastics.

6.13.1 *Thermoplastics*

Acetal and polyacetal

These combine very high strength, good temperature and abrasion resistance, exceptional dimensional stability and low coefficient of thermal expansion. They compete with nylon (but with many better properties) and with diecastings (but are lighter). Chemical resistance is good except for strong acids. Typical applications are water-pump parts, pipe fittings, washing machines, car instrument housings, bearings and gears.

Acrylics (methylmethacrylate, PMMA)

These are noted for their optical clarity and are available as sheet, rod, tubing, etc., as Perspex (UK) and Plexiglas (USA, Germany, etc.). They are hard and brittle and resistant to discolouring and weathering. Applications include optical lenses and prisms, transparent coverings, draughting instruments, reflectors, control knobs, baths and washbasins. They are available in a wide range of transparent and opaque colours.

Acrylonitrile–butadiene–styrene (ABS)

This combination of three materials gives a material which is strong, stiff and abrasion resistant with good properties, except out of doors, and ease of processing. The many applications include pipes, refrigerator liners, car-instrument surrounds, radiator grills, telephones, boat shells, and radio and television parts. Available in medium, high and very high impact grades.

Cellulose

'Cellulose nitrate' is inflammable and has poor performance in heat and sunlight. Its uses are currently limited. *Cellulose acetate* has good strength, stiffness and hardness and can be made self-extinguishing. Glass-filled grades are made. *Cellulose acetobutyrate* (CAB) has superior impact strength, dimensional stability and service temperature range and can be weather stabilized. *Cellulose proprionate* (CP) is similar to CAB, but has better dimensional stability and can have higher strength and stiffness. *Ethyl cellulose* has better low-temperature strength and lower density than the others. Processing of cellulose plastics is by injection moulding and vacuum forming. Applications include all types of mouldings, electrical insulation, and toys.

Ethylene–vinyl acetate (EVA)

This material gives tough flexible mouldings and extrusions suitable for a wide temperature range. The material may be stiffened by the use of fillers and is also used for adhesives. Applications include all types of mouldings, disposable liners, shower curtains, gloves, inflatables, gaskets, and medical tubing. The material is considered competitive with polyvinyl chloride (PVC), polythene and synthetic rubbers, and is also used for adhesives and wax blends.

Fluorocarbons

These have outstanding chemical, thermal and electrical properties. The four main types are described below.

Polytetrafluoroethylenes (PTFE) 'Teflon' or 'Fluon', these are the best known types of PTFEs. PTFEs resist all known chemicals, weather and heat, have extremely low coefficients of friction, and are 'non-stick'. They are inert, with good electrical properties. They are non-toxic, non-flammable and have a working temperature range of $-270\,°C$ to $260\,°C$. They may be glass filled for increased strength.

Applications include chemical, mechanical and electrical components, bearings (plain or filled with glass and/or bronze), tubing, and vessels for 'aggressive' chemicals.

Fluoroethylenepropylene (FEP) Unlike PTFE, this can be processed on conventional moulding machines and extruded, but the thermal and chemical properties are slightly less good.

Ethylenetetrafluoroethylene (ETFE) The properties are similar to those of PTFE, with a thermoplasticity similar to that of polyethylene.

Perfluoroalcoxy (PFA) This has the same excellent properties as PTFE, but is melt processable and, therefore, suitable for linings for pumps, valves, pipes and pipe fittings.

Ionomers

These thermoplastics based on ethylene have high melt strength which makes them suitable for deep forming, blowing, etc. They are used for packaging, bottles, mouldings for small components, tool handles, trim, etc. They have a high acceptance of fillers.

Methylpentene (TPX)

This is a high clarity resin with excellent chemical and electrical properties and the lowest density of all thermoplastics. It has the best resistance of all transparent plastics to distortion at high temperature – it compares well with acrylic for optical use, but has only 70% of its density. It is used for light covers, medical and chemical ware, high frequency electrical insulation, cables, microwave-oven parts, and radar components. It can withstand soft soldering temperatures.

Polyethylene terephthalate (PETP)

This has good strength, rigidity, chemical and abrasion resistance and a very low coefficient of friction. It is attacked by acetic acid and strong nitric and sulphuric acids. It is used for bearings, tyre reinforcement, bottles, car parts, gears, and cams.

Polyamides (nylons)

These are a range of thermoplastics, e.g. Nylon 6, Nylon 66 and Nylon 610, which are among the toughest engineering plastics with high vibration-damping capacity, abrasion resistance and high load capacity for high-speed bearings. They have low coefficient of friction and good flexibility. Pigment-stabilized types are not affected by ultraviolet radiation and chemical resistance is good. Unfilled nylon is prone to swelling due to moisture absorption. Nylon bearings may be filled with molybdenum disulphide or graphite. Applications include bearings, electrical insulators, gears, wheels, screw fasteners, cams, latches, fuel lines and rotary seals.

Polyethylene

Low density polyethylene is generally called 'polythene' and is used for films, coatings, pipes, domestic mouldings, cable sheathing and electrical insulation. The high-density type is used for larger mouldings and is available in the form of sheet, tube, etc. Polyethylene is limited as an engineering material because of its low strength and hardness. It is attacked by many chemicals.

Polyethersulphone

This is a high-temperature engineering plastic – useful up to $180\,°C$ and some grades up to $200\,°C$. It is resistant to most chemicals and may be extruded or injection moulded to close tolerances. The properties are similar to those of nylons. Applications are as a replacement for glass for medical needs and food handling, circuit boards, general electrical components, and car parts requiring good mechanical properties and dimensional stability.

Polypropylene oxide (PPO)

This is a useful engineering plastic with excellent mechanical, thermal and fatigue properties, low creep, and low moisture absorption. Filled grades can be used as alternatives to thermosets and some metals. Applications are light engineering parts, and car, aircraft and business components (especially for heat and flame resistance).

Polystyrene

This plastic is not very useful as an engineering material, but used for toys, electrical insulation, refrigerator linings, packaging and numerous commercial articles. It is available in unmodified form, in transparent form and opaque colours, high-impact form and extra-high-impact form, as well as in a heat-resistant grade. It can be stabilized against ultraviolet radiation and also made in expanded form. It is attacked by many chemicals and by ultraviolet light.

Polysulphone

This has similar properties to nylon but they are maintained up to 180 °C (120 °C for nylon). Its optical clarity is good and its moisture absorption lower than that of nylon. Applications are replacement for glass for medical needs and chemistry equipment, circuit boards, and many electrical components.

Polyvinyl chloride (PVC)

This is one the most widely used of all plastics. With the resin mixed with stabilizers, lubricants, fillers, pigments and plasticizers, a wide range of properties is possible from flexible to hard types, in transparent or opaque-colour form. It is tough, strong, with good resistance to chemicals, good low-temperature characteristics and flame-retardant properties. It is used for electrical conduit and trunking, junction boxes, rainwater pipes and gutters, decorative profile extrusions, tanks, guards, ducts, etc.

Polycarbonate

This is tough thermoplastic with outstanding strength, dimensional stability, and electrical properties, high heat distortion temperature and low temperature resistance (down to −100 °C). It is available in optical, translucent and opaque grades (many colours). Polycarbonates have good chemical resistance and weathering properties and can be stabilized against ultraviolet radiation. They are used for injection mouldings and blow extrusions for glazing panels, helmets, face shields, dashboards, window cranks, and gears. Polycarbonate is an important engineering plastic.

Polypropylene

This is a low density, hard, stiff, creep-resistant plastic with good resistance to chemicals, good wear resistance, low water absorption and of relatively low cost. Produced as filaments, weaves and in many other forms, polypropylene may be glass filled. It is used for food and chemical containers, domestic appliances, furniture, car parts, twine, toys, tubing, cable sheath, and bristles.

Polyphenylene sulphide

This is a high-temperature plastic useful up to 260 °C with room temperature properties similar to those of nylon. It has good chemical resistance and is suitable for structural components subject to heat. Glass filler improves strength and heat resistance. Uses are similar to those of nylon, but for high temperatures.

Polyphenylene oxide

This is a rigid engineering plastic similar to polysulphone in uses. It can be injection moulded and has mechanical properties the same as those for nylon. It is used for car parts, domestic appliances, and parts requiring good dimensional stability.

6.13.2 Thermosets

Alkyds

There are two main groups of alkyds: diallylphthalate (DAP) and diallylisophthalate (DIAP). These have good dimensional stability and heat resistance (service temperature 170 °C; intermittent use 260 °C), excellent electrical properties, good resistance to oils, fats and most solvents, but restricted resistance to strong acids and alkalis. The mechanical properties are improved by filling with glass or minerals. The main uses are for electrical components and encapsulation. A wide range of colours and fast-curing grades are available.

Amino resins

These are based on formaldehyde with urea or melamine formulated as coatings and adhesives for laminates, impregnated paper and textiles. Moulding powder is compounded with fillers of cellulose and wood flour, and extenders, etc. Composites with open-weave fabric are used for building panels. Uses include domestic electrical appliances and electric light fittings; the melamine type is used for tableware. The strength is high enough for use in stressed components, but the material is brittle. Electrical, thermal and self-extinguishing properties are good.

Epoxies

These resins are used extensively. They can be cold cured without pressure using a 'hardener', or be heat cured. Inert fillers, plasticizers, flexibilizers, etc., give a wide range of properties from soft flexible to rigid solid materials. Bonding to wood, metal, glass, etc., is good and the mechanical, electrical and chemical properties are excellent. Epoxies are used in all branches of engineering, including large castings, electrical parts, circuit boards, potting, glass and carbon fibre structures, flooring, protective coatings and adhesives.

Epon resins

These can be formulated for surface coatings and have excellent adhesion, chemical resistance and flexibility. They are used for casting and potting materials, adhesives, structural laminates and foams.

Phenolics (phenol formaldehyde, PF)

PF is the original Bakelite and is usually filled with 50–70% wood flour for moulded non-stressed or lightly stressed parts. Other fillers are: mica for electrical parts; asbestos for heat resistance; glass fibre for strength and electrical properties; nylon; and graphite. Phenolics represent one of the best thermosets for low creep. Mouldings have good strength, good gloss and good temperature range (150 °C wood filled; intermittent use 220 °C), but are rather brittle. Applications include electrical circuit boards, gears, cams, and car brake linings (when filled with asbestos, glass, metal powder, etc.). The cost is low and the compressive strength very high.

Polyester

This can be cured at room temperature with a hardener or alone at 70–150 °C. It is used unfilled as a coating, for potting, encapsulation, linings, thread locking, castings, and industrial mouldings. It is used mostly for glass-reinforced-plastic (GRP) mouldings.

Polyimides

These are noted for their high resistance to oxidation and service temperatures of up to 250 °C (400 °C for intermittent use). The low coefficient of friction and high resistance to abrasion makes them ideal for non-lubricated bearings. Graphite or molybdenum disulphide filling improves these properties. They are used for high density insulating tape. Polyimides have high strength, low moisture absorption, and resist most chemicals, except strong alkalis and ammonia solutions.

Silicones

These may be cold or heat cured and are used for high-temperature laminates and electrical parts resistant to heat (heat distortion temperature 450 °C). Unfilled and filled types are used for special-duty mouldings. Organosilicones are used for surface coatings and as an adhesive between organic and non-organic materials.

6.13.3 Laminated plastics

These consist of layers of fibrous material impregnated with and bonded together by a thermosetting resin to produce sheet, bars, rods, tubes, etc. The laminate may be 'decorative' or 'industrial', the latter being of mechanical or electrical grade.

Phenolics

Phenolic plastics can be reinforced with paper, cotton fabric, asbestos paper fabric or felt, synthetic fabric, or wood flour. They are used for general-purpose mechanical and electrical parts. They have good mechanical and electrical properties.

Epoxies

These are used for high-performance mechanical and electrical duties. Fillers used are paper, cotton fabric and glass fibre.

Tufnol

'Tufnol' is the trade name for a large range of sheet, rod and tube materials using phenolic resin with paper and asbestos fabric, and epoxy resin with glass or fabric.

Polyester

This is normally used with glass fabric (the cheapest) filler. The mechanical and electrical properties are inferior to those of epoxy. It can be rendered in self-colours.

Melamine

Fillers used for melamine are paper, cotton fabric, asbestos paper fabric, and glass fabric. Melamines have a hard non-scratch surface, superior electrical properties and can be rendered in self-colours. They are used for insulators, especially in wet and dirty conditions, and for decorative and industrial laminates.

Silicone

This is used with asbestos paper and fabric and glass fabric fillers for high-temperature applications (250°C; intermittent use 300 °C). It has excellent electrical but inferior mechanical properties.

Polyimide

This is used with glass fabric as filler. Polyimides have superior thermal and electrical properties with a service temperature as for silicones but with two to three times the strength and flexibility.

6.13.4 Foam and cellular plastics

Thermoplastics

Polyurethane foams The 'flexible' type is the one most used. It is 'open cell' and used for upholstery, underlays, thermal and vibration insulation, and buoyancy. It can be used *in situ*. The rigid type has 'closed cells' and is used for sandwich construction, insulation, etc. Moulded components are made from rigid and semi-rigid types.

Expanded polystyrene This is made only in rigid form with closed cells. It can be used *in situ*. The density is extremely low, as is the cost. Chemical resistance is low and the service temperature is only 70 °C. It is used for packaging, thermal and acoustic insulation and buoyancy applications.

High-density polystyrene foam This has a porous core with a solid skin. It is used for structural parts.

Cellular polyvinyl chlorides (PVC) The low-density type is closed cell and flexible. It is used for sandwich structures, thermal insulation, gaskets, trim, buoyancy, and insulating clothing. The moderate to high density open-cell type is similar to latex rubber and is used as synthetic leather cloth. The rigid closed-cell type is used for structural parts, sandwich construction, thermal insulation and buoyancy. Rigid open-cell PVC (microporous PVC) is used for filters and battery separators. In general, cellular PVC has high strength and good fire resistance and is easy to work.

Polyethylene foams The flexible type is closed cell and has low density with good chemical resistance and colour availability, but is a poor heat insulator and costly. The flexible foams are used for vibration damping, packaging and gaskets. The rigid type has high density and is used for filters, cable insulation. A structural type has a solid skin and a foam core.

Ethylene vinyl acetates (EVA) These are microcellular foams similar to microcellular rubber foam, but are much lighter with better chemical resistance and colour possibilities.

Other types Other types of thermoplastics include: *cellular acetate* which is used as a core material in constructions; *expanded acrylics*, which have good physical properties, thermal insulation and chemical resistance; *expanded nylon* (and *expanded ABS*) which are low-density, solid-skin constructions; *expanded PVA* which has similar properties to expanded polystyrene; and *expanded polypropylene* which gives high-density foams.

Thermosets

Phenolids These can be formed *in situ*. They have good rigidity, thermal insulation and high service temperature. They are brittle.

Urea formaldehyde (UF) foam This is readily formed *in situ* and has good thermal insulation. It has open pores and is used for cavity-wall filling.

Expanded epoxies These have limited use due to their high cost. They give a uniform texture and good dimensional stability, and are used for composite foams, e.g. with polystyrene beads.

Silicon foams These are rigid and brittle with a high service temperature (300 °C; 400 °C intermittent use). Their use is limited to high-temperature-resistant sandwich constructions. The flexible closed-cell type is costly but will operate up to 200 °C and is used for high-temperature seals and gaskets.

Elastomers

Cellular rubbers There are three types: 'sponge', solid rubber blown to give an open-cell structure; 'foam', a liquid rubber expanded to form open or closed cells and stiffer than sponge; and 'expanded', a solid rubber blown with mainly closed cells – it is stiffer than sponge. Uses include gaskets, seals, thermal insulation, cushioning, shock absorption, sound and vibration damping, buoyancy and sandwich constructions.

6.13.5 *Properties of plastics*

Typical physical properties of plastics

Properties of plastics	ρ (kg m^{-3})	Tensile strength (N mm^{-2})	Elongation (%)	E (GN m^{-2})	BHN	Machinability
Thermoplastics						
PVC rigid	1330	48	200	3.4	20	Excellent
Polystyrene	1300	48	3	3.4	25	Fair
PTFE	2100	13	100	0.3	—	Excellent
Polypropylene	1200	27	200–700	1.3	10	Excellent
Nylon	1160	60	90	2.4	10	Excellent
Cellulose nitrate	1350	48	40	1.4	10	Excellent
Cellulose acetate	1300	40	10–60	1.4	12	Excellent
Acrylic (Perspex)	1190	74	6	3.0	34	Excellent
Polythene (high density)	1450	20–30	20–100	0.7	2	Excellent
Thermosetting plastics						
Epoxy resin (glass filled)	1600–2000	68–200	4	20	38	Good
Melamine formaldehyde (fabric filled)	1800–2000	60–90	—	7	38	Fair
Urea formaldehyde (cellulose filled)	1500	38–90	1	7–10	51	Fair
Phenol formaldehyde (mica filled)	1600–1900	38–50	0.5	17–35	36	Good
Acetals (glass filled)	1600	58–75	2–7	7	27	Good

BHN = Brinell hardness number, ρ = density, E = Young's modulus.

Relative properties of plastics

Material	Tensile strength	Compressive strength	Machining properties	Chemical resistance
Thermoplastics				
Nylon	E	G	E	G
PTFE	F	G	E	O
Polypropylene	F	F	E	E
Polystyrene	E	G	F	F
Rigid PVC	E	G	E	G
Flexible PVC	F	P	P	G

Relative properties of plastics (*continued*)

Material	Tensile strength	Compressive strength	Machining properties	Chemical resistance
Thermosetting plastics				
Epoxy resin (glass-fibre filled)	O	E	G	E
Formaldehyde (asbestos filled)	G	G	F	G
Phenol formaldehyde (Bakelite)	G	G	F	F
Polyester (glass-fibre filled)	E	G	G	F
Silicone (asbestos filled)	O	G	F	F

O = outstanding, E = excellent, G = good, F = fair, P = poor.
Tensile strength (typical): $E = 55 \, N \, mm^{-2}$; $P = 21 \, N \, mm^{-2}$.
Compressive strength (typical): $E = 210 \, N \, mm^{-2}$; $P = 35 \, N \, mm^{-2}$.

6.14 Elastomers

Elastomers, or rubbers, are essentially amorphic polymers with linear chain molecules with some cross-linking which ensures elasticity and the return of the material to its original shape when a load is removed. They are characterized by large strains (typically 100%) under stress. The synthetic rubber styrene butadiene is the most used elastomer, with natural rubber a close second. The following describes the commonly used elastomers and gives some applications and properties.

6.14.1 Natural rubbers (polyisoprene, NR)

These have high strength, flexibility and resilience, but have poor resistance to fuels, oils, flame and sunlight ageing. They are more costly than synthetic rubbers which replace them. 'Soft rubber' contains 1–4% sulphur. Wear resistance is increased by inclusion of fillers such as carbon black, silicon dioxide, clay, and wood flour. 'Hard rubber' contains over 25% sulphur. Full vulcanization of 45% produces ebonite. Applications include vehicle tyres and tubes, seals, anti-vibration mountings, hoses and belts.
　　Shore hardness: 30–90. Temperature range: $-55\,°C$ to $82\,°C$.

6.14.2 Synthetic rubbers

Styrene butadiene rubbers (SBR, GRS, BUNA S)

These are similar to natural rubbers in application, but are inferior in mechanical properties, although cheaper. They are used in car brake hydraulic systems and for hoses, belts, gaskets and anti-vibration mountings.
　　Shore hardness: 40–80. Temperature range: $-50\,°C$ to $82\,°C$.

Butadiene rubbers (polybutadiene, BR)

These are used as substitutes for natural rubber, but are generally inferior. They have similar applications as natural rubber.

Shore hardness: 40–90. Temperature range: −100 °C to 93 °C.

Butyl rubbers (isobutylene isoprene, GR 1)

These are extremely resistant to water, silicon fluids and grease, and gas permeation. They are used for puncture-proof tyres, inner tubes and vacuum seals.

Shore hardness: 40–90. Temperature range: −45 °C to 150 °C.

Nitrile rubbers (butadiene acrylonitrile, BUNA N.NBR)

These have good physical properties and good resistance to fuels, oils, solvents, water, silicon fluids and abrasion. They are used for O rings and other seals, petrol hoses, fuel-pump diaphragms, gaskets and oil-resistant shoe soles.

Shore hardness: 40–95. Temperature range: −55 °C to 82 °C.

Neoprene rubbers (polychloroprene, chloroprene)

These are some of the best general-purpose synthetic rubbers. They have excellent resistance to weather ageing, moderate resistance to oils, and good resistance to refrigerants and mild acids.

Shore hardness: 30–95. Temperature range: −40 °C to 115 °C.

Chlorosulphonated polyethylene rubbers (CSM)

These have poor mechanical properties but good resistance to acids and heat with complete resistance to ozone. They are used for chemical plant, tank linings, and high-voltage insulation.

Shore hardness: 45–100. Temperature range: −100 °C to 93 °C.

Ethylene propylene rubbers (EP.FPM)

These are specialized rubbers especially resistant to weather ageing, heat, many solvents, steam, hot water, dilute acids and alkalis, and ketones, but not petrol or mineral oils. They are used for conveyor belts, limited car applications, silicone fluid systems, and electrical insulation.

Shore hardness: 40–90. Temperature range: −50 °C to 177 °C.

Fluorocarbon rubbers

These comprise a wide range of rubbers with excellent resistance to chemical attack, heat, acids, fuels, oils, aromatic compounds, etc. They have a high service temperature. They are particularly suitable for vacuum duties.

Shore hardness: 60–90. Temperature range: −23 °C to 260 °C.

Isoprenes (polyisoprene, IR)

These are chemically the same as natural rubber but are more costly. The properties and applications are similar to those of natural rubber.

Shore hardness: 40–80. Temperature range: −50 °C to 82 °C.

Polyacrylic rubbers (ACM, ABR)

This is a group of rubbers midway between nitrile and fluorocarbon rubbers with excellent resistance to mineral oils, hypoid oils and greases, and good resistance to hot air and ageing. The mechanical strength is low. They are used for spark-plug seals and transmission seals.

Shore hardness: 40–90. Temperature range: −30 °C to 177 °C.

Polysulphide rubbers

These have poor physical properties and heat resistance but good resistance to oils, solvents and weather ageing and are impermeable to gases and moisture. They are used for caulking and sealing compounds and as a casting material.

Shore hardness: 40–85. Temperature range: −50 °C to 121 °C.

Polyurethane rubbers

These have exceptional strength and tear and abrasion resistance (the best of all rubbers), low-temperature flexibility and good resistance to fuels, hydrocarbons, ozone and weather. Resistance to solutions of acids and alkalis, hot water, steam, glycol and ketones is poor. They are used for wear-resistant applications such as floor coverings.

Shore hardness: 35–100. Temperature range: −53 °C to 115 °C.

Silicone rubbers (SI)

These have exceptionally high service temperature ranges, but the mechanical properties and chemical resistance are poor. They cannot be used for fuels, light mineral oils, or high-pressure steam. They are used for high- and low-temperature seals, high-temperature rotary seals, cable insulation, hydraulic seals, and aircraft door and canopy seals.

Shore hardness: 30–90. Temperature range: $-116\,°C$ to $315\,°C$ ($380\,°C$ for intermittent use).

Fluorosilicone rubbers

These are similar to silicone rubbers but have better oil resistance and a lower temperature range.

Shore hardness: 40–80. Temperature range: $-64\,°C$ to $204\,°C$.

6.15 Wood

Permitted stresses in structural timbers ($N\,mm^{-2}$)

Timber	Bending			Compression			
	Stress in extreme fibre		Horizontal shear stress	Stress parallel to grain		Stress perpendicular to grain	
	Outside location	Dry location	All locations	Outside locations	Dry location	Outside location	Dry location
Oak	8.3	9.7	0.9	6.0	6.9	1.6	3.5
Douglas fir	7.6	9.0	0.6	6.0	6.9	1.6	2.1
Norway spruce	6.9	7.6	0.6	5.5	5.5	1.2	2.1

Mechanical properties of timbers

Wood	Moisture (%)	Density, ρ ($kg\,m^{-3}$)	Fibre stress at elastic limit ($N\,mm^{-2}$)	Modulus of elasticity, E ($N\,mm^{-2}$)	Modulus of rupture ($N\,mm^{-2}$)	Compressive strength parallel to grain ($N\,mm^{-2}$)	Shear strength ($N\,mm^{-2}$)
Ash	15	657	60	10 070	103	48	10
Beech	—	740	60–110	10 350	—	27–54	8.3–14
Birch	9–10	710	85–90	15 170	130–135	67–74	13–18.5
Elm, English	—	560	40–54	11 790	—	17–32	8–11.3
Elm, Dutch	—	560	42–60	7 720	—	18–32	7.2–10
Elm, Wych	—	690	65–100	7 860	—	29–47	7.3–11.4
Fir, Douglas	6–9	530	45–73	10 340–15 170	71–97	49–74	7.4–8.8
Mahogany	15	545	60	8 690	80	45	6.0
Oak	—	740	56–87	14 550	—	27–50	8–12
Pine, Scots	—	530	41–83	8 550–10 340	—	21–42	5.2–9.7
Poplar	—	450	40–43	7 240	—	20	4.8
Spruce, Norway	—	430	36–62	7 380–8 620	—	18–39	4.3–8
Sycamore	—	625	62–106	8 970–13 450	—	26–46	8.8–15

6.16   Adhesives

Adhesives are materials which are used to join solids (adherents) by means of a thin layer which adheres to the solids. At some stage the adhesive is liquid or plastic and sets to form a solid. In the final stage it may be rigid or flexible.

In engineering, joining by adhesives has in many cases replaced other methods such as soldering, brazing, welding, riveting and bolting.

Advantages of adhesive bonding
Dissimilar materials may be joined, e.g. plastics to metal.
Large bonding areas are possible.
Uniform stress distribution and low stress concentration is obtained.
Bonding is usually carried out at low temperature.

The bond is generally permanent.
A smooth finish is usually obtained.

Disadvantages of adhesive bonding
A curing time, which may be long, is required for optimum strength.
The adhesive may be flammable or toxic.
The bond may be affected by the environment, e.g. heat, cold, or humidity.

Adhesives may be classified as follows:

(1) natural adhesives,
(2) elastomers,
(3) thermoplastics,
(4) thermosets, and
(5) Other adhesives.

6.16.1   *Natural adhesives*

These are set by solvent evaporation. They are generally of low strength and are weakened by moisture and mould. They are restricted to joining low-strength materials.

Animal glues

These are made from collagen (from the bones and skins of animals) with sugar and glycerol added for increased flexibility. They are available in sheet (Scotch Glue), bead and powder forms, all of which dissolve in water at 60 °C, and also as a liquid with gelling inhibitors. Degradation occurs at about 100 °C. These glues have a long 'pot life' a long dry life and a 'tacky' stage useful for 'initial set'. They will join wood, paper, leather, cloth and most porous materials.

Fish glues

These have similar applications to animal glues but are usually liquid at room temperature and have better resistance to water and a better recovery of strength on drying.

Vegetable glues

These are based on starch or dextrine from starch and are available either as a powder to be mixed with water or ready mixed. The shear strength is low but they are only used for paper and cardboard. Resistance to water and high temperatures is low.

Casein

This is a protein glue made from milk precipitated with acid. It is supplied as a powder to be mixed with water and is used for joining wood, paper, cloth and asbestos. Latex/casein is used for foil/paper laminations. Casein has better resistance to water and better strength than animal and fish glues. Other protein glues are made from blood, soya bean residue, etc.

6.16.2   *Elastomer adhesives*

These adhesives are based on natural and synthetic rubbers set by solvent evaporation or heat curing. They have relatively low shear strength and suffer from creep and are therefore used for unstressed joints. They are useful for flexible bonds with plastics and rubbers.

'Contact adhesives' use rubber in a solvent and will join many materials.

Natural rubbers

Solvent-type natural rubber adhesives have service temperatures up to 60 °C, and hot-curing types are serviceable up to 90 °C. The former may incorporate resin for improved strength (see later). Resistance to water is good, but resistance to oils and solvents is poor. Adherents include: natural rubber; some plastics such as acrylics and PTFE; expanded natural rubber, polystyrene and polyurethane; aluminium alloy, iron and steel; fabrics, card, leather, paper, wood; and glass and ceramics. Solutions are used for car upholstery, paper, fabric-backed PVC to hardboard, and floor coverings. The latex type is also used to adhere paper to plastics and metal. Reclaim rubber adhesives are used for car sound-proofing, draught excluding and undersealing. Pressure-sensitive adhesive is used for tapes, labels and gluing polythene sheet to metals.

Polychloroprenes (neoprene)

These synthetic-rubber-based adhesives have good resistance to water, oils and solvents and are either solvent setting or vulcanizing by heat curing or catalyst with or without resin modifiers. They are used for bonding metal, wood, leather, synthetic leather and plastics (except PVC) with applications in car, aircraft and ship-building industries.

Acrylonitride butadienes (nitrile)

These adhesives are similar to neoprene types and are supplied in the form of solutions for joining rubber to rubber, unbacked PVC to itself, and metal, wood, leather and PVC sheet to metals. The latex type is used for PVC film to paper, textiles, aluminium foil to plastics, paper and wood, etc. The shear strength is up to $7 \, MN \, m^{-2}$.

Butyl rubber adhesives

These are used in the car and building industries and are applied by gun or tape. Resistance to water is good, but that to oils is poor.

Styrene butadiene rubber adhesives

These are based on the synthetic rubber used for car tyres and are used in the car industry for bonding felt carpets and for gluing metal to rubber trims. The pressure-sensitive type is used for tapes and labels.

Polyurethane adhesives

These are used for many plastics including PVC, polystyrene, and melamine. They have good strength at room temperature, excellent resistance to oils, acids, alkalis and many solvents, but poor resistance to water. They give a flexible bond suitable for resisting shock and vibration.

Polysulphide rubber adhesives

These have outstanding resistance to oils, solvents, light, air and heat, and will bond steel, aluminium, glass, concrete, ceramics and wood. Uses include sealants for fuel tanks, aircraft pressure cabins and windscreens, lights and pipe joints. With epoxy resins they are used for filling and sealing aluminium roof panels and car body panels.

Silicone rubber adhesives

These vulcanize at room temperature and bond a wide range of materials, including silicone rubber. The shear strength is up to $1.4 \, MN \, m^{-2}$ at the maximum service temperature of 316 °C. Although the strength is not high, they have excellent resistance to high temperatures. Formulation with epoxy resin gives good strength up to 340 °C.

6.16.3 *Thermoplastic adhesives*

In general, these have a low shear strength and suffer from creep at high loading. They are therefore used in low-stress conditions. Resistance is good to oils and poor to good for water.

Polyvinyl acetate (PVA)

This is the well-known 'white glue' used for woodworking. It also bonds metals, glass, ceramics, leather and many plastics. The shear strength is good and the resistance high to oils and mould, but poor to heat and limited to water. Emulsion types are used for ceramic tiles. A fast-setting type is available.

Polyvinyl alcohol (PVA)

This is made from PVA and is similar to it. It is used for paper in a re-sealable form. Resistance to oils and greases is good, but poor to water.

Polyacrylates

These are generally used for textiles and the pressure-sensitive types are used for labels. Water-based acrylic sealants are available.

Polyester acrylics

These cure in the absence of air (anaerobic) and give an extremely strong bond for metals, glass, ceramic and many other materials. The shear strength may be as high as 14 MN m^{-2}.

Acrylic solvent cement

This consists of polymethyl methacrylate (PMMA) dissolved in methyl chloride and is used for bonding PMMA to itself and to cellulosics, styrene, polycarbonate and rigid PVC. The shear strength is about 7 MN m^{-2} at 38 °C.

Cyanoacrylates

These set in the presence of moisture (from the adherents) in several seconds to give an extremely high strength (up to 20 MN m^{-2}). They are used for the rapid assembly of small components, metal to metal, and metal to non-metal joints, but not for porous materials since voids are not filled.

Silicone resins

These will bond fluorocarbons. They have low strength but a high service temperature. They can be formulated with other adhesives to give higher strength.

Polyamides

These are applied hot and set on cooling. They bond metals, wood, plastics, leather and laminates. The chemical resistance is the same as that for nylon.

Acrylic acid diesters

These are anaerobic adhesives used for e.g. nut locking and as a gasket cement. Their performance is satisfactory up to 150 °C.

6.16.4 *Thermoset adhesives*

These adhesives set as a result of the build-up of molecular chain length to give rigid cross-linked matrices. They include epoxy resins, which are some of the most widely used adhesives.

Phenolic formaldehyde (PF) resins

These are widely used in woodworking especially for plywood, and have excellent resistance to water, oils, solvents, etc. They will bond fluorocarbons, nylons and epoxy resin. Engineering adhesives are based on mixtures with other resins.

Phenolic neoprene

This is a heat-curing adhesive good for bonding metal to metal and metal to wood with a strength of 20 MN m^{-2}.

Phenolic nitrile

This is a hot-curing adhesive with a shear strength of 28 MN m^{-2} at a service temperature of 175 °C. It is used for metal to non-metal joints such as car brake linings.

Phenolic polyamides

These are usually available as a thermoplastic polyamide film and phenolic resin solution. The shear strength is up to 35 NM m^{-2}.

Phenolic vinyls

These have a high strength (up to 35 NM m^{-2}), but are not very useful above 100 °C. They are used for bonding honeycomb sandwich constructions, metal to metal and rubber to metal.

Resorcinol formaldehydes (RF)

These are used for wood and have superior strength, water resistance and temperature resistance compared with PF adhesives. They bond acrylics, nylons, phenolics and urea plastics.

Polyesters (unsaturated)

These have limited use and are unsuitable for glass-reinforced plastic. They bond copper, copper alloys, most fabrics, PVC, polyester films and polystyrene (in certain cases).

Polyimides

These cure at 260–370 °C and require post-curing for maximum strength which is retained up to 400 °C. These structural adhesives will bond metals, but the cost is high.

Epoxy resins

These adhesives are available as a two-part mixture (resin and hardener) for self-curing at room temperature or as one part for heat curing. Curing can take from 5 min (two part) to 24 h (one-part). They bond metal, glass, ceramics, wood, many rubbers and some plastics. They have excellent resistance to oils and good resistance to water and most solvents. The shear strength is up to $35 \, \text{MN m}^{-2}$.

Epoxy phenolics

These have an increased service temperature with 50% strength at 200 °C and are useful up to 565 °C, with low creep. They are useful in the car industry.

Epoxy polyamides

These have improved flexibility and peel strength, but relatively low shear strength.

Epoxy polysulphides

These have improved peel strength and flexibility, with a shear strength of $28 \, \text{MN m}^{-2}$.

Epoxy silicones

These have the best heat resistance (up to 300 °C) and a shear strength of $14 \, \text{MN m}^{-2}$. They are used for bonding metals and laminates.

Phenolic polyvinylacetates

These set under pressure and at elevated temperatures. They have good strength and good resistance to water, oils and solvents.

Redux adhesive

This is a mixture of polyvinyl formal powder and phenol formaldehyde liquid resin which gives a strong metal-to-metal joint that is better than riveting and spot-welding. It is normally useful up to 80 °C, but can be formulated to 250 °C.

6.16.5 Other adhesives

Sodium silicate

Known as 'water glass', this is a cheap, colourless adhesive used for bonding aluminium foil to paper, insulating materials to walls and for dry-mould bonding.

Ceramic adhesive

This is typically borosilicate glass compounded with alkaline earths and oxides of alkaline metals set by firing at 700–1200 °C. It is used for metal-to-metal joints.

Bitumen

This is a substance derived from coal and lignite. It is used in solution or as a hot melt in the car industry and for roofing and tiles.

6.16.6 Maximum and minimum service temperatures for adhesives

Adhesive	Temperature (°C)	
	Minimum	Maximum
Cyanoacrylate	—	80
Epoxy	—	90
Epoxy phenolic	—	200
Epoxy polyamide	—	100
Epoxy polysulphide	—	90
Epoxy silicone	—	300
Natural rubber	−40	65
Natural rubber (vulcanized)	−30	90
Neoprene	−50	90
Nitrile	−50	150
Polyurethane	−200	150

6.16.7 *Complementary adhesives and adherents**

Adhesive

Adherends	Animal glues	Starch	Dextrine	Casein	Acrylonitrile butadiene	Polychloroprene	Polyurethane	Silicone rubber	Polybutadiene	Natural rubber	Butyl	Cellulose nitrate	Polyvinyl alcohol	Polyvinyl acetate	Polyacrylate	Silicone resin	Cyanoacrylate	Phenolic formaldehyde	Urea formaldehyde	Resorcinol formaldehyde	Melamine formaldehyde	Polyesters (unsaturated)	Epoxy resins	Polyamides	Phenolic-vinyl formal	Phenolic-polyvinylacetal	Phenolic nitrile	Phenolic epoxy	Sodium silicate
Metals					×	×				×				×			×						×	×	×	×	×		
Glass ceramics	×					×						×		×			×						×		×		×		×
Wood			×							×		×		×				×	×	×	×		×						
Paper	×	×	×	×								×	×	×															×
Leather	×				×	×				×		×																	
Textiles, felt	×					×				×		×		×															
Elastomers																													
Polychloroprene (neoprene)						×																							
Nitrile					×												×												
Natural					×					×							×									×			
Silicone								×																					
Butyl					×						×																		
Polyurethane					×	×	×																						
Thermoplastics																													
Polyvinyl chloride (flexible)					×	×	×																						
Polyvinyl chloride (rigid)					×	×	×																×						
Cellulose acetate						×						×					×												
Cellulose nitrate						×						×					×												
Ethyl cellulose												×					×						×				×		
Polyethylene (film)							×		×						×														
Polyethylene (rigid)																							×				×		
Polypropylene (film)							×		×						×														
Polypropylene (rigid)																							×				×		
Polycarbonate						×																	×						
Fluorocarbons									×							×	×						×						
Polystyrene						×										×							×						
Polyamides (nylon)					×														×		×		×				×		
Polyformaldehyde (acetals)						×											×					×	×						
Methylpentene						×											×												
Thermosets																													
Epoxy																	×	×		×			×						
Phenolic						×											×		×				×				×		
Polyester																						×	×						
Melamine					×	×																	×						
Polyethylene terephthalate					×	×									×								×						
Diallylphthalate					×																	×	×						
Polyimide																						×	×						

*From Shields, J. *Adhesive Bonding*, The Design Council.

Note: in general, any two adherends may be bonded together if the chart shows that they are compatible with the same adhesive.

6.16.8 *Typical shear strength of adhesives*

Adhesive	Shear strength ($N\,mm^{-2}$)
Epoxy	35
Filled epoxy	14–21
Epoxy polyamide	25
Epoxy nylon	42
Epoxy polysulphide	20–28
Epoxy silicone	10–14
Neoprene	2
Nitrile	7
Phenolic neoprene	14–20
Phenolic nitrile	28
Phenolic polyamide	35
Phenolic vinyl	35
Polyvinyl acetate	20
Polyimide	14–18
Polyurethane	4–10
Silicone (unmodified)	14

6.16.9 *Joints for adhesives*

Lap joints

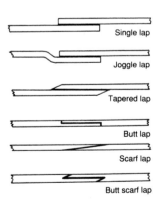

Single lap

Joggle lap

Tapered lap

Butt lap

Scarf lap

Butt scarf lap

Joints with increased bond area

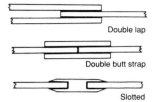

Double lap

Double butt strap

Slotted

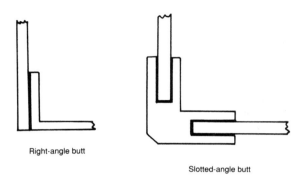

Right-angle butt

Slotted-angle butt

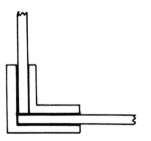

Right-angle-butt support

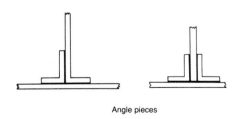

Angle pieces

Angle pieces increase the bonded area and thus reduce the cleavage stress.

6.17 Composites

A composite is a material consisting of two (or more) different materials bonded together, one forming a 'matrix' in which are embedded fibres or particles that increase the strength and stiffness of the matrix material.

A natural composite is wood in which cellulose fibres are embedded in a lignin matrix. Concrete is a composite in which particles of stone add strength with a further increase in strength provided by steel rein-forcing rods. Vehicle tyres consist of rubber reinforced with woven cords.

Plastics are reinforced with glass, carbon and other fibres. The fibres may be unidirectional, woven or random chopped. Metals, carbon and ceramics are also used as matrix materials.

So-called 'whiskers', which are single crystals of silicon carbide, silicon nitride, sapphire, etc., give extremely high strength.

6.17.1 *Elastic modulus of a composite (continuous fibres in direction of load)*

Let:

E_f = modulus of fibres
E_m = modulus of matrix
E_c = modulus of composite

r = (cross-sectional area of fibres)/(total cross-sectional area)

$$E_c = rE_f + (1-r)E_m$$

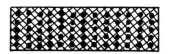

Matrix with fibres

6.17.2 *Acronyms for composites*

FRP   Fibre-reinforced plastic
FRT   Fibre-reinforced thermoplastic
GRP   Glass-reinforced plastic
GRC   Glass-reinforced composite
CFC   Carbon fibre composite
CFRP   Carbon-fibre-reinforced plastic
CFRT   Carbon-fibre-reinforced thermoplastic

6.17.3 *Forms of fibres for composites*

Fibre: length over 10 times the diameter; diameter less than 0.25 mm.
Filament: a continuous fibre.
Wire: a metallic fibre
Whisker: a fibre consisting of a single crystal.

Arrangement of fibres in composites

Type	Arrangement	Remarks
Unidirectional		Load taken in direction of fibres. Weak at right angles to fibres
Bidirectional		Takes equal load in both directions. Weaker since only half the fibres used in each direction

Arrangement of fibres in composites (*continued*)

Type	Arrangement	Remarks
Multidirectional		Load capacity much reduced but can take load in any direction in plane of fibres
Woven mat		Similar to bidirectional type but easy to handle
Random, chopped		Low in strength but multidirectional. Has handling advantages

6.17.4 Matrix materials for composites

Polymers: epoxies, polyesters, phenolics, silicones, polyimides, and other high-temperature polymers.

Thermoplastics: Perspex, nylon, etc.

Miscellaneous: metals, carbon, ceramics.

6.17.5 Properties of some fibres, wires and whiskers

Material	Type	Density, ρ (kg m^{-3})	Young's modulus, E (GN m^{-2})	Tensile strength (N mm^{-2})	Filament diameter, α (μm)
E glass	Fibre	2 500	62	3 500	2.5
Carbon	Fibre	2 000	415	1 750	7.5
Silica	Fibre	2 500	72	6 000	5.0
18/8 Stainless steel	Wire	7 900	205	2 100	150
Tungsten	Wire	19 300	350	2 900	150
Tungsten	Wire	19 300	350	3 800	25
Graphite	Whisker	2 200	675	21 000	—
Sapphire (Al$_2$O$_3$)	Whisker	4 000	525	6 000	—
Silicon carbide	Whisker	3 200	690	21 000	—
Silicon nitride	Whisker	3 100	380	14 000	—

6.18   Ceramics

Aluminium oxide (alumina)

	% Al_2O_3			
	75	86–94	94–98	>98
Density ($kg\,m^{-3}$)	3200	3300	3500	3700
Hardness (Moh scale)	8.5	9.0	9.0	9.0
Compressive strength ($N\,mm^{-2}$)	1250	1750	1750	1750
Flexural strength ($N\,mm^{-2}$)	270	290	350	380
Max. working temperature (°C)	800	1100	1500	1600

Silicon nitride

	Type	
	Reaction sintered	Hot pressed
Density ($kg\,m^{-3}$)	2 300–2 600	3 120–3 180
Open porosity (%)	18–28	0.1
Hardness (Moh scale)	9	9
Young's modulus ($N\,mm^{-2}$)	160 000	290 000
Flexural strength ($N\,mm^{-2}$)		
at 20 °C	110–175	550–680
at 1200 °C	210	350–480

6.19   Cermets

Cermets consist of powdered ceramic material in a matrix of metal, combining the hardness and strength of ceramic with the ductility of the metal to produce a hard, strong, yet tough, combination; the process involves compaction and sintering.

Typical cermets and applications

Ceramic	Matrix	Applications
Tungsten carbide	Cobalt	Cutting-tool bits
Titanium carbide	Molybdenum, cobalt or tungsten	
Molybdenum carbide	Cobalt	Dies
Silicon carbide	Cobalt or chromium	

Typical cermets and applications (*continued*)

Ceramic	Matrix	Applications
Aluminium oxide	Cobalt, iron or chromium	High-temperature components
Magnesium oxide	Magnesium, aluminium, cobalt, iron or nickel	Rocket and jet engine parts
Chromium oxide	Chromium	Disposable tool bits
Uranium oxide	Stainless steel	Nuclear fuel elements
Titanium boride	Cobalt or nickel	
Chromium boride	Nickel	Mainly as cutting tool tips
Molybdenum boride	Nickel or nickel–chromium alloy	

6.20   Materials for special requirements

High-strength metals

High carbon steel
Tool steel, carbon or alloy
Spring steel
Nickel steel
High tensile steel
Chrome–molybdenum steel
Nickel–chrome–molybdenum steel
18% nickel maraging steels
Phosphor bronze
Aluminium bronze
Beryllium copper
High-strength aluminium alloys

High temperature metals

Tungsten
Tantalum
Molybdenum
Chromium
Vanadium
Titanium
Nimonic alloys
Stellite
Hastelloy
Inconel
Stainless steel
Nichrome
Heat-resisting alloy steels

Malleable metals

Gold
Silver
Lead
Palladium
Rhodium
Tantalum
Vanadium

Corrosion-resistant metals

Stainless steels (especially austenitic)
Cupronickel
Monel
Titanium and alloys
Pure aluminium
Nickel
Lead
Tin
Meehanite (cast iron)

Solders

Lead–tin
Pure tin
Lead–tin–cadmium
Lead–tin–antimony
Silver solder
Aluminium solder

Coating metals

Copper
Cadmium
Chromium
Nickel
Gold
Silver
Platinum
Tin
Zinc
Brass
Bronze
Lead

Brazing metals

Copper, zinc (spelter)
Copper, zinc, tin
Silver, copper, zinc, cadmium (Easy-flo)
Silver, copper eutectic
Silver, copper, zinc
Silver, copper, phosphorus
Gold alloys
Palladium alloys
Pure gold, silver, palladium and platinum

Good conductors of electricity

Silver
Copper
Gold
Aluminium
Magnesium
Brass
Copper
Phosphor bronze
Beryllium copper

Permanent-magnet materials

Alnico I
Alnico II
Alnico V
Cobalt steel 35%
Tungsten steel 6%
Chrome steel 3%
Electrical sheet steel 1% Si
Barium ferrite

Metals with high electrical resistance

Advance (Cu, Ni)
Constantan or Eureka (Cu, Ni)
Manganin (Cu, Mn, Ni)
Nichrome (Ni, Cr)
Platinoid
Mercury
Bismuth

Good electrical insulators

Thermoplastics
Thermosetting plastics
Glass
Mica
Transformer oil
Quartz
Ceramics
Soft natural and synthetic rubber
Hard rubber
Silicone rubber
Shellac
Paxolin
Tufnol
Ebonite
Insulating papers, silks, etc.
Gases

Semiconductors

Silicon
Germanium
Gallium arsenide
Gallium phosphide
Gallium arsenide phosphide
Cadmium sulphide
Zinc sulphide
Indium antimonide

Low-loss magnetic materials

Pure iron
Permalloy
Mumetal
Silicon sheet steel 4.5%
Silicon sheet steel 1%
Permendur
Annealed cast iron
Ferrite

Good conductors of heat

Aluminium
Bronze
Copper
Duralumin
Gold
Magnesium
Molybdenum
Silver
Tungsten
Zinc

Good heat insulators

Asbestos cloth
Balsa wood
Calcium silicate
Compressed straw
Cork
Cotton wool
Diatomaceous earth
Diatomite
Expanded polystyrene
Felt
Glass fibre and foam
Glass wool
Hardboard
Insulating wallboard
Magnesia
Mineral wool
Plywood
Polyurethane foam
Rock wool
Rubber
Sawdust
Slag wool
Urea formaldehyde foam
Wood
Wood wool

Sound-absorbing materials

Acoustic tiles and boards:
    Cellulose
    Mineral

Acoustic plasters

Blanket materials:
    Rock wool
    Glass wool
    Wood wool

Perforated panels with absorbent backing

Suspended absorbers

Bearing materials

Tin based alloy	Silver overlay on
Lead based alloy	lead–tin
Lead–tin–antimony	Porous bronze
alloy	Porous leaded bronze
Copper–lead alloy	Porous iron
Leaded bronze	Chrome plating
Tin bronze	Carbon
Aluminium bronze	Carbon (graphite)
Cast iron (Meehanite)	Rubber
Cadmium–nickel alloy	Phenolics
Cadmium–silver alloy	Nylon
Cadmium–copper–silver	Teflon (PTFE)
alloy	Cermets
Silver overlay on	Lignum vitae
lead–indium	Jewels

High strength-to-weight ratio materials

Magnesium alloys
High strength aluminium alloys
Titanium
Titanium alloys
Nylon
Glass-reinforced nylon
Glass-reinforced plastics
Carbon-fibre-reinforced plastics
Ceramic-whisker-reinforced metals
Duralumin

Lubricants

Mineral oils
Vegetable oils
Mineral grease
Tallow
Silicone oil
Silicone grease
Flaked graphite
Colloidal graphite
Graphite grease
Molybdenum disulphide
Water
Gases

6.21 Miscellaneous information

6.21.1 *Densities*

In the following tables the densities ρ are given for normal pressure and temperature.

Metals				**Wood (15% moisture)**	
Metal	ρ (kg m^{-3})	Metal	ρ (kg m^{-3})	Wood	ρ (kg m^{-3})
Aluminium	2 700	Monel	18 900	Ash	660
Aluminium bronze (90%Cu, 10%Al)	7 700	Nickel	8 900	Balsa	100–390
		Nimonic (average)	8 100	Beech	740
Antimony	6 690	Palladium	12 160	Birch	720
Beryllium	1 829	Phosphor bronze	8 900	Elm: English	560
Bismuth	9 750	(typical)		Dutch	560
Brass (60/40)	8 520	Platinum	21 370	wych	690
Cadmium	8 650	Sodium	971	Fir, Douglas	480–550
Chromium	7 190	Steel: mild	7 830	Mahogany	545
Cobalt	8 900	stainless	8 000	Pine: Parana	550
Constantan	8 920	Tin: grey	5 750	pitch	640
Copper	8 930	rhombic	6 550	Scots	530
Gold	19 320	tetragonal	7 310	Spruce, Norway	430
Inconel	8 510	Titanium	4 540	Teak	660
Iron: pure	7 870	Tungsten	19 300		
cast	7 270	Uranium	18 680		
Lead	11 350	Vanadium	5 960		
Magnesium	1 740	Zinc	7 140		
Manganese	7 430				
Mercury	13 546				
Molybdenum	10 200				

Miscellaneous solids

Solid	ρ $(\mathrm{kg\,m^{-3}})$	Solid	ρ $(\mathrm{kg\,m^{-3}})$
Acrylic	1180	Polyethylene	910–965
Asbestos	2450 (average)	Polypropylene	900 (approx.)
Brickwork, common	1600–2000	Polystyrene	1030
Compressed straw slab	260	Polyurethane foam	30
Concrete: lightweight	450–1000	PTFE	2170
medium	1300–1700	PVC	1390
dense	2000–2400	Rock wool	220–390
Epoxy resin	1230	Rubber: butadiene	910
Epoxy/glass fibre	1500	natural	920
Expanded polystyrene	15–30	neoprene	1250
Glass: flint	3500	nitrile	1000
Pyrex	2210	Stone	2300–2800
window	2650	Urea formaldehyde foam	8
Glass-wool mat/quilt	25	Wood wool slab	500–800
Ice	917		
Mineral wool quilt	50		
Nylon	1130		

Liquids and gases

Liquid	ρ $(\mathrm{kg\,m^{-3}})$	Gas	ρ $(\mathrm{kg\,m^{-3}})$	Gas	ρ $(\mathrm{kg\,m^{-3}})$
Amyl alcohol	812	Air	1.293	Oxygen	1.43
Ethanol	794	Argon	1.78	Propane	2.02
Methanol	769	Carbon dioxide	1.98	Smoke	0.13
Lubricating oil	910	Carbon monoxide	1.25	(average)	
Paraffin (kerosene)	800	Ethane	1.36	Steam (100 °C)	0.63
Petrol	700	Helium	0.177	Sulphur dioxide	2.92
Pure water	1000	Hydrogen	0.0899	Xenon	5.89
Sea water	1030	Krypton	3.73		
Heavy water (11.6 °C)	1105	Methane	0.72		
		Neon	0.90		
		Nitrogen	1.25		

6.21.2 *Thermal expansion*

Let:

α = coefficient of linear expansion $(°\mathrm{C}^{-1})$
β = coefficient of superficial expansion $(°\mathrm{C}^{-1})$
γ = coefficient of cubical expansion $(°\mathrm{C}^{-1})$
θ = temperature change $(°\mathrm{C})$
L = initial length
A = initial area
V = initial volume

L' = final length
A' = final area
V' = final volume

Then:

$L' = L(1 + \alpha\theta)$
$A' = A(1 + \beta\theta)$
$V' = V(1 + \gamma\theta)$

Approximately:

$\beta = 2\alpha$
$\gamma = 3\alpha$

Coefficients of linear expansion $\alpha(\times 10^6\,°C^{-1})$ at normal temperature (unless otherwise stated)

Material	α	Material	α	Material	α
Aluminium	23	Gold	14	Rubber: natural, soft	150–220
	29 (0–600 °C)		15 (0–500 °C)	natural, hard	80
Antimony	11	Granite	8.3	nitrile	110
Brass	19	Graphite	7.9	silicone	185
Brick	5	Gunmetal	18	Sandstone	12
Bronze	18	Ice	50	Silver	19
Cadmium	30	Iron: cast	11		20.5 (0–900 °C)
Cement	11	Wrought	12	Slate	10
Chromium	7		15 (0–700 °C)	Solder (2 lead: 1 tin)	25
	11 (0–900 °C)	Lead	29	Steel: hardened	12.4
Cobalt	12		33 (0–320 °C)	mild	11
	18 (23–350 °C)	Magnesium	25	stainless	10.4
Concrete	13		30 (0–400 °C)	Tin	21
Copper	16.7	Nickel	12.8	Titanium	9
	20 (0–1000 °C)		18 (0–1000 °C)	Tungsten	4.5 (20 °C)
Diamond	1.3	Phosphor bronze	16.7		6
Duralumin	23	Plaster	17		(600–1400 °C)
Ebonite	70	Platinum	8.9		7
German silver	18.4		11 (0–800 °C)		(1400–2200 °C)
Glass	8.6 (0–100 °C)	Porcelain	4	Vanadium	8
	9.9 (100–200 °C)	Quartz	8–14	Zinc	30
	11.9 (200–300 °C)				

6.21.3 *Freezing mixtures*

Ammonium nitrate (parts)	Crushed ice or snow in water (parts)	Temperature (°C)
1	0.94	−4
1	1.20	−14
1	1.31	−17.5
1	3.61	−8

Calcium chloride (parts)	Crushed ice or snow in water (parts)	Temperature (°C)
1	0.49	−20
1	0.61	−39
1	0.70	−55
1	1.23	−22
1	4.92	−4

Solid carbon dioxide with alcohol −72
Solid carbon dioxide with   −77
  chloroform or ether

6.21.4 *Coefficients of cubical expansion of liquids at normal temperature (unless otherwise stated)*

$\gamma(\times 10^6\,°C^{-1})$

Liquid	γ	Liquid	γ
Acetic acid	107	Olive oil	70
Aniline	85	Paraffin	90
Benzene	124	Sulphuric acid	51
Chloroform	126	(20%)	
Ethanol	110	Turpentine	94
Ether	163	Water	41.5
Glycerine	53		(0–100 °C)
Mercury	18		100
			(100–200 °C)
			180
			(200–300 °C)

6.21.6 *Anti-freeze mixtures*

	Freezing point (°C)				
Concentration (%vol.)	10	20	30	40	50
Ethanol (ethyl alcohol)	−3.3	−7.8	−14.4	−22.2	−30.6
Methanol (methyl alcohol)	−5.0	−12.1	−21.1	−32.2	−45.0
Ethylene glycol	−3.9	−8.9	−15.6	−24.4	−36.7
Glycerine	−1.7	−5.0	−9.4	−15.6	−22.8

Engineering measurements

7.1 Length measurement

7.1.1 Engineer's rule

These are made from hardened and tempered steel marked off with high accuracy in lengths from about 10–30 cm with folding rules up to 60 cm. They are used for marking off, setting callipers and dividers, etc. When used directly, the accuracy is ±0.25 mm, and when used to set a scribing block the accuracy is ±0.125 mm.

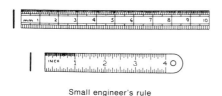

Small engineer's rule

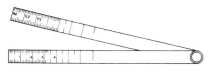

Folding rule

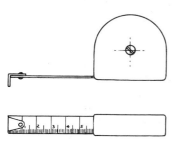

Spring tape rule

7.1.2 Feeler gauge (thickness gauge)

These consist of a number of thin blades of spring steel of exact, various thicknesses. They are used for measuring small gaps between parts.

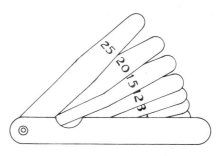

Thickness gauge

7.1.3 Micrometers

Micrometers are used for the measurement of internal and external dimensions, particularly of cylindrical shape. Measurement is based on the advance of a precision screw. The 'outside micrometer' is made in a variety of sizes, the most popular being 25 mm in 0.01-mm steps. It has a fixed 'barrel' graduated in

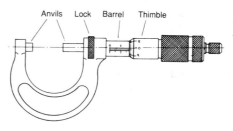

Outside micrometer

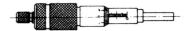

Micrometer head

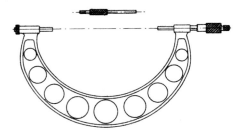

Large outside micrometer with extension rod

Inside micrometer

1-mm and 0.5-mm divisions screwed with a 0.5 mm pitch thread and a 'thimble' graduated around its circumference with 50–0.01 mm divisions.

An 'inside micrometer' has the fixed anvil projecting from the thimble; extensions may be attached. A 'micrometer head' is available consisting of the barrel and thimble assembly for use in any precision measuring device.

Reading a micrometer

Reading shown:

Reading on barrel = 5.5 mm
Reading on thimble = 0.28 mm
Total reading = 5.78 mm

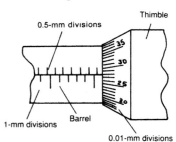

Micrometer

7.1.4 *Vernier calliper gauge*

This is used for internal and external measurement. It has a long flat scale with a fixed jaw and a sliding jaw, with a scale, or cursor, sliding along the fixed scale and read in conjunction with it. Two scales are provided to allow measurement inside or outside of the jaws.

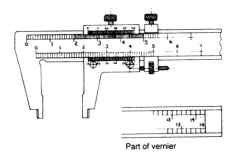

Part of vernier

Vernier calliper guage

Reading a vernier calliper gauge

Reading shown:

Reading on main scale = 43.5 mm
Reading on cursor = 0.18 mm
Total reading = 43.68 mm

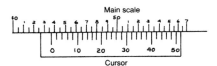

7.1.5 *Dial test indicator (dial gauge)*

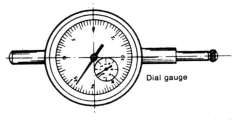

Dial gauge

The linear movement of a spring-loaded plunger is magnified by gears and displayed on a dial. Various sensitivities are available and a smaller scale shows complete revolutions of the main pointer. A typical indicator has a scale with 100–0.01 mm divisions and a small dial reading up to 25 revolutions of the pointer, i.e. a total range of 25 mm.

7.1.6 Gauge blocks (slip gauges)

These are hardened, ground and lapped rectangular blocks of steel made in various thicknesses of extreme accuracy and with a high degree of surface finish so that they will 'wring' together with a slight twist and pressure and remain firmly attached to one another. They are made in a number of sets; BS 888 recommends metric sets, two of which are given in the table below.

Gauge block sets (BS 888)

	No. blocks
Set M78	
1.01–1.49 mm in 0.01-mm steps	49
0.05–9.50 mm in 0.50-mm steps	19
10, 20, 30, 40, 50, 75, 100 mm	7
1.0025 mm	1
1.005 mm	1
1.0075 mm	1
Set M50	
1.01–1.09 mm in 0.01-mm steps	9
1.10–1.90 mm in 0.01-mm steps	9
1–25 mm in 1-mm steps	25
50, 75, 100 mm	3
1.0025, 1.0050, 1.0075 mm	3
0.05 mm	1

Protective slips are provided for use at the ends of the combinations.

7.1.8 Accuracy of linear measurement

The following table gives the accuracy of different methods of linear measurement.

7.1.7 Measurement of large bores

The size of very large bores may be measured by means of a gauge rod of known length slightly less than the bore. The rod is placed in the bore and the 'rock' noted. The bore can be determined from the amount of rock and the rod length.

Bore diameter $D = L + \dfrac{a^2}{8L}$

where: L = gauge length, a = 'rock'.

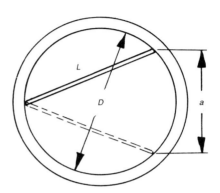

Instrument	Use	Accuracy (mm)
Steel rule	Directly	± 0.25
	To set a scribing block	± 0.125
Vernier calipers	External	± 0.03
	Internal	± 0.05
25-mm micrometer	Directly	± 0.007
	Preset to gauge blocks	± 0.005
Dial gauge	Over complete range	$\pm 0.003–0.03$
Dial gauge	As comparator over small range	$\pm 0.0001–0.0025$

7.2 Angle measurement

7.2.1 Combination angle slip gauges

Precision angle blocks are available with faces inclined to one another at a particular angle accurate to one second of arc. The gauges may be wrung together as with slip gauges, and angles may be added or subtracted to give the required angle. Details of a 13-block set are given.

13-Block set:

Degrees: 1, 3, 9, 9, 27, 41.
Minutes: 1, 3, 9, 27.
Seconds: 3, 9, 27.
Plus 1 square block.

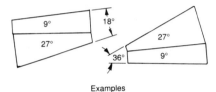

Examples

7.2.2 Measurement of angle of tapered bores

The method of measuring the angle of internal and external bore tapers is shown using precision balls, rollers and slip gauges.

External taper (using rollers and slip gauges)

$$\theta = \tan^{-1} \frac{(b_2 - b_1)}{2h}$$

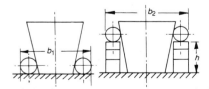

Internal taper (using two balls)

$$\theta = \sin^{-1} \frac{(D_1 - D_2)}{(2h_2 + D_2) - (2h_1 + D_1)}$$

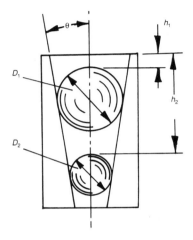

7.2.3 Sine bar

This is used to measure the angle of one surface relative to another. It consists of a precision bar with rollers, a precise distance apart. The angle of tilt is determined from the size of slip gauge used.

Angle of surface $\theta = \sin^{-1}\left(\dfrac{h}{L}\right)$

where: $L =$ distance between rollers, $h =$ height of slip gauges.

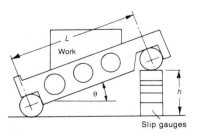

Slip gauges

7.3 Strain measurement

In carrying out strength tests on materials it is necessary to measure the strain. This is defined as the extension divided by the original length. In the case of mechanical extensometers, the original length is a 'gauge length' marked on the specimen. A typical gauge length is 2 cm and the magnification is up to 2000.

7.3.1 *Extensometer*

A typical extensometer (the Huggenberger) is shown. The knife edges A and B are held on to the specimen by a clamp with gauge length L. There are pivots at C and D and knife edges E and F are held in contact by a tension spring. The magnified increase in L is indicated by a pointer H on a scale J.

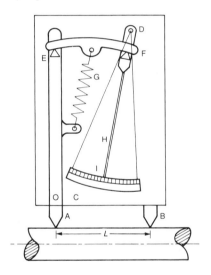

7.3.2 *Strain gauges*

The commonest type of strain gauge is the electrical resistance strain gauge ('strain gauge' for short). These are devices which produce an electrical signal proportional to the mechanical strain of the surface to which they are bonded. They can be made extremely small and can be attached to components of any shape which may be moving, e.g. an engine con-rod.

The gauge consists of a grid of resistance wire or, more usually, foil mounted on an insulating backing cemented to the component. Leads are connected to a bridge circuit and the strain is measured by a galvanometer or calibrated resistor. Dynamic strains may be indicated on an oscilloscope or suitable recorder. It is usually necessary to use 'dummy' gauges mounted on an unstressed surface at the same temperature to compensate for temperature effects.

Electrical resistance strain gauge

The sensitivity of a strain gauge is given by the 'gauge factor', i.e. the ratio of change in resistance to gauge resistance divided by the strain. Various arrangements are used, depending on the type of stress being measured, e.g. tension, compression, bending and torsion. For two-dimensional stress situations a 'strain gauge rosette' consisting of three gauges at different angles is used. The principal stresses and their direction can be calculated from the three strains.

7.3.3 *Strain-gauge applications*

Symbols used:
  R = resistance
  R_g = gauge resistance
  R_d = dummy gauge resistance
  dR = change in resistance
  e = strain
  E = Young's modulus
  σ = direct stress
  V = voltage applied to bridge
  $\bar{V}$ = galvanometer voltage
  I_g = gauge current
  F_g = gauge factor

Gauge factor $F_g = \dfrac{dR/R}{e}$

Direct stress $\sigma = eE$

Tension or compression (one active gauge, one dummy gauge)

Galvanometer voltage $\bar{V} = F_g e \dfrac{V}{2}$

Gauge current $I_g = \dfrac{V}{2R_g}$

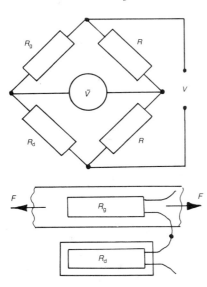

Bending (two active gauges: one in tension, one in compression)

$\bar{V} = F_g e V;\ I_g = \dfrac{V}{2R_g}$

Bending (four active gauges: two in tension, two in compression)

$\bar{V} = 2F_g e V;\ I_g = \dfrac{V}{2R_g}$

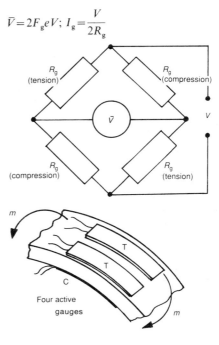

Four active gauges

Tension or compression (two active gauges and two dummy gauges in series)

This arrangement eliminates the effect of bending.

$\bar{V} = F_g e \dfrac{V}{2};\ I_g = \dfrac{V}{4R_g}$

Torque measurement

Two gauges are mounted on a shaft at 45° to its axis and perpendicular to one another. Under torsion one gauge is under tension and the other under compression, the stresses being numerically equal to the shear stress. The gauges are connected in a bridge circuit, as for bending. To eliminate bending effects four gauges may be used, two being on the opposite side of the shaft. In this case:

$$\bar{V} = 2F_g eV$$

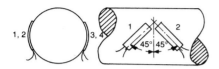

7.3.4 Strain gauge rosette

In the case of two-dimensional stress, it is necessary to use three gauges. If the gauges are at 45° to one another, then the principal stresses may be found as follows.

Let:

e_a, e_b, e_c = measured strains
E = Young's modulus
v = Poisson's ratio

7.3.5 Characteristics of some strain gauges

Principal stresses

$$\sigma_1 = E\left[\frac{K_1}{1-v} + \frac{K_2}{1+v}\right]$$

$$\sigma_2 = E\left[\frac{K_1}{1-v} - \frac{K_2}{1+v}\right]$$

Angle between σ_1 and e_a

$$\theta = \tfrac{1}{2}\tan^{-1}\left(\frac{2e_b - e_a - e_c}{e_a - e_c}\right)$$

where: $K_1 = \dfrac{(e_a + e_c)}{2}$ and $K_2 = \sqrt{\dfrac{(e_a - e_b)^2 + (e_b + e_c)^2}{2}}$

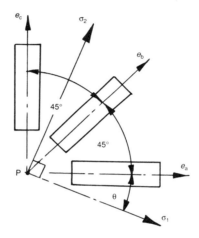

Material	Gauge factor, F_g	Resistance, R_g (Ω)	Temperature coefficient of resistance (°C^{-1})	Remarks
Advance (57%Cu, 43%Ni)	2.0	100	0.11×10^{-4}	F_g constant over wide range of strain; low-temperature (<250 °C) use
Platinum alloys	4.0	50	0.22×10^{-2}	For high-temperature (>500 °C) use
Silicon semiconductor	−100 to +100	200	0.09	Brittle, but high F_g. Not suitable for large strains

7.4 Temperature measurement

7.4.1 *Liquid-in-glass thermometers*

Mercury

The commonest type of thermometer uses mercury which has a freezing point of $-39\,°C$ and a boiling point of $357\,°C$, although it can be used up to $500\,°C$ since the thermometer may contain an inert gas under pressure.

The advantages of this thermometer are: good visibility; linear scale; non-wetting; good conductor of heat; and pure mercury is easily available.

The disadvantages are: it is fragile; slow cooling of glass; long response time; and errors arise due to non-uniform bore and incorrect positioning.

Alcohol

Alcohol can be used down to $-113\,°C$, but its boiling point is only $78\,°C$. The alcohol needs colouring. It is cheaper than mercury, and its low-temperature operation is an advantage in a number of applications.

Mercury in steel

This thermometer employs a mercury filled capillary tube connected to a Bourdon-type pressure gauge which deflects as the mercury expands with temperature. It is extremely robust and can give a remote indication.

7.4.2 *Thermocouples*

When a junction is made of two dissimilar metals (or semi-conductors) a small voltage, known as a 'thermal electromotive force (e.m.f.)' exists across it, which increases, usually linearly, with temperature. The basic circuit includes a 'cold junction' and a sensitive measuring device, e.g. a galvanometer, which indicates the e.m.f. The cold junction must be maintained at a known temperature as a reference, e.g. by an ice bath or a thermostatically controlled oven. If two cold junctions are used then the galvanometer may be connected by ordinary copper leads. A number of

thermocouples connected in series, known as a 'thermopile', gives an e.m.f. proportional to the number of thermocouples. Practical thermocouples are protected by a metal sheath with ceramic beads as insulation.

The advantages of thermocouples are: they are simple in construction, compact, robust and relatively cheap; they are suitable for remote control, automatic systems and recorders since they have a short response time.

The disadvantages are that they suffer from errors due to voltage drop in the leads, variation in cold-junction e.m.f. and stray thermoelectric effects in leads.

7.4.3 *Thermocouple circuits*

Basic thermocouple circuit

$V = $ Constant $\times$ Temperature (usually)

Galvanometer e.m.f. $V = V_h - V_c$

where: $V_h = $ e.m.f. for 'hot' junction, $V_c = $ e.m.f. for 'cold' junction

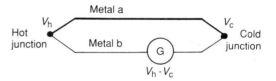

Thermocouple circuit with ice bath

A bath of melting ice is used for the cold junction. Temperature is given relative to $0\,°C$.

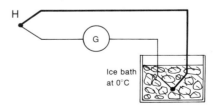

$G = $ galvanometer, $C = $ cold junction, $H = $ hot junction

Thermocouple circuit with extension leads

Two cold junctions at the same temperature are used and copper extension leads to the measuring instrument.

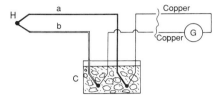

Practical thermocouple

The wires pass through ceramic beads inside a protective metal sheath.

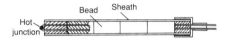

Thermopile

This consists of a number of thermocouples connected in series to give a higher e.m.f.

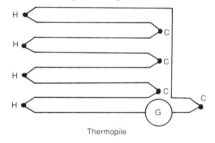

Thermopile

7.4.4 Thermocouple pairs and temperature limit

Materials	Temperature (°C)		Applications
	Minimum	Maximum	
Copper/constantan (57%Cu, 43%Ni)	−250	400	Flue gases, food processes, sub-zero temperatures
Iron/constantan	−200	850	Paper pulp mills, chemical reactors, low-temperature furnaces
Chromel (90%Ni, 10%Cr)/Alumel (94%Ni, 3%Mn, 2%Al, 1%Si)	0	1100	Blast-furnace gas, brick kilns, glass manufacture
Platinum/platinum rhodium	0	1400	Special applications
Tungsten/molybdenum	1250	2600	Special applications

7.4.5 Thermoelectric sensitivity of materials

Thermoelectric sensitivity of thermocouple materials relative to platinum (reference junction at 0 °C)

Metal	Sensitivity $(\mu V\,°C^{-1})$	Metal	Sensitivity $(\mu V\,°C^{-1})$
Bismuth	−72	Silver	6.5
Constantan	−35	Copper	6.5
Nickel	−15	Gold	6.5
Potassium	−9	Tungsten	7.5
Sodium	−2	Cadmium	7.5
Platinum	0	Iron	18.5
Mercury	0.6	Nichrome	25
Carbon	3	Antimony	47
Aluminium	3.5	Germanium	300
Lead	4	Silicon	440
Tantalum	4.5	Tellurium	500
Rhodium	6	Selenium	900

7.4.6 Thermal e.m.f. for thermocouple combinations

Thermal e.m.f. for common thermocouple combinations (reference junction at 0 °C)

Temperature		E.m.f. (mV)				
°F	°C	Copper/ constantan	Chromel/ constantan	Iron/ constantan	Chromel/ alumel	Platinum 10% rhodium
−300	−184	−5.284	−8.30	−7.52	−5.51	—
−250	−157	−4.747	—	−6.71	−4.96	—
−200	−129	−4.111	−6.40	−5.76	−4.29	—
−150	−101	−3.380	—	−4.68	−3.52	—
−100	−73	−2.559	−3.94	−3.49	−2.65	—
−50	−46	−1.654	—	−2.22	−1.70	—
0	−18	−0.670	−1.02	−0.89	−0.68	—
50	10	0.389	—	0.05	0.04	—
100	38	1.517	2.27	1.94	1.52	0.221
150	66	2.711	—	3.41	2.66	0.401
200	93	3.967	5.87	4.91	3.82	0.595
250	121	5.280	—	6.42	4.97	0.800
300	149	6.647	9.71	7.94	6.09	1.017
350	177	8.064	—	9.48	7.20	1.242
400	204	9.525	13.75	11.03	8.31	1.474
450	232	11.030	—	12.57	9.43	1.712
500	260	12.575	17.95	14.12	10.57	1.956
600	316	15.773	22.25	17.18	12.86	2.458
700	371	19.100	26.65	20.26	15.18	2.977
800	427	—	31.09	23.32	17.53	3.506
1000	538	—	40.06	29.52	22.26	4.596
1200	649	—	49.04	36.01	26.98	5.726
1500	816	—	62.30	—	33.93	7.498
1700	927	—	70.90	—	38.43	8.732
2000	1093	—	—	—	44.91	10.662
2500	1371	—	—	—	54.92	13.991
3000	1649	—	—	—	—	17.292

7.4.7 Electronic thermocouple thermometer

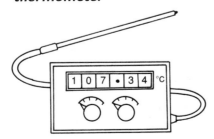

This has a robust sheathed thermocouple connected to a voltmeter which gives a digital or analogue readout of temperature. It avoids many of the usual disadvantages of thermocouples.

7.4.8 Resistance thermometers

Resistance thermometers are based on the fact that the electrical resistance of a metal wire varies with temperature. The metals most used are platinum and nickel, for which the resistance increases with temperature in a linear manner.

If R_o is the resistance at 0 °C, then the resistance R_t at T°C is:

$$R_t = R_o(1 + \alpha T)$$

or $T = \dfrac{(R_t - R_o)}{\alpha R_o}$

where: α = temperature coefficient of resistance.

The value of α is given for a number of metals as well as electrolytes and semi-conductors in the table below.

Resistance temperature coefficients (at room temperature) °C^{-1}

Material	α (°C^{-1})	Material	α (°C^{-1})
Nickel	0.0067	Gold	0.004
Iron	0.002–0.006	Platinum	0.00392
Tungsten	0.0048	Mercury	0.00099
Aluminium	0.0045	Manganin	±0.00002
Copper	0.0043	Carbon	−0.0007
Lead	0.0042	Electrolytes	−0.02 to −0.09
Silver	0.0041	Semi-conductor (thermistor)	−0.068 to +0.14

The construction of a typical resistance thermometer is shown in the figure. It consists of a small resistance coil enclosed in a metal sheath with ceramic insulation beads. The temperature range is 100 °C to 300 °C for nickel and 200 °C to 800 °C for platinum.

With other metals it is possible to reach 1500 °C. The small resistance change is measured by means of a Wheatstone bridge and dummy leads eliminate temperature effects on the element leads.

The resistance thermometer is used for heat treatment and annealing furnaces and for calibration of other thermometers.

The main disadvantages are fragility and slow response.

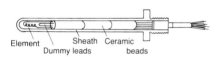

Resistance thermometer

7.4.9 Thermistors

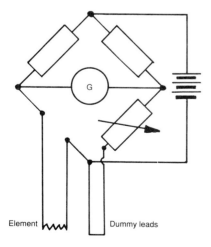

Resistance thermometer measuring bridge

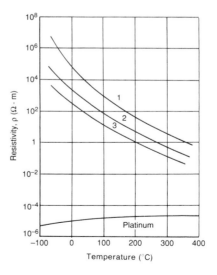

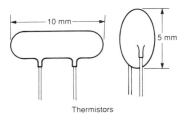

Thermistors

Most metals have a positive temperature coefficient of resistance, i.e. resistance increases with temperature. Semi-conductors may have a very large negative coefficient which is non-linear. A 'thermistor' is a bead of such material, e.g. oxides of copper, manganese and cobalt, with leads connected to a measuring circuit. They are extremely sensitive; for example, a change from $400\,\Omega$ at $0\,°C$ to $100\,\Omega$ at $140\,°C$. They are inexpensive and suitable for very small changes in temperature. The graph shows curves of resistivity for three thermistor materials compared with platinum.

7.4.10 Pyrometers

Total radiation pyrometer

At very high temperatures where thermometers and thermocouples are unsuitable, temperature can be deduced from the measurement of radiant energy from a hot source. The radiation is passed down a tube and focused, using a mirror, onto a thermocouple or thermopile which is shielded from direct radiation.

Disappearing-filament pyrometer

The brightness and colour of a hot body varies with temperature and in the case of the disappearing filament pyrometer it is compared with the appearance of a heated lamp filament. The radiation is focused

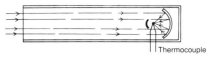

Total radiation pyrometer

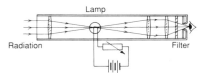

Disappearing-filament pyrometer

onto the filament the brightness of which is varied by means of a calibrated variable resistor until the filament appears to vanish. A red filter protects the eye.

7.4.11 *Bimetallic thermometer*

The deflection of a bimetallic strip or coil may be used to indicate temperature. This type is not very accurate but is simple and cheap. These thermometers are used for alarms and temperature controllers when connected to a mechanical system.

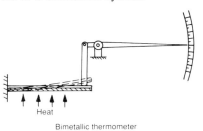

Bimetallic thermometer

7.4.12 *Temperature-sensitive paints*

Kits are available of paints and crayons made of chemicals which change colour at definite temperatures. The range is from about $30\,°C$ to $700\,°C$, with an accuracy of about 5%. Several paints are required to cover the range. Crayons are the easiest to use. The method is suitable for inaccessible places.

7.4.13 *Fixed-point temperatures*

The table below gives fixed-point temperatures known to a high degree of accuracy from which instruments can be calibrated.

	Temperature (°C)
Boiling point of liquid oxygen	−182.97
Melting point of ice	0.00
Triple point of water	0.01
Boiling point of water	100.00
Freezing point of zinc	419.505
Boiling point of liquid sulphur	444.60
Freezing point of liquid antimony	630.50
Melting point of silver	960.80
Melting point of gold	1063.00

7.5   Pressure measurement

7.5.1   *Pressure units*

1 newton per square metre ($1 \, \mathrm{N \, m^{-2}}$) = 1 pascal (1 Pa)
1 bar = 100 000 (10^5) Pa = 1000 millibar (mbar)
1 mbar = 100 Pa
1 bar = 760 mm Hg (approximately)

7.5.2   *Barometers*

Mercury barometers

The basic barometer consists of a vertical glass tube closed at the top, filled with mercury and standing in a mercury bath. There is a space at the top of the tube in which a vacuum exists and the height of the column is a measure of atmospheric pressure. The so-called 'Fortin barometer' is a mercury barometer with a Vernier scale.

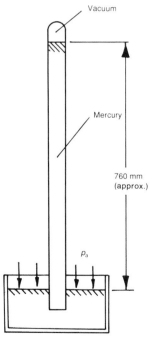

Mercury barometer

Aneroid barometer

A sealed flexible metal bellows or capsule with a very low internal pressure is connected to a lever with pointer and scale. Atmospheric-pressure variations cause a corresponding deflection of the capsule and movement of the pointer. The pointer usually carries a pen which records the temperature on a rotating chart.

Mercury barometer

Atmospheric pressure supports a column of mercury of approximately 760 mm Hg.

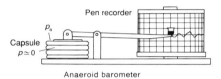

Anaeroid barometer

Standard atmospheric pressure = 1.0135 bar ≡ 1013.25 mbar ≡ 101 325 Pa.
Gauge pressure $p_g = p - p_a$

where: p = absolute pressure, p_a = atmospheric pressure.

7.5.3   *Manometers*

The U-tube manometer may be used to measure a pressure relative to atmospheric pressure, or the difference between two pressures. If one 'leg' is much larger in diameter than the other, a 'single-leg manometer' is obtained and only a single reading is required (as for the barometer). The inclined single-leg manometer gives greater accuracy. When the manometer fluid is less dense than the fluid, the pressure of which is to be measured, an inverted manometer is used. When pressure is measured relative to atmospheric pressure the air density is assumed to be negligible compared with that of the manometer fluid.

U-tube manometer – pressure relative to atmosphere (gauge pressure)

Let:

ρ_m = density of manometer fluid
h = manometer reading
g = acceleration due to gravity

Measured pressure $p = \rho_m g h$

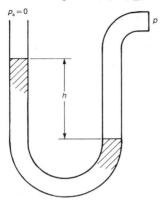

U-tube manometer – differential pressure

Pressure difference $p_1 - p_2 = (\rho_m - \rho_f)gh$

where: ρ_f = density of measured fluid.

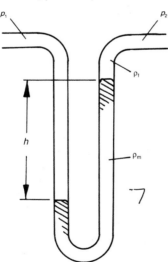

Single-leg manometer – gauge pressure

Measured pressure $p = \rho_m g h$

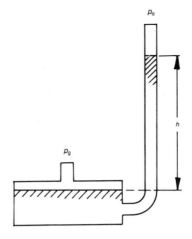

Inclined single-leg manometer

Measured pressure $p = \rho_m g L \sin \theta$

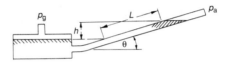

Inverted U-tube manometer

Pressure difference $(p_1 - p_2) = (\rho_f - \rho_m)gh$

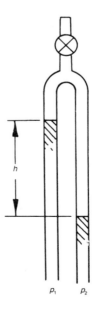

7.5.4 *Bourdon pressure gauge*

In the Bourdon gauge a curved flattened metal tube is closed at one end and connected to the pressure source

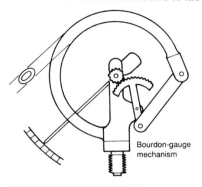

Bourdon-gauge mechanism

at the other end. Under pressure the tube tends to straighten and causes a deflection of a pointer through a lever and rack and pinion amplifying system. This gauge can be used for liquids or gases from a fraction of a bar pressure up to 10 000 bar. Calibration is by means of a 'dead-weight tester'.

7.5.5 *Pressure transducers*

A wide range of transducers is available which convert the deflection of a diaphragm or Bourdon tube into an electrical signal which gives a reading on an indicator or is used to control a process, etc. Transducers cover a wide range of pressure and have a fast response. Types include, piezo-crystal, strain gauge, variable capacity, and variable inductance.

7.6 Flow measurement

The simplest method of measuring the mass flow of a liquid is to collect the liquid in a bucket or weigh tank over a given time and divide the mass by the collection time. For gases, a volume can be collected in a gasometer over a known time to give the volume flow rate.

7.6.1 *Measurement by weight*

Bucket

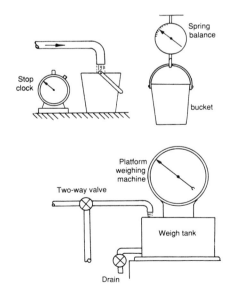

Spring balance

Stop clock

bucket

Platform weighing machine

Two-way valve

Weigh tank

Drain

$$\dot{m} = \text{mass per second} = \frac{\text{Mass collected}}{\text{Collection time}}$$

$$\text{Volume per second} = \frac{\text{Mass per second}}{\text{Density}}$$

Weigh tank

$$\dot{m} = \frac{\text{Mass collected}}{\text{Collection time}}$$

7.6.2 *Measurement by gas tank (gasometer)*

$$\text{Volume per second} = \frac{\text{Volume collected}}{\text{Collection time}}$$

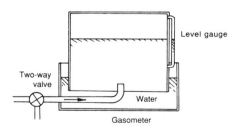

Level gauge

Two-way valve

Water

Gasometer

7.6.3 *Rotameter*

This is a type of variable-orifice meter consisting of a vertical glass tapered tube containing a metal 'float'. The fluid, which may be a liquid or gas, flows through the annular space between the float and the tube. As the flow is increased the float moves to a greater height. The movement is roughly proportional to flow, and calibration is usually carried our by the supplier. Angled grooves in the rim of the float cause rotation and give the float stability.

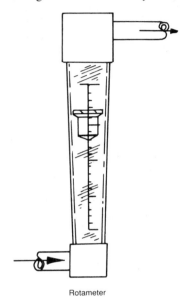

Rotameter

7.6.4 *Turbine flow meters*

An axial or tangential impeller mounted in a pipe rotates at a speed roughly proportional to the velocity, and hence the flow, of the fluid in the pipe. The rotational speed is measured either mechanically or electronically to give flow or flow rate.

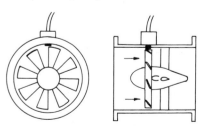

Axial-impeller flowmeter

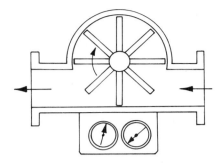

Tangential-impeller flowmeter

7.6.5 *Differential pressure flowmeters*

These depend on the pressure difference caused by a change in section or obstruction in a pipe or duct. British Standard BS 1042 deals with the design of the 'venturi-meter' the 'orifice plate' and the 'nozzle'. Pressure difference is measured by a manometer or transducer; the position of the pressure tappings is important. Flow is proportional to the square root of the pressure difference and calibration is therefore necessary. Of the three types the venturi-meter is the most expensive but gives the least overall pressure loss. The orifice plate is the simplest and cheapest type and occupies the least space, but has an appreciable overall pressure loss. The nozzle type is a compromise between the other two.

Venturi meter

(See Section 4.3.3)

Flow $Q = \text{Constant} \sqrt{(p_1 - p_2)}$
Pressure difference $(p_1 - p_2) = (\rho_m - \rho_f)gh$

Symbols are as for manometers (see above).

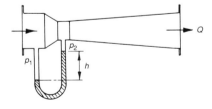

Orifice meter

The flow formula is as for the Venturi meter.

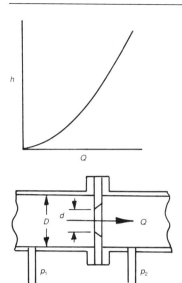

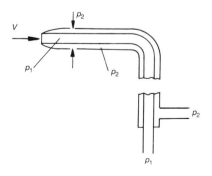

Nozzle meter

The flow formula is as for the Venturi meter.

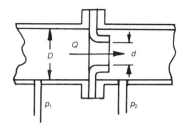

7.7 Velocity measurement

7.7.1 *Pitot-static tube*

The pitot-static tube consists of two concentric tubes, the central one with an open end pointing upstream of the fluid flow and the other closed at the end but with small holes drilled at right angles to the direction of flow. The central tube pressure is equal to the static pressure plus the 'velocity pressure', whereas the outer tube pressure is the static pressure only.

A manometer or other differential pressure measuring device measures the pressure difference between the tubes which is equal to the 'velocity pressure'. For large pipes or ducts, traversing gear is used and an average value of velocity calculated.

Fluid velocity $V = \sqrt{\dfrac{2(p_2 - p_1)}{\rho_f}}$

$\qquad\qquad = \sqrt{\dfrac{2\rho_m g h}{\rho_f}}$

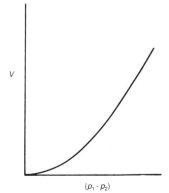

7.7.2 Anemometers

Various types of anemometer are used to measure the velocity, usually of air. The 'cup type' is used for free air and has hemispherical cups on arms attached to a rotating shaft. The shape of the cups gives a greater drag on one side than the other and results in a speed of rotation approximately proportional to the air speed. Velocity is found by measuring revolutions over a fixed time. The 'vane anemometer' has an axial impeller attached to a handle with extensions and an electrical pick-up which measures the revolutions. A meter with several ranges indicates the velocity.

The 'hot-wire anemometer' is used where it is necessary to investigate the change in velocity over a small distance, e.g. in a boundary layer. A probe terminating in an extremely small heated wire element is situated in the fluid stream and cools to an extent which depends on the velocity. The resulting change in resistance of the element is measured by a bridge circuit and is related to velocity by calibration. The response is rapid.

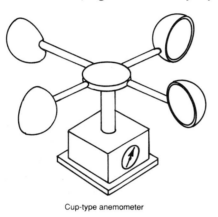

Cup-type anemometer

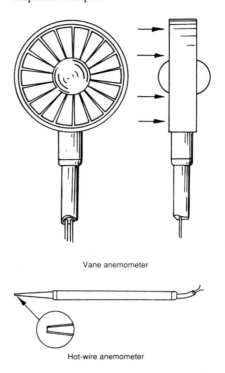

Vane anemometer

Hot-wire anemometer

7.8 Rotational-speed measurement

7.8.1 Mechanical tachometers

These may be permanently mounted on a machine or hand-held. The hand-held type has several shaft attachments with rubber ends (see figure), including a conical end for use with a shaft centre hole, a wheel to run on a cylindrical surface, and a cup end for use where there is no centre hole.

7.8.2 Electrical tachometers

The tachogenerator is driven by the shaft and gives an output voltage proportional to speed which is indicated as rotational speed on a meter. Alternatively, a toothed wheel passing an inductive pick-up generates pulses which are counted over a fixed time and displayed on a meter as the speed of rotation.

7.8.3 Stroboscope

This has an electronic flash tube which flashes at a variable rate and which is adjusted to coincide with the rotational speed so that the rotating object, or a suitable mark on it, appears to stand still. The flash-rate control is calibrated in rotational speed.

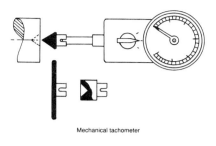

Mechanical tachometer

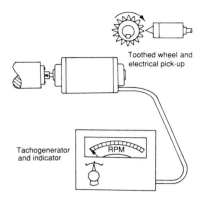

Toothed wheel and electrical pick-up

Tachogenerator and indicator

RPM

7.9 Materials-testing measurements

7.9.1 *Hardness testing*

Hardness tests on materials consist of pressing a hardened ball or point into a specimen and measuring the size of the resulting indentation. The two methods shown are the Brinell method, which utilizes a ball, and the Vicker's pyramid method which utilizes a pyramidal point.

Other methods in use are the Rockwell method which uses a ball or diamond cone, and the Shore scleroscope, a portable instrument which measures the height of rebound of a hammer falling on the surface.

Measurement of Brinnel hardness number (BHN)

The ball size is 10 mm for most cases or 1 mm for light work.

Let:
D = diameter of indentation (mm)
D_b = diameter of ball (mm)
F = force on ball (kg)

Values of F: steel, $F = 30D_b^2$; copper, $F = 10D_b^2$; aluminium, $F = 5D_b^2$

$$\text{Hardness BHN} = \frac{F}{1.57 D_b (D_b - \sqrt{D_b^2 - D^2})}$$

Vicker's pyramid number (VPN)

Let:
F = load (kg)
b = diagonal of indentation (mm)

$$\text{VPN} = 1.854 \frac{F}{b^2}$$

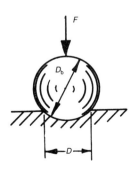

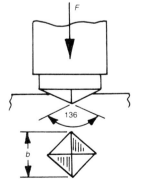

7.9.2 Toughness tests

Toughness testing consists of striking a notched test piece with a hammer and measuring the energy required to cause fracture. The energy is indicated on the dial of the test machine and the force is produced by a swinging mass.

$$\text{Toughness} = \text{Constant} \times \frac{\text{Energy to fracture specimen}}{\text{Energy of the swinging mass}}$$

The energy of the swinging mass is 163 J for the Izod impact test and 294 J for the Charpy test.

Izod impact test machine and test piece

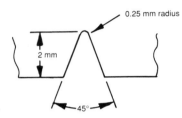

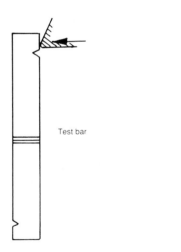

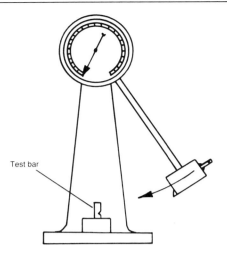

Charpy test piece

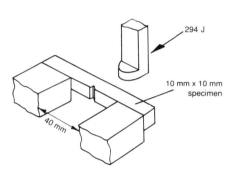

7.9.3 Tensile test on steel

Testing machines are used to determine the mechanical properties of materials under tension, compression, bending, shear and torsion.

One of the most important tests is the tensile test, especially that for steel. Typical curves are shown for ductile steel and hard steel. In the case of a ductile steel such as 'mild steel', there is a definite yield point above which the steel is no longer elastic. In the case of hard steel the load–extension curve becomes non-linear and it is necessary to specify a 'proof stress' for a specified strain, e.g. 0.1%.

Load–extension curves for steel

Symbols used:
W = load
W_e = elastic limit
W_y = yield load
W_f = fracture load
W_m = maximum load
W_p = proof load
e = strain
x = extension
σ = stress
E = Young's modulus

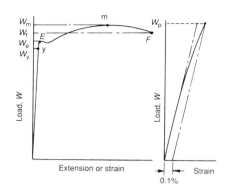

Tensile strength $\text{TS} = \dfrac{W_m}{\text{Original area of cross-section}}$ (N mm^{-2})

Yield stress $\text{YS} = \dfrac{W_y}{\text{Area of cross-section}}$ (N mm^{-2})

Proof stress $\text{PS} = $ Stress for a specified strain (e.g. 0.1%), (N mm^{-2})

Strain $e = \dfrac{\text{Extension at load } W \text{ (mm)}}{\text{Original gauge length (mm)}}$

Young's modulus $E = \dfrac{\text{Stress in elastic region}}{\text{Corresponding strain}}$ (N mm^{-2})

Percentage elongation (Elong. %) $= \dfrac{\text{Extension at failure}}{\text{Original gauge length}} \times 100\%$

Percentage reduction in area $= \dfrac{\text{Original area of cross-section} - \text{Area at fracture}}{\text{Original area}} \times 100\%$

8.1 Units and symbols

8.1.1 Symbols and units for physical quantities

Quantity	Symbol	Unit
Acceleration: gravitational	g	$m\,s^{-2}$
linear	a	$m\,s^{-2}$
Admittance	Y	S
Altitude above sea level	z	m
Amount of substance	n	mol
Angle: plane	$\alpha, \beta, \theta, \phi$	rad
solid	Ω, ω	steradian
Angular acceleration	α	$rad\,s^{-2}$
Angular velocity	ω	$rad\,s^{-1}$
Area	A	m^2
Area, second moment of	I	m^4
Bulk modulus	K	$N\,m^{-2}$, Pa
Capacitance	C	μF
Capacity	V	l, m^3
Coefficient of friction	μ	No unit
Coefficient of linear expansion	α	$°C^{-1}$
Conductance: electrical	G	S
thermal	h	$kW\,m^{-2}\,K^{-1}$
Conductivity: electrical	α	$kS\,mm^{-1}$
thermal	λ	$W\,m^{-1}\,K^{-1}$
Cubical expansion, coefficient of	β	$°C^{-1}$
Current, electrical	I	A
Current density	J	$A\,mm^{-2}$
Density	ρ	$kg\,m^{-3}$
Density, relative	d	No unit
Dryness fraction	x	No unit
Dynamic viscosity	η	$Ns\,m^{-2}$, cP
Efficiency	η	No unit
Elasticity, modulus of	E	$N\,m^{-2}$, Pa
Electric field strength	E	$V\,m^{-1}$
Electric flux	ϕ	C

Quantity	Symbol	Unit
Electric flux density	D	$C\,m^{-2}$
Energy	W	J
Energy: internal	U, E	J
specific internal	u, e	$kJ\,kg^{-1}$
Enthalpy	H	J
Enthalpy, specific	h	$kJ\,kg^{-1}$
Entropy	S	$kJ\,K^{-1}$
Expansion, coefficient of cubical	β	$°C^{-1}$
Expansion, coefficient of linear	α	$°C^{-1}$
Field strength: electric	E	$V\,m^{-1}$
magnetic	H	$A\,m^{-1}$
Flux density: electric	D	$C\,m^{-2}$
magnetic	B	T
Flux: electric	ψ	C
magnetic	Φ	Wb
Force	F	N
Force, resisting	R	N
Frequency	f	Hz
Frequency, resonant	f_r	Hz
Gravitational acceleration	g	$m\,s^{-2}$
Gibbs' function	G	J·
Gibbs' function, specific	g	$kJ\,kg^{-1}$
Heat capacity, specific	c	$kJ\,kg^{-1}\,K^{-1}$
Heat flow rate	q, ϕ	W
Heat flux intensity	ϕ	$kW\,m^{-2}$
Illumination	E	lux
Impedance	Z	Ω
Inductance: self	L	H
mutual	M	H
Internal energy	U, E	J
Internal energy, specific	u, e	$kJ\,kg^{-1}$

Quantity	Symbol	Unit
Inertia, moment of	I, J	$kg\,m^2$
Kinematic viscosity	v	$m^2\,s^{-1}$, St
Length	l	m
Light: velocity of	c	$m\,s^{-1}$
Light, wavelength of	λ	m
Linear expansion, coefficient of	α	$°C^{-1}$
Luminance	L	$cd\,m^{-2}$
Luminous flux	ϕ	lm
Luminous intensity	I	cd
Magnetic field strength	H	$A\,m^{-1}$
Magnetic flux	Φ	Wb
Magnetic flux density	B	T
Magnetomotive force	F	A
Mass:	m	kg
Mass: rate of flow	$\dot{m}$	$kg\,s^{-1}$
Modulus, bulk	K	$N\,m^{-2}$
Modulus of elasticity	E	$N\,m^{-2}$
Modulus of rigidity	G	$N\,m^{-2}$
Modulus of section	Z	m^3
Molar mass of gas	M	$kg\,K^{-1}\,mol^{-1}$
Molar volume	V_m	$m^3\,K^{-1}\,mol^{-1}$
Moment of force	M	N-m
Moment of inertia	I, J	kg-m^2
Mutual inductance	M	H
Number of turns in a winding	N	No unit
Periodic time	T	s
Permeability: absolute	μ	$\mu H\,m^{-1}$
absolute of free space	μ_0	$\mu H\,m^{-1}$
relative	μ_r	
Permeance	Λ	H
Permittivity, absolute	ε	$pF\,m^{-1}$
Permittivity of free space	ε_0	$pF\,m^{-1}$
Permittivity, relative	ε_r	No unit
Poisson's ratio	v	No unit
Polar moment of area	J	m^4
Power: apparent	S	V-A
active	P	W
reactive	Q	V-A_r
Pressure	p	$N\,m^{-2}$, Pa
Quantity of heat	Q	J
Quantity of electricity	Q	A-h, C
Reactance	X	Ω

Quantity	Symbol	Unit
Reluctance	S	H, $A\,Wb^{-1}$
Relative density	d	No unit
Resistance, electrical	R	Ω
Resisting force	R	N
Resistance, temperature coefficients of	α, β, γ	$°C^{-1}$
Resistivity: conductors	ρ	$M\Omega$-mm
insulators	ρ	$M\Omega$-mm
Resonant frequency	f_r	Hz
Second moment of area	I	m^4
Self-inductance	L	H
Shear strain	γ	No unit
Shear stress	τ	$N\,m^{-2}$, Pa
Specific gas constant	R	$kJ\,kg^{-1}\,K^{-1}$
Specific heat capacity	c	$kJ\,kg^{-1}\,K^{-1}$
Specific volume	v	$m^3\,kg^{-1}$
Strain, direct	ε	No unit
Stress, direct	σ	$N\,m^{-2}$, Pa
Shear modulus of rigidity	G	$N\,m^{-2}$, Pa
Surface tension	γ	$N\,m^{-1}$
Susceptance	B	S
Temperature value	θ	$°C$
Temperature coefficients of resistance	α, β, γ	$°C^{-1}$
Thermodynamic temperature value	T	K
Time	t	s
Torque	T	Nm
Vapour velocity	C	$m\,s^{-1}$
Velocity	v	$m\,s^{-1}$
Velocity, angular	ω	$rad\,s^{-1}$
angular	N	rev/s rev/min
Velocity of light	c	$m\,s^{-1}$
Velocity of sound	c_s	$m\,s^{-1}$
Voltage	V	V
Volume	V	m^3
Volume, rate of flow	V	$m^3\,s^{-1}$
Viscosity: dynamic	μ, η	$Ns\,m^{-2}$, cP
kinematic	v	$m^2\,s^{-1}$, cSt
Wavelength	λ	m
Work	W	J
Young's modulus of elasticity	E	$N\,m^{-2}$, Pa

8.1.2 *Abbreviations for technical terms*

Term	Abb.	Term	Abb.
Absolute	abs.	High tension	h.t.
Alternating current	a.c.	High voltage	h.v.
Aqueous	aq.	Horse power	h.p.
Atomic number	at. no.	Indicated mean effective pressure	i.m.e.p.
Atomic weight	at. wt.	Infra-red	i.r.
Audio frequency	a.f.	Intermediate frequency	i.f.
Boiling point	b.p.	Internal combustion	i.c., IC
Bottom dead centre	b.d.c., BDC	Internal combustion engine	i.c.e.
Brake mean effective pressure	b.m.e.p.	Kinetic energy	k.e.
Calculated	calc.	Lower calorific value	l.c.v., LCV
Calorific value	c.v., CV	Low pressure	l.p.
Cathode-ray oscilloscope	c.r.o.	Low tension	l.t.
Cathode-ray tube	c.r.t.	Low voltage	l.v.
Centre of gravity	c.g.	Magnetomotive force	m.m.f.
Compare	cf.	Maximum	max.
Computer-aided design	CAD	Mean effective pressure	m.e.p.
Computer-aided manufacture	CAM	Melting point	m.p.
Concentrated	conc.	Minimum	min.
Constant	const.	Moment	mom.
Corrected	corr.	Numerical control	n.c.
Critical	crit.	Pitch circle diameter	p.c.d.
Cross-sectional area	c.s.a.	Potential difference	p.d.
Decomposition	decomp.	Potential energy	p.e.
Degree	deg.	Pressure	press.
Diameter	dia.	Proof stress	p.s.
Differential coefficient	d.c.	Radian	rad.
Dilute	dil.	Radio frequency	r.f.
Direct current	d.c.	Radius	rad.
Dry flue gas	d.f.g.	Relative density	r.d.
Elastic limit	e.l.	Relative humidity	r.h.
Electromotive force	e.m.f.	Root mean square	r.m.s.
Equation	eqn.	Specific	spec.
Equivalent	equiv.	Specific gravity	s.g.
Example	ex.	Standard temperature and pressure	s.t.p.
Experiment(al)	expt.	Strain energy	s.e.
Freezing point	f.p.	Temperature	temp.
Frequency	freq.	Tensile strength	t.s., TS
Higher calorific value	h.c.v., HCV	Thermocouple	t/c
High frequency	h.f.	Top dead centre	t.d.c., TDC
High pressure	h.p.	Ultraviolet	u.v.
High speed steel	h.s.s.	Ultra-high frequency	u.h.f.
High tensile	h.t.	Very high frequency	v.h.f.
		Yield stress	y.s., YS

8.1.3 *Abbreviations for units*

Unit	Abb.	Unit	Abb.	Unit	Abb.	Unit	Abb.
metre	m	steradian	sr	newton	N	mole	mol
angström	A	radian per	rad s^{-1}	bar	bar	watt	W
square metre	m^2	second		millibar	mb	decibel	dB
cubic metre	m^3	hertz	Hz	standard	atm	kelvin	K
litre	l	revolution per	rev. min^{-1}	atmosphere		centigrade	°C
second	s	minute		millimetre of	mm Hg	coulomb	C
minute	min	kilogramme	kg	mercury		ampere	A
hour	h	gramme	g	poise	P	volt	V
lumen	lm	tonne	t	stokes	S, St	ohm	Ω
candela	cd	(= 1 Mg)		joule	J	farad	F
lux	lx	seimen	S	kilowatt hour	kW-h	henry	H
day	d	atomic mass	u	electron volt	eV	weber	Wb
year	a	unit		calorie	cal	tesla	T
radian	rad	pascal	Pa				

8.1.4 *Multiples and submultiples*

Multiplying factor	Prefix	Symbol
10^{12}	tera	T
10^{9}	giga	G
10^{6}	mega	M
10^{3}	kilo	k
10^{-3}	milli	m
10^{-6}	micro	μ
10^{-9}	nano	n
10^{-12}	pico	p
10^{-15}	femto	f
10^{-18}	atto	a

8.1.5 *SI equivalents for Imperial and US customary units*

Abbreviations used

m = metre
km = kilometre
in. = inch
ft = foot
yd = yard
m, mi = mile
Pa = pascal (N m^2)
psi = pounds per square inch

Tsi = tons per square inch
atm = atmosphere
l = litre
cc = cubic centimetre
gal = gallon
lb = pound
lbm = pound mass
lbf = pound force
k, kip = kilopound
t, T = ton
tnf, tonf = ton force
mph = miles per hour
fpm = feet per minute
kt = knot (nautical mile per hour)
gpm = gallons per minute
cfs = cubic feet per second
cfm = cubic feet per minute
N = newton
s, sec = second
min = minute
h = hour
hp = horsepower
kW = kilowatt
Btu = British thermal unit
J = joule

Length

1 in. = 25.4 mm = 0.0254 m.
1 ft = 305 mm = 0.305 m.

1 yard = 914 mm = 0.914 m.
1 mile = 1609 m = 1.609 km.
1 nautical mile = 1.835 km = 1.14 miles.
1 μm = 10^{-6} m.
1 Å = 10^{-10} m.

Area

1 in.^{2} = 645 mm^2 = 0.645 × 10^{-3} m^2.
1 ft^2 = 9.29 × 10^4 mm^2 = 0.0929 m^2.
1 yard^2 = 0.836 m^2.
1 acre = 4047 m^2.
1 mile^2 = 2.59 × 10^4 m^2 = 2.59 km^2.
1 hectare = 10 000 m^2

Volume (capacity)

1 in.^{3} = 16.4 × 10^3 mm^3 = 16.4 × 10^{-6} m^3.
1 ft^3 = 0.0283 m^3.
1 yard^3 = 0.765 m^3.
1 pint (UK) = 0.568 l.
1 pint (US) = 0.456 l.
1 quart (UK) = 1.137 l.
1 quart (US) = 0.9464 l.
1 gallon (UK) = 1.201 gallon (US) = 4.546 l.
1 gallon (US) = 3.785 l.
1 barrel = 42 gallons (US) = 159 l.
1 cm^3 = 1000 mm^3.
1 l. = 1000 cm^3.
1 m^2 = 1000 l.

Mass

1 lbm = 0.454 kg.
1 slug = 32.17 lbm = 14.6 kg.
1 ton (US or 'short') = 2000 lbm = 907.2 kg.
1 ton (UK or 'long') = 2240 lbm = 1016 kg.
1 tonne (metric ton) = 1000 kg

Density

1 lb in.$^{-3}$ = 27 680 kg m^{-3}.
1 lb ft^{-3} = 16.02 kg m^{-3}.
1 slug ft^{-3} = 515.4 kg m^{-3}.

Velocity

1 in. s^{-1} = 0.0254 m s^{-1}.
1 ft s^{-1} = 0.3048 m s^{-1}.
1 ft min^{-1} = 0.00508 m s^{-1}.

1 mile h^{-1} = 0.447 m s^{-1} = 1.61 km h^{-1}.
1 km h^{-1} = 0.719 m s^{-1}.
1 knot = 1 nautical mile/hour = 0.515 m s^{-1}.

Mass flow rate

1 lbm s^{-1} = 0.454 kg s^{-1}.
1 lbm h^{-1} = 1.26 × 10^{-4} kg s^{-1}.
1 ton h^{-1} = 0.282 kg s^{-1}.
1 slug s^{-1} = 14.6 kg s^{-1}.

Volume flow rate

1 ft^3 s^{-1} = 0.283 m^3 s^{-1}.
1 UK gallon sec^{-1} = 0.00455 m^3 s^{-1}.
1 US gallon s^{-1} = 0.00379 m^3 s^{-1}.
1 UK gallon min^{-1} = 7.58 × 10^{-5} m^3 s^{-1}.
1 US gallon min^{-1} = 6.31 × 10^{-5} m^3 s^{-1}.

Force

1 lbf = 4.45 N.
1 kip (1000 lbf) = 4.45 kN.
1 tonf = 9964 N.
1 poundal = 0.138 N.
1 dyne = 10^{-5} N.

Stress or pressure

1 lbf in.^{2} (psi) = 6895 N m^{-2} (Pa).
1 lbf ft^{-2} (psf) = 47.9 N m^{-2}.
1 kip in.$^{-2}$ (ksi) = 6895 kN m^{-2} (kPa).
1 kip ft^{-2}. (ksf) = 47.9 kN m^{-2} (kPa).
1 poundal ft^{-2} = 1.49 N m^{-2}.
1 tonf in.$^{-2}$ = 15.44 × 10^6 N m^{-2}.
1 tonf ft^{-2} = 1.073 × 10^5 N m^{-2}.
1 in. water (39.2 °F) = 249 N m^{-2}.
1 ft water (39.2 °F) = 2989 N m^{-2}.
1 in. mercury = 3386 N m^{-2}.
1 atmos = 14.7 psi = 1.01325 × 10^5 N m^{-2}.
1 MPa = 10^6 N m^{-2} = 1 N mm^{-2}.
1 bar = 10^5 N m^{-2}.

Work and energy

1 in. lbf = 0.113 J (Nm).
1 ft. lbf = 1.365 J.
1 Btu = 778 ft lbf = 252 calories = 1055 J.
1 cal = 4.186 J.
1 kcal = 4.186 kJ.

1 ft poundal $= 0.0421$ J.
1 horsepower-hour $= 2.685$ MJ.
1 kW-h $= 3.6$ MJ.
1 erg $= 10^{-7}$ J.

Power

1 ft lbf s^{-1} $= 1.356$ W.
1 ft lbf min^{-1} $= 0.0226$ W.
1 horsepower $(550\,\text{ft lbf s}^{-1}) = 746$ W $= 0.746$ kW.

1 ft poundal sec $= 0.0421$ W

Acceleration

1 ft s^{-2} $= 0.305$ m s^{-2}.
1 g $= 32.174$ ft s^{-2} $= 9.807$ m s^{-2}.

Fuel consumption

1 mile per gallon (mpg) $= 0.425$ km l^{-1}.

8.2 Fasteners

8.2.1 *Bolt and screw types*

Bolts

Bolts are used for fastening machine parts together often in conjunction with nuts and washers to form non-permanent connections. The bolt head is usually hexagonal, but may be square or round. The 'shank' may be screwed for part or the whole of its length, in the latter case it is sometimes called a 'screw' or 'machine screw'.

Most bolts are made of low or medium carbon steel by forging or machining with threads cut or rolled. Forged bolts are called 'black' and machined bolts 'bright'. They are also made in high tensile, alloy and stainless steels as well as non-ferrous metals and alloys, and plastics. Bolts may be plated or galvanized to prevent corrosion.

In the UK, metric threads (ISOM) have largely replaced BSW and BSF threads. For small sizes British Association (BA) threads are used. In the USA, the most used threads are 'unified fine' (UNF) and 'unified coarse' (UNC).

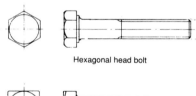

Hexagonal head bolt

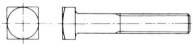

Square head bolt

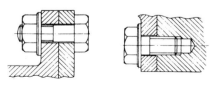

Bolted joint
(through bolt)
application

Tap bolt–
application

Stud (stud bolt)

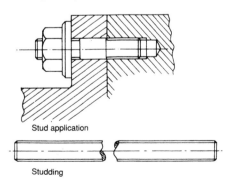

Stud application

Studding

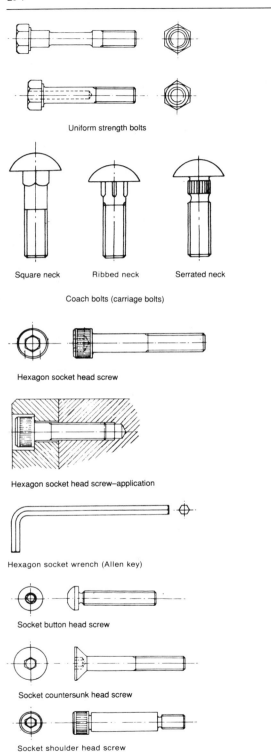

Uniform strength bolts

Square neck　　Ribbed neck　　Serrated neck

Coach bolts (carriage bolts)

Hexagon socket head screw

Hexagon socket head screw–application

Hexagon socket wrench (Allen key)

Socket button head screw

Socket countersunk head screw

Socket shoulder head screw

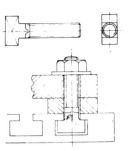

T bolt and application

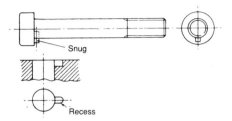

Snug

Recess

Cheese head bolt

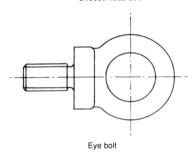

Eye bolt

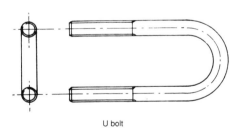

U bolt

Indented foundation bolt

Rag bolt

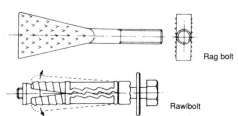

Rawlbolt

Screws

The term 'screw' is applied to a wide range of threaded fasteners used with metal, wood, plastics, etc. Screws have a variety of types of head and are made in many materials (steel, brass, nylon, etc.), some are plated. Small screws usually have BA threads and special threads are used for wood and self-tapping screws.

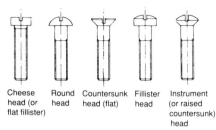

Cheese head (or flat fillister)   Round head   Countersunk head (flat)   Fillister head   Instrument (or raised countersunk) head

Slotted head machine screws

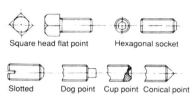

Square head flat point   Hexagonal socket

Slotted   Dog point   Cup point   Conical point

Set screws

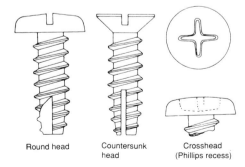

Wood screws

Round head   Countersunk head   Crosshead (Phillips recess)

Self-tapping screws

Drive screw

8.2.2 Nuts and washers

Nuts are usually hexagonal, but may be square or round. Steel hexagon nuts may be 'black' or 'bright' and have one or both faces chamfered. Washers are used to distribute load and prevent damage to a surface. They are mostly of steel, but brass, copper, aluminium, fibre, leather and plastics are used.

A wide variety of lock washers and locking devices are available, including adhesives such as 'Loctite'.

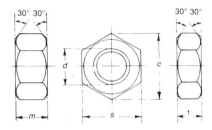

ISO metric precision hexagon nut and twin nut

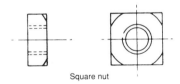

Square nut

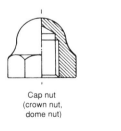

Cap nut (crown nut, dome nut)

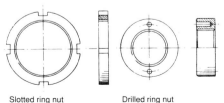

Slotted ring nut   Drilled ring nut

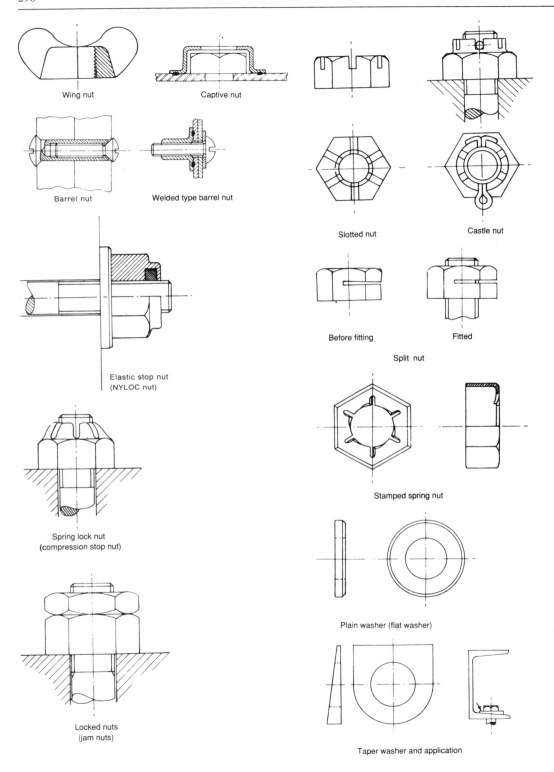

Wing nut

Captive nut

Barrel nut

Welded type barrel nut

Slotted nut

Castle nut

Before fitting

Fitted

Split nut

Elastic stop nut
(NYLOC nut)

Stamped spring nut

Spring lock nut
(compression stop nut)

Plain washer (flat washer)

Locked nuts
(jam nuts)

Taper washer and application

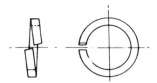

Helical spring lock washer

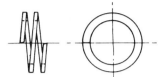

Two-coil spring lock washer

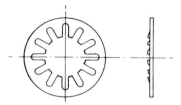

Internally serrated lock washer (tooth lock washer)

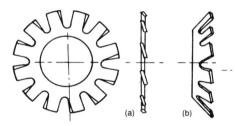

Externally serrated lock washer: (a) flat and (b) for countersunk hole

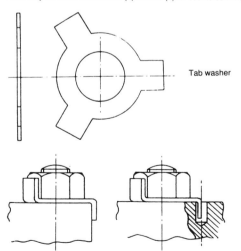

Tab washer

Tab washer–application

8.2.3 *Rivets and pins*

Rivets

Rivets are used to make permanent joints between two or more plates. Steel rivets may be closed when red hot; rivets of softer metals such as aluminium and copper may be closed cold. There are a number of types of riveted joint configurations for plates, two of which are shown in the figure.

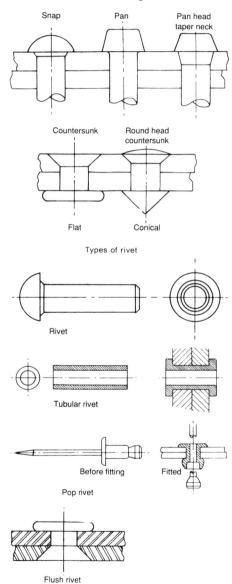

Types of rivet

Rivet

Tubular rivet

Before fitting   Fitted

Pop rivet

Flush rivet

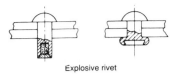

Explosive rivet

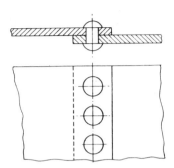

Riveted lap joint

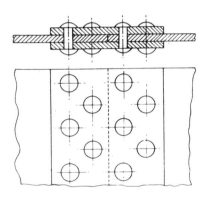

Double riveted butt joint with two straps

Pins

The term 'pin' refers to a large number of components used for fixing, locating and load carrying. Dowel pins are used to locate accurately one part relative to another. Taper pins fit into taper holes and are often used for light shaft couplings. A grooved pin has grooves with raised edges to give a tight fit in a hole. The roll pin is a spring steel tube which closes to give a

tight fit. Split pins are used mainly for locking nuts. Cotter pins are used to connect rods in tension and fits into mating slots.

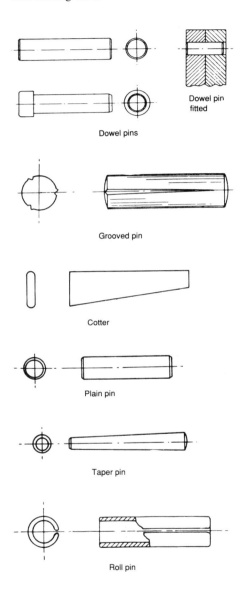

Dowel pins

Dowel pin fitted

Grooved pin

Cotter

Plain pin

Taper pin

Roll pin

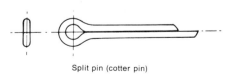

Split pin (cotter pin)

8.2.4 *ISO metric nut and bolt sizes*

ISO metric precision hexagon nuts and bolts (all quantities) (in mm)

D	p_f	f_{max}	c_{max}	h_{max}	L_{min}	t_{1max}	t_{2max}	p_c	A_b	D_t
M1.6	0.35	3.2	3.7	1.225	9.2	1.3	—	0.35	0.795	1.25
M2	0.4	4	4.6	1.525	10	1.6	—	0.4	1.53	1.6
M2.5	0.45	5	5.8	1.825	11	2	—	0.45	2.61	2.05
M3	0.5	5.5	6.4	2.125	12	2.4	—	0.5	4.0	2.5
M4	0.7	7	8.1	2.925	14	3.2	—	0.7	6.82	3.3
M5	0.8	8	9.2	3.65	16	4	—	0.8	11.3	4.2
M6	1	10	11.5	4.15	18	5	—	1	15.8	5
M8	1.25	13	15	5.65	22	6.5	5.0	1.25	30.0	6.8
M10	1.5	17	19.6	7.18	26	8	6.0	1.5	48	8.5
M12	1.75	19	21.9	8.18	30	10	7.0	1.75	70.5	10.2
M16	2	24	27.7	10.18	38	13	8.0	2	136	14
M20	2.5	30	34.6	13.215	46	16	9.0	2.5	212	17.5
M24	3	36	41.6	15.215	54	19	10.0	3	305	21
M30	3.5	46	53.1	19.26	66	24	12.0	3.5	492	26.5
M36	4	55	63.5	23.26	78	29	14.0	4	722	32
M42	4.5	65	75.1	26.26	90	34	16.0	4.5	1007	37.5
M48	5	75	86.6	30.26	102	38	18.0	5	1330	43
M56	5.5	85	98.1	35.31	118	46	—	5.5	1830	50.5
M64	6	95	109.7	40.31	134	51	—	6	2430	58

D = nominal diameter $\qquad$ L_{min} = minimum length of thread
p_f = pitch (fine series) $\qquad$ t_1 = thickness of normal nut
p_c = pitch (coarse series) $\qquad$ t_2 = thickness of thin nut
f = width across flats $\qquad$ A_b = area at bottom of thread
c = width across corners $\qquad$ D_t = tapping drill diameter for coarse thread
h = height of head $\qquad$ L = bolt length

Standard bolt lengths (L)

20, 25, 30, 35, 40, 45, 50, 55, 60, 65, 70, 75, 80, 90, 100, 110, 120, 130, 140, 150

Standard screw lengths

10, 12, 16, 18, 20, 22, 25, 30, 35, 40, 45, 50, 55, 60, 70.

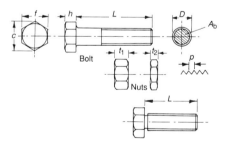

8.2.5 *Clearance holes for bolts*

Clearance holes for metric bolts

Bolt size, D (mm)	Clearance hole diameter, (mm)		
	Fine	Medium	Coarse
1.6	1.7	1.8	2
2	2.2	2.4	2.6
2.5	2.7	2.9	3.1
3	3.2	3.4	3.6
4	4.3	4.5	4.8
5	5.3	5.5	5.8
6	6.4	6.6	7
7	7.4	7.6	8
8	8.4	9	10
10	10.5	11	12
12	13	14	15
14	15	16	17
16	17	18	19
18	19	20	21
20	21	22	24
22	23	24	26
24	25	26	28
27	28	30	32
30	31	33	35
33	34	36	38
36	37	39	42
39	40	42	45

8.2.6 *British Association (BA) screw threads*

No.	Major diameter (mm)	Pitch (mm)	Core diameter (mm)	Area at bottom of thread (mm^2)
0	6.0	1.0	4.80	18.10
1	5.3	0.9	4.22	13.99
2	4.7	0.81	3.73	10.93
3	4.1	0.73	3.22	8.14
4	3.6	0.66	2.81	6.20
5	3.2	0.59	2.49	4.87
6	2.8	0.53	2.16	3.66
7	2.5	0.48	1.92	2.89
8	2.2	0.43	1.68	2.22
9	1.9	0.39	1.43	1.61
10	1.7	0.35	1.28	1.29
11	1.5	0.31	1.13	1.00
12	1.3	0.28	0.96	0.72
13	1.2	0.25	0.90	0.64
14	1.0	0.23	0.72	0.41
15	0.9	0.21	0.65	0.33
16	0.79	0.19	0.56	0.25
17	0.70	0.17	0.50	0.20
18	0.62	0.15	0.44	0.15
19	0.54	0.14	0.37	0.11
20	0.48	0.12	0.34	0.091
21	0.42	0.11	0.29	0.066
22	0.37	0.10	0.25	0.049
23	0.33	0.09	0.22	0.038
24	0.29	0.08	0.19	0.028
25	0.25	0.07	0.17	0.023

8.2.7 *Unified screw threads*

Size designation	Nominal major diameter		Coarse series (UNC)			Fine series (UNF)		
			No. threads per inch	Area at bottom of threads		No. threads per inch	Area at bottom of threads	
	in.	mm		in.^{2}	mm^2		in.^{2}	mm^2
0	0.0600	1.524				80	0.00151	0.974
1	0.0730	1.854	64	0.00218	1.406	72	0.00237	1.529
2	0.0860	2.184	56	0.00310	2.000	64	0.00339	2.187
3	0.0990	2.515	48	0.00406	2.619	56	0.00451	2.910
4	0.1120	2.845	40	0.00496	3.200	48	0.00566	3.652
5	0.1250	3.175	40	0.00672	4.335	44	0.00716	4.619
6	0.1380	3.505	32	0.00745	4.806	40	0.00874	5.639
8	0.1640	4.166	32	0.01196	7.716	36	0.01285	8.290
10	0.1900	4.826	24	0.01450	9.355	32	0.0175	11.29
12	0.2160	5.486	24	0.0206	13.29	28	0.0226	14.58
1/4	0.2500	6.350	20	0.0269	17.35	28	0.0326	21.03
5/16	0.3125	7.938	18	0.0454	29.29	24	0.0524	33.81
3/8	0.375	9.525	16	0.0678	43.74	24	0.0809	52.19
7/16	0.4375	11.11	14	0.0933	60.19	20	0.1090	70.32
1/2	0.5000	12.70	13	0.1257	81.10	20	0.1485	95.87
9/16	0.5625	14.29	12	0.162	104.5	18	0.1890	121.9
5/8	0.6250	15.88	11	0.202	130.3	18	0.240	154.8
3/4	0.7500	19.05	10	0.302	194.8	16	0.351	226.5
7/8	0.8750	22.23	9	0.419	270.3	14	0.480	309.7
1	1.0000	25.40	8	0.551	355.5	12	0.625	403.2
$1\frac{1}{4}$	1.2500	31.75	7	0.890	574.2	12	1.024	660.6
$1\frac{1}{2}$	1.5000	38.10	6	1.294	834.8	12	1.260	812.9

8.2.8 *Pipe threads*

BSP pipe threads (BS 2779: 1973) – Whitworth thread form

Nominal size (in.)	Threads per inch	Pitch (mm)	Major diameter (mm)	Minor diameter (mm)
1/16	28	0.907	7.723	6.561
1/8	28	0.907	9.728	8.566
1/4	19	1.337	13.157	11.445
3/4	19	1.337	16.662	14.950
1/2	14	1.814	20.955	18.631
5/8	14	1.814	22.911	20.587
3/4	14	1.814	26.441	24.117
7/8	14	1.814	30.201	27.877
1	11	2.309	33.249	30.291
$1\frac{1}{8}$	11	2.309	37.897	34.939
$1\frac{1}{4}$	11	2.309	41.910	38.952

BSP pipe threads (BS 2779: 1973) – Whitworth thread form (*continued*)

Nominal size (in.)	Threads per inch	Pitch (mm)	Major diameter (mm)	Minor diameter (mm)
$1\frac{1}{2}$	11	2.309	47.803	44.845
$1\frac{3}{4}$	11	2.309	53.746	50.788
2	11	2.309	59.614	56.656
$2\frac{1}{4}$	11	2.309	65.710	62.752
$2\frac{1}{2}$	11	2.309	75.189	72.226
$2\frac{3}{4}$	11	2.309	81.534	78.576
3	11	2.309	87.884	84.926
$3\frac{1}{2}$	11	2.309	100.330	97.372
4	11	2.309	113.030	110.072
$4\frac{1}{2}$	11	2.309	125.73	122.772
5	11	2.309	138.43	135.472
$5\frac{1}{2}$	11	2.309	151.13	148.172
6	11	2.309	163.83	160.372

8.2.9 Rectangular BS keys

Dimensions (mm)

Shaft diameter, D	Key $b \times d$	Depth in shaft, d_1	Depth in hub, d_2	Radius, r Max.	Min.
6–8	2 × 2	1.2	1	0.16	0.08
8–10	3 × 3	1.8	1.4	0.16	0.08
10–12	4 × 4	2.5	1.8	0.16	0.08
12–17	5 × 5	3	2.3	0.25	0.16
17–22	6 × 6	3.5	2.8	0.25	0.16
22–30	8 × 7	4	3.3	0.25	0.16
30–38	10 × 8	5	3.3	0.40	0.25
38–44	12 × 8	5	3.3	0.40	0.25
44–50	14 × 9	5.5	3.8	0.40	0.25
50–58	16 × 10	6	4.3	0.40	0.25
58–65	18 × 11	7	4.4	0.40	0.25
65–75	20 × 12	7.5	4.9	0.60	0.40
75–85	22 × 14	9	5.4	0.60	0.40
85–95	25 × 14	9	5.4	0.60	0.40
95–110	28 × 16	10	6.4	0.60	0.40
110–130	32 × 18	11	7.4	0.60	0.40
130–150	36 × 20	12	8.4	1.00	0.70
150–170	40 × 22	13	9.4	1.00	0.70
170–200	45 × 25	15	10.4	1.00	0.70
200–230	50 × 28	17	11.4	1.00	0.70
230–260	56 × 32	20	12.4	1.60	1.20
260–290	63 × 32	20	12.4	1.60	1.20
290–330	70 × 36	22	14.4	1.60	1.20
330–380	80 × 40	25	15.4	2.50	2.00
380–440	90 × 45	28	17.4	2.50	2.00
440–500	100 × 50	31	19.5	2.50	2.00

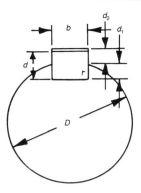

8.2.10 *ISO straight-sided splines*

Dimensions (mm)

Light series				Medium series			
D_o	D_i	n	b	D_o	D_i	n	b
26	23	6	6	14	11	6	3
30	26	6	6	16	13	6	3.5
32	28	6	7	20	16	6	4
36	32	8	6	22	18	6	5
40	36	8	7	25	21	6	5
46	42	8	8	28	23	6	6
50	46	8	9	32	26	6	6
58	52	8	10	34	28	6	7
62	56	8	10	38	32	8	6
68	62	8	12	42	36	8	7
78	72	10	12	48	42	8	8
83	82	10	12	54	46	8	9
98	92	10	14	60	52	8	10
103	102	10	16	65	56	8	10
120	112	10	18	72	62	8	12
				82	72	10	12
				92	82	10	12
				102	92	10	14
				112	102	10	16
				125	112	10	18

n = number of splines.

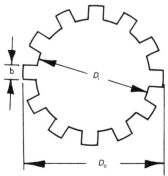

8.3 Engineering stock

8.3.1 *Circular, square and rectangular hollow steel sections*

M = mass per unit length
A = cross-sectional area
I_X = second moment of area about axis XX
I_Y = second moment of area about axis YY

Circular hollow steel sections (BS 4848: Part 2)

D_o (mm)	t (mm)	M (kg m^{-1})	A (cm^2)	I_X (cm^4)	D_o (mm)	t (mm)	M (kg m^{-1})	A (cm^2)	I_X (cm^4)
21.3	3.2	1.43	1.82	0.77	139.7	5.0	16.6	21.2	481
26.9	3.2	1.87	2.38	1.70		6.3	20.7	26.4	589
33.7	2.6	1.99	2.54	3.09		8.0	26.0	33.1	720
	3.2	2.41	3.07	3.60		10.0	32.0	40.7	862
	4.0	2.93	3.73	4.19	168.3	5.0	20.1	25.7	856
42.4	2.6	2.55	3.25	6.46		6.3	25.2	32.1	1053
	3.2	3.09	3.94	7.62		8.0	31.6	40.3	1297
	4.0	3.79	4.83	8.99		10.0	39.0	49.7	1564
48.3	3.2	3.56	4.53	11.60	193.7	5.4	25.1	31.9	1417
	4.0	4.37	5.57	13.8		6.3	29.1	37.1	1630
	5.0	5.34	6.80	16.2		8.0	36.6	46.7	2016
60.3	3.2	4.51	5.74	23.5		10.0	45.3	57.7	2442
	4.0	5.55	7.07	28.2		12.5	55.9	71.2	2934
	5.0	6.82	8.69	33.5		16.0	70.1	89.3	3554
76.1	3.2	5.75	7.33	48.8	219.1	6.3	33.1	42.1	2386
	4.0	7.11	9.06	59.1		8.0	41.6	53.1	2960
	5.0	8.77	11.2	70.9		10.0	51.6	65.7	3598
88.9	3.2	6.76	8.62	79.2		12.5	63.7	81.1	4345
	4.0	8.38	10.7	96.3		16.0	80.1	102	5297
	5.0	10.3	13.2	116		20.0	98.2	125	6261
114.3	3.6	9.83	12.5	192					
	5.0	13.5	17.2	257					
	6.3	16.8	21.4	313					

Hollow square steel sections (BS 4848: Part 2)

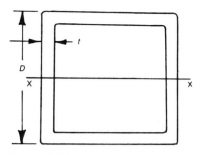

D (mm)	t (mm)	M (kg m^{-1})	A (cm^2)	I_X (cm^4)	D (mm)	t (mm)	M (kg m^{-1})	A (cm^2)	I_X (cm^4)
20	2.0	1.12	1.42	0.76	120	5.0	18.0	22.9	503
	2.6	1.39	1.78	0.88		6.3	22.3	28.5	610
25	2.0*	1.43	1.82	1.59		8.0	27.9	35.5	738
	2.6*	1.80	2.30	1.90		10	34.2	43.5	870
	3.2*	2.15	2.74	2.14	140	5.0*	21.1	26.9	814
30	2.6	2.21	2.82	3.49		6.3*	26.3	33.5	994
	2.9*	2.44	3.10	3.76		8.0*	32.9	41.9	1 212
	3.2	2.65	3.38	4.00		10*	40.4	51.5	1 441
40	2.4*	2.81	3.58	8.39	150	5.0	22.7	28.9	1 009
	2.6	3.03	3.86	8.94		6.3	28.3	36.0	1 236
	2.9	3.35	4.26	9.71		8.0	35.4	45.1	1 510
	3.2	3.66	4.66	10.4		10	43.6	55.5	1 803
	4.0	4.46	5.68	12.1		12.5	53.4	68.0	2 125
50	2.5*	3.71	4.72	17.7		16	66.4	84.5	2 500
	2.9*	4.26	5.42	19.9	180	6.3	34.2	43.6	2 186
	3.2	4.66	5.94	21.6		8.0	43.0	54.7	2 689
	4.0	5.72	7.28	25.5		10	53.0	67.5	3 237
	5.0	6.97	8.88	29.6		12.5	65.2	83.0	3 856
60	2.9*	5.17	6.58	35.6		16	81.4	104	4 607
	3.2	5.67	7.22	38.7	200	6.3	38.2	48.6	3 033
	4.0	6.97	8.88	46.1		8.0	48.0	61.1	3 744
	5.0	8.54	10.90	54.4		10	59.3	75.5	4 525
70	2.9*	6.08	7.74	57.9		12.5	73.0	93.0	5 419
	3.6	7.46	9.50	69.5		16	91.5	117	6 524
	5.0	10.10	12.90	90.1	250	6.3	48.1	61.2	6 049
80	2.9*	6.99	8.90	88.0		8.0	60.5	77.1	7 510
	3.6	8.59	10.90	106		10	75.0	95.5	9 141
	5.0	11.70	14.90	139		12.5	92.6	118	11 050
	6.3	14.40	18.40	165		16	117	149	13 480
90	3.6	9.72	12.4	154	300	10	90.7	116	16 150
	5.0	13.30	16.9	202		12.5	112	143	19 630
	6.3	16.40	20.9	242		16	142	181	24 160
100	4.0	12.00	15.3	234	350	10.0	106	136	26 050
	5.0	14.80	18.9	283		12.5	132	168	31 810
	6.3	18.40	23.4	341		16.0	167	213	39 370
	8.0	22.90	29.1	408	400	10.0	122	156	39 350
	10.0	27.90	35.5	474		12.5	152	193	48 190

*Not to BS 4848: Part 2.

Rectangular hollow steel sections (BS 4848:Part 2)

$D \times B$ (mm × mm)	t (mm)	M (kg m^{-1})	A (cm^2)	I_X (cm^4)	I_Y (cm^4)	$D \times B$ (mm × mm)	t (mm)	M (kg m^{-1})	A (cm^2)	I_X (cm^4)	I_Y (cm^4)
50 × 30	2.4*	2.91	3.58	11.6	5.14	200 × 100	5.0	22.7	28.9	1 509	509
	2.6	3.03	3.86	12.4	5.45		6.3	28.3	36.0	1 851	618
	2.9*	3.35	4.26	13.3	5.90		8.0	35.4	45.1	2 269	747
	3.2	3.66	4.66	14.5	6.31		10.0	43.6	55.5	2 718	881
60 × 40	2.5*	3.71	4.72	23.1	12.2		12.5	53.4	68.0	3 218	1 022
	2.9*	4.26	5.42	26.2	13.7		16.0	66.4	84.5	3 808	1 175
	3.2	4.66	5.94	28.3	14.8	250 × 150	6.3	38.2	48.6	4 178	1 886
	4.0	5.72	7.28	33.6	17.3		8.0	48.0	61.1	5 167	2 317
80 × 40	2.9*	5.17	6.58	53.5	17.7		10.0	59.3	75.5	6 259	2 784
	3.2	5.67	7.22	58.1	19.1		12.5	73.0	93.0	7 518	3 310
	4.0	6.97	8.88	69.6	22.6		16.0	91.5	117	9 089	3 943
90 × 50	2.9*	6.08	7.74	82.9	32.8	300 × 200	6.3	48.1	61.2	7 880	4 216
	3.6	7.46	9.50	99.8	39.1		8.0	60.5	77.1	9 798	5 219
	5.0	10.1	12.9	130	50.0		10.0	75.0	95.5	11 940	6 331
100 × 50	2.9*	6.53	8.32	108	36.1		12.5	92.6	118	14 460	7 619
	3.2	7.18	9.14	117	39.1		16.0	117	149	17 700	92 931
	4.0	8.86	11.3	142	46.7	400 × 200	10.0	90.7	116	24 140	8 138
	5.0	10.9	13.9	170	55.1		12.5	112	143	29 410	9 820
	6.3*	13.4	17.1	202	64.2		16.0	142	181	36 300	11 950
100 × 60	2.9*	6.99	8.90	121	54.6	450 × 250	10.0	106	136	37 180	14 900
	3.6	8.59	10.9	147	65.4		12.5	132	168	45 470	18 100
	5.0	11.7	14.9	192	84.7		16.0	167	213	56 420	22 250
	6.3	14.4	18.4	230	99.9						
120 × 60	3.6	9.72	12.4	230	76.9						
	5.0	13.3	16.9	304	99.9						
	6.3	16.4	20.9	366	118						
120 × 80	5.0	14.8	18.9	370	195						
	6.3	18.4	23.4	447	234						
	8.0	22.9	29.1	537	278						
	10.0	27.9	35.5	628	320						
150 × 100	5.0	18.7	23.9	747	396						
	6.3	23.3	29.7	910	479						
	8.0	29.1	37.1	1106	577						
	10.0	35.7	45.5	1312	678						
160 × 80	5.0	18.0	22.9	753	251						
	6.3	22.3	28.5	917	302						
	8.0	27.9	35.5	1 113	361						
	10.0	34.2	43.5	1 318	419						

*Not to BS 4848: Part 2.

8.3.2 *ISO metric metal sheet, strip and wire sizes*

Preference is given in the order: R 10, R20, R40.
Sizes (mm)

R 10	R 20	R 40	R 10	R 20	R 40	R 10	R 20	R 40
0.020	0.020	0.020	0.250	0.250	0.250	3.15	3.15	3.15
		0.021			0.265			3.35
	0.022	0.022		0.280	0.280		3.55	3.55
		0.024			0.300			3.75
0.025	0.025	0.025	0.315	0.315	0.315	4.00	4.00	4.00
		0.026			0.335			4.25
	0.028	0.028		0.355	0.355		4.50	4.50
		0.030			0.375			4.75
0.032	0.0320	0.032	0.400	0.400	0.400	5.00	5.00	5.00
		0.034			0.425			5.30
	0.036	0.036		0.450	0.450		5.60	5.60
		0.038			0.475			6.00
0.040	0.040	0.040	0.500	0.500	0.500	6.30	6.30	6.30
		0.042			0.530			6.70
	0.045	0.045		0.560	0.560		7.10	7.10
		0.048			0.600			7.50
0.050	0.050	0.050	0.630	0.630	0.630	8.00	8.00	8.00
		0.053			0.670			8.50
	0.056	0.056		0.710	0.710		9.00	9.00
		0.060			0.750			9.50
0.063	0.063	0.063	0.800	0.800	0.800	10.00	10.00	10.00
		0.067			0.850			10.6
	0.071	0.071		0.900	0.900		11.2	11.2
		0.075			0.950			10.6
0.080	0.080	0.080	1.000	1.000	1.000	12.5	12.5	12.5
		0.085			1.06			13.2
	0.090	0.090		1.12	1.12		14.0	14.0
		0.095			1.18			15.0
0.100	0.100	0.100	1.25	1.25	1.25	16.0	16.0	16.0
		0.106			1.32			17.0
	0.112	0.112		1.40	1.40		18.0	18.0
		0.118			1.50			19.0
0.125	0.125	0.125	1.60	1.60	1.60			
		0.132			1.70	20.0	20.0	20.0
	0.140	0.140		1.80	1.80			21.2
		0.150			1.90		22.4	22.4
								23.6
0.160	0.160	0.160	2.00	2.00	2.00	25.0	25.0	25.0
		0.170			2.12			
	0.18	0.180		2.24	2.24			
		0.190			2.36			
0.200	0.200	0.200	2.50	2.50	2.50			
		0.212			2.65			
	0.224	0.224		2.80	2.80			
		0.236			3.00			

8.3.3 Copper pipe sizes for domestic water pipes, etc.

Size are given in BS 2871: Part 1.

Size of pipe* (mm)	Nominal thickness (mm)		
	Table X: Half-hard, light gauge	Table Y: half-hard, annealed	Table Z: hard drawn, thin wall
6	0.6	0.8	0.5
8	0.6	0.8	0.5
10	0.6	0.8	0.5
12	0.6	0.8	0.5
15	0.7	1.0	0.5
18	0.8	1.0	0.6
22	0.9	1.2	0.6
28	0.9	1.2	0.6
35	1.2	1.5	7.0
42	1.2	1.5	8.0
54	1.2	2.0	9.0
76.1	1.5	2.0	1.2
108	1.5	2.5	1.2

*Outer diameter.

8.4 Miscellaneous data

8.4.1 Factors of safety

Factor of safety $FS = \dfrac{\text{Tensile strength or Proof stress}}{\text{Permissible working stress}}$ (sometimes based on yield stress)

Typical factors of safety for various materials

Material	Type of load			
	Steady	Varying, of same kind	Alternating	Shock
Grey cast iron	4	6	10	15
Malleable cast iron	4	6	8	12
Carbon steel	4	6	8	12
Brittle alloys	5	6	10	15
Soft alloys	5	6	8	12
Timber	6	10	14	20
Brick	15	20	25	30
Stone	15	20	25	30

Components

Component	FS	Component	FS
Boilers	4.5–6	Gears: static load	1.25
Shafts for flywheels, armatures, etc.	7–9	fatigue load	2.0
Lathe spindles	12	Wire rope: general hoists	5–7
Shafting	24	guys	3.5
Steelwork: buildings	4	mine shafts	5–8
bridges	5	lifts	7–12
small-scale	6	Springs: small, light duty	2
Cast-iron wheels	20	small, heavy duty	3
Welds not subject to fatigue	3–6	large, light duty	3
Turbine blades and rotors	3–5	large, heavy duty	4.5
Bolts	8.5		

8.4.2 *Velocity of sound in various media*

Solid	Velocity $(\mathrm{m\,s^{-1}})$	Liquid	Velocity $(\mathrm{m\,s^{-1}})$	Gas	Velocity $(\mathrm{m\,s^{-1}})$
Aluminium	5280	Water: fresh	1430	Air	331
Copper	3580	sea	1510	Oxygen	315
Iron	3850	Alcohol	1440	Hydrogen	1263
Steel	5050	Mercury	1460	Carbon monoxide	336
Lead	1200			Carbon dioxide	258
Glass	45–5600				
Rubber	30				
Wood	4–5000				

8.4.3 *Loudness of sounds*

Source	Intensity (db)	Source	Intensity (db)
Threshold of hearing	0	Loud conversation	70
Virtual silence	10	Door slamming	80
Quiet room	20	Riveting gun	90
Average home	30	Loud motor horn	100
Motor car	40	Thunder	110
Ordinary conversation	50	Aero engine	120
Street traffic	60	Threshold of pain	130

8.4.4 *Greek alphabet*

Upper case	Lower case	Name	Upper case	Lower case	Name	Upper case	Lower case	Name
A	α	alpha	I	ι	iota	P	ρ	rho
B	β	beta	K	κ	kappa	Σ	σ	sigma
Γ	γ	gamma	Λ	λ	lambda	T	τ	tau
Δ	δ	delta	M	μ	mu	Y	u	upsilon
E	ε	epsilon	N	ν	nu	Φ	ϕ	phi
Z	ζ	zeta	Ξ	ξ	xi	X	χ	chi
H	η	eta	O	o	omicron	Ψ	ψ	psi
Θ	θ	theta	Π	π	pi	Ω	ω	omega

Glossary of terms

abrasion The process of rubbing, grinding or wearing away by friction using an abrasive such as emery, corundum, diamond, etc.

absolute pressure Pressure measured from absolute zero pressure as opposed to 'gauge pressure'.

absolute temperature Temperature measured with respect to 'absolute zero temperature', units are 'kelvin' (symbol K). $K = °C + 273.15$.

acceleration The rate of change of velocity with respect to time, (d^2x/dt^2) or $\ddot{x}$ metres per second per second $(m\, s^{-2})$.

a.c. machines machines producing or using alternating current, e.g. alternator and a.c. generator. a.c. motors

Addendum The radial distance between the pitch circle and the major diameter of a gear.

adhesive Substances used for joining materials, usually without the necessity for heat, based on natural substances (animal bone, casein, rubber, etc.) or synthetic resins.

adiabatic process A thermodynamic process in which there is no transfer of heat between the working substance and the surroundings.

aerofoil A body shaped so as to produce an appreciable 'lift', i.e. a force normal to the direction of fluid flow relative to the body, and a small 'drag' force in the same direction as the flow. Aerofoil sections are used for turbine blades, wing sections, etc.

air–fuel ratio The ratio of the mass of air to mass of fuel entering an internal combustion engine, gas turbine or boiler furnace.

air motor A motor which converts the energy of compressed air into mechanical energy, usually as a rotation. The main types are axial or radial piston, and vane.

alloy A substance with metallic properties composed of two or more chemical elements, at least one of which is a metal.

alloy steel Steel containing significant quantities of alloying elements other than carbon and commonly accepted amounts of manganese, sulphur, silicon and phosphorus, added to change the mechanical and physical properties.

alternating current Abbreviation a.c. Electric current whose flow changes direction cyclicly. The normal waveform is sinusoidal.

alternator A type of a.c. generator driven at constant speed to generate the desired frequency.

anemometer A mechanical or electrical instrument for measuring the velocity of a fluid stream, particularly wind velocity. The main types are, cup, vane and hot wire.

aneroid barometer A barometer with a partially evacuated bellows chamber connected to a pointer with a pen recording atmospheric pressure on a drum chart. The bellows responds to atmospheric pressure.

angle gauges Sets of metal blocks with two opposite faces at various angles to one another, used separately or jointly to measure angles to a high degree of accuracy.

angular acceleration The rate of change of angular velocity expressed in radians per second squared: $d^2\theta/dt^2$ or $\ddot{\theta}$ $(rad\, s^{-2})$.

angular momentum The product $I\omega$ of the moment of inertia, I and the angular velocity ω of a body moving in a curve, e.g. a flywheel.

angular velocity The rate of change of angular displacement with respect to time, expressed in radians per second, $d\theta/dt$ or $\dot{\theta}$ $(rad\, s^{-1})$.

annealing Heating a metal to, and holding at, a suitable temperature and cooling at a suitable rate so as to reduce hardness, improve machineability, ease cold working, etc.

Archimedes principle States that a body wholly or partially submerged suffers an apparent loss of weight equal to the weight of fluid displaced.

arc welding A process for joining metals by fusion in which heat is produced by an electric arc.

arithmetic mean The sum of n numbers divided by n.

arithmetic progression A series of numbers where each number is obtained by adding a fixed quantity to the previous number.

atomic weight   Relative atomic mass where one unit is 1.660×10^{-27} kg.

axial flow machines   Pumps, fans, compressors, turbines, etc., in which the fluid flows generally parallel to the axis of rotation.

balancing   Measuring the static or dynamic out-of-balance forces in a rotating part and adding or subtracting mass to cancel them out.

barometer   Instrument for measuring atmospheric pressure, the main types being the aneroid and Fortin barometers.

beams   Bars, rods, etc., of metal or other material carrying transverse loads with various types of support, e.g. simple supports, built-in ends, continuous supports.

bearing   A fixed support for a rotating shaft or sliding part with minimum friction.

belt drive   The transmission of power from one shaft to another by means of an endless belt which may be flat or of vee section, etc.

bending moment   The algebraic sum of the moments of all the forces to either side of a transverse section of a beam, etc.

bending modulus   A property of a section equal to the bending moment divided by the maximum bending stress.

bend loss   The loss of pressure in a fluid flowing around a bend in a pipe or duct.

Bernoulli equation   States that in a pipe or duct in which a fluid flows, the sum of the pressure, potential and kinetic energies is equal at any point.

bevel gear   A toothed wheel with teeth formed on a conical surface used for transmitting rotation from a shaft to one at an angle to it in the same plane, usually at right angles.

binary numbers   A scale of numbers with 'radix' equal to 2 as opposed to the usual scale radix of 10 (decimal numbers). Only two symbols are used: 0 and 1.

binomial coefficients   Coefficients of terms of the expansion of $(1+x)^n$ using the binomial theorem.

binomial distribution   A distribution used in statistics based on the binomial theorem which gives the probability of an event taking place.

black body   In the study of radiation of heat, a body which completely absorbs heat or light falling on it.

black-body radiation   The quantity or quality of radiation from a black body, e.g. from the inside of a cavity.

blade   A curved plate often of aerofoil section used to deflect a fluid flow, e.g. airscrew or propeller blade, turbine blade, impeller vane.

blank   A piece of sheet metal cut to a suitable shape to be subject to further pressing processes. A pressed sintered component requiring further machining, etc.

blower   A rotating, usually air, compressor for supplying relatively large flows at a low pressure.

boiling point   The temperature at which a liquid boils at standard atmospheric pressure of 101.325 kN m^{-2}.

bolt   A cylindrical partly screwed bar with a (usually) hexagonal head used in conjunction with a 'nut' to fasten two or more parts together.

bore   Hole or cavity produced by a single- or multi-point tool, usually cylindrical.

boundary layer   A thin layer of fluid adjacent to a surface over which the fluid flows, which exerts a viscous drag on the surface due to the large velocity gradient.

boundary lubrication   A state of partial lubrication in a plain bearing where there is no oil film, only an adsorbed monomolecular layer of lubricant in the surfaces.

Bourdon tube pressure gauge   A gauge in which fluid pressure tends to straighten a curved, flattened tube connected to a pointer mechanism; pressure is read from a circular scale. A differential form is available having two tubes connected to a single pointer.

Boyle's law   States that, for a 'perfect gas' the volume of a given mass varies inversely as the pressure at constant temperature.

brake   A device for applying resistance to the motion of a body, either to retard it or to absorb power (dynamometer).

brazing   The joining of metals by a thin capillary layer of non-ferrous metal filler in the space between them. Carried out above about 800 °C.

brittle fracture   Fracture of a material with little or no plastic deformation.

broaching   The cutting of holes of various shapes or cutting of an outside surface, with a 'broach' consisting of a tapered bar with cutting edges. The broach moves in a reciprocating axial manner.

buckling   Sudden large-scale deformation of a strut, thin cylinder, etc., due to instability when loaded, e.g. an axial load on a strut.

bulk modulus   The ratio of pressure (three-dimensional stress) to volumetric strain of a material.

buoyancy   The apparent loss of weight experienced by a submerged or floating body due to the upthrust caused by fluid pressure.

butt welding The welding together of abutting members lying in the same plane.

cam A sliding mechanical device used to convert rotary to linear (usually) motion, and vice versa.

capacitance The 'charge' on a conducting body divided by its 'potential'. Unit the 'farad'.

capacitor An electrical component having capacitance usually consisting of two conducting surfaces of large area separated by a very thin (usually) dielectric.

carbide tools High-speed machine tools of tungsten, titanium or tantalum carbide, or combinations of these in a matrix of cobalt or nickel.

carbon steel Steel containing carbon up to about 2% and only residual quantities of other elements, except for small amounts of silicon and manganese.

carburizing Introducing carbon into solid ferrous alloys by heating in the presence of a carbonaceous material.

Carnot cycle An ideal heat engine cycle having the maximum thermal efficiency, called the 'Carnot efficiency'.

case hardening The production of a hard surface on steel by heating in a carbonaceous medium to increase the carbon content, and then quenching.

casting An object at or near-finished shape obtained by the solidification of a molten substance in a 'mould'. The name of the process.

cast iron Iron containing carbon suitable for casting, e.g. grey, white, malleable, nodular.

cavitation The formation and sudden collapse of bubbles in a liquid due to local reduction in pressure. Cavitation erosion may be caused on local metal surfaces.

centre drilling Drilling of a conical hole in the end of a workpiece to support it while being rotated. A 'centre drill' is used.

centreless grinding The grinding of cylindrical or conical surfaces on workpieces running in rollers instead of centres.

centre of buoyancy The 'centroid' of the immersed portion of a floating body.

centre of gravity (centre of mass) The imaginary point in a body at which the mass may be assumed to be concentrated.

centre of percussion The point on a compound pendulum whose distance from the centre of oscillation is the same as the length of a simple pendulum with the same periodic time.

centre of pressure The point on a submerged surface at which the resultant pressure may be taken to act.

centrifugal casting A casting made by pouring molten material into a rotating mould. This improves the quality of the casting.

centrifugal compressor A machine similar to the centrifugal pump used for increasing the pressure of gases such as air. It may have several stages.

centrifugal force A body constrained to move in a curved path reacts with a force (centrifugal force) directed away from the centre of curvature. It is equal and opposite to the force deviating the body from a straight line called the 'centripetal force'. Both are equal to the mass multiplied by the 'centripetal acceleration'.

centrifugal pump A pump, usually for liquids, which has a rotating 'impeller' which increases the pressure and kinetic energy of the fluid.

centripetal force See 'centrifugal force'.

centroid The centre of gravity of a lamina. Centre of area.

ceramics Non-organic, non-metallic materials of brittle nature, e.g. alumina, carbides.

cermet A body of ceramic particles bonded with a metal.

chain drive A device consisting of an endless chain (usually a 'roller chain') connecting two wheels (sprockets) on parallel shafts.

chamfer A corner bevelled to eliminate a sharp edge.

charge A quantity of unbalanced electricity in a body, i.e. an excess or deficiency of electrons.

Charles' law States that for a 'perfect gas' at constant pressure the volume increases by 1/273 of its volume at 0 °C for each degree celsius rise in temperature.

chip A piece of metal removed by a cutting tool or abrasive.

chip breaker A groove in a cutting tool used to break continuous chips for safety and handling reasons.

chuck A device for holding work or tools during machining operations.

clearance The gap or space between two mating components.

closed cycle gas turbine A gas turbine unit in which the working fluid continuously circulates without replenishment.

clutch A device used to connect or disconnect two rotating shafts, etc., either while rotating or at rest.

cold working Plastic deformation of metal below the recrystallization temperature.

column A vertical member with a compressive load; a strut.

combined stress A state of stress combining tensile (or compressive), shear, and bending stresses.

combustion equations Chemical equations used in the study of combustion of fuels for engines, boilers, etc.

combustion products Chemical products resulting from the combustion of fuels in air.

complex number A number of the form $(a+ib)$ having a 'real' part a and an imaginary part ib where $i=\sqrt{-1}$. The symbol j is also used.

composite A material consisting of a mixture of two or more materials, e.g. glass or carbon fibres in a plastic matrix.

compressibility The reciprocal of 'bulk modulus'.

compression ignition engine An engine in which ignition takes place as the result of temperature rise in the air/fuel mixture due to compression.

compression ratio In an internal combustion engine, the ratio of the total volume in a cylinder at outer dead centre to the clearance volume. In powder metallurgy, the ratio of the volume of loose powder to the volume of the 'compact' made from it.

compressive strength The maximum compressive stress a material will withstand, based on the original cross-sectional area.

compressive stress Compressive force divided by area of cross-section.

compressor A rotary or reciprocating machine which compresses air or other gases.

condenser A heat exchanger in which a vapour, e.g. steam, is condensed, usually by water flowing in tubes over which the vapour passes.

conductance The property of a substance which makes it conduct electricity. The unit is the 'siemens' (symbol G). The reciprocal of resistance.

conduction of heat Heat transferred from one part of a medium to another without motion, the heat being passed from one molecule to another.

conductivity (electrical) Conduction (reciprocal of resistance) between opposite faces of a 1 m cube at a specified temperature. The unit is the 'ohm metre' (symbol Ω-m).

conductivity (thermal) A measure of the rate at which heat flows through a wall by conduction. The unit is watt per metre per kelvin ($W\,m^{-1}\,K^{-1}$).

conservation of angular momentum In a closed system the sum of the angular momenta $\Sigma I\omega$ is a constant, where I = moment of inertia, ω = angular velocity.

conservation of energy The energy in a closed system cannot be changed but only interchanged, e.g. potential to kinetic energy.

conservation of matter Matter is neither created nor destroyed during any physical or chemical change.

conservation of momentum In a closed system the sum of the momenta Σmv, is constant, where: m = mass, v = velocity.

constant-pressure cycle (Diesel cycle) An ideal engine cycle in which combustion is assumed to take place at constant pressure.

constant volume cycle (Otto cycle) An ideal cycle in which combustion is assumed to take place at constant volume. The basis for the petrol engine cycle.

contact stresses The localized stress between contacting curved surfaces and between a curved and a flat surface, such as occurs in ball and roller bearings.

continuous beam A beam supported on three or more supports.

continuous casting A process in which an ingot, billet or tube is produced continuously.

convection of heat The transfer of heat from one part of a fluid to another due to 'convection currents' often due to gravity (natural convection) or by induced flow (forced convection).

convergent–divergent nozzle A nozzle for fluid flow which decreases in area to a throat and then increases in area to the exit; the flow may be supersonic at outlet.

convergent nozzle A nozzle for fluid flow which decreases in area to a 'throat' at outlet.

core A formed object inserted into a mould to shape an internal cavity.

core box In casting, a box in which cores are formed in sand, etc.

corrosion The deterioration of a metal by chemical or electrochemical reaction with its environment.

cosine rule A mathematical rule for solving triangles: $a^2=b^2+c^2-2bc\cos A$, where a, b, c = lengths of the sides, A = angle opposite side a.

counterboring Drilling or boring a flat-bottomed hole, often concentric with other holes.

counterflow heat exchanger A heat exchanger in which the two fluids flow in opposite directions.

countersinking Forming a conical depression at the entrance to a hole for deburring, and for countersunk screw heads.

couple Two equal and opposite forces parallel to one another. The distance between them is the 'arm'. Its magnitude is the product of one force and the arm.

crank An arm on a shaft with a pin used to produce reciprocating motion with a connecting rod.

crankshaft A shaft carrying several cranks, usually at different angular positions, to which connecting rods are fitted in an engine, reciprocating pump, etc.

creep Slow plastic deformation of metals under stress, particularly at high temperatures.

creep resistance Resistance of metals to creep.

critical speed A rotational speed corresponding to a natural frequency of transverse vibrations of the member. Also called 'whirling speed'.

crossflow heat exchanger A heat exchanger in which the two fluids flow at right angles to one another.

cutting fluid A fluid used in metal cutting to improve finish, tool life, and accuracy. It acts as a chip remover and a coolant.

cutting speed The linear or peripheral speed of relative motion between a cutting tool and workpiece in the principal direction of cutting.

cyaniding The introduction of carbon and nitrogen into a solid ferrous alloy by holding it at a suitable high temperature in contact with molten cyanide.

cycloidal gears Gears with teeth whose flank profile consists of a cycloidal curve.

cylindrical grinding Grinding the outer cylindrical surfaces of a rotating part.

damped vibration Vibrations reduced in amplitude due to energy dissipation.

damping The reduction in amplitude of vibrations due to mechanical friction in a mechanical system or by electrical resistance in an electrical one.

deceleration Negative acceleration. The rate of diminution of velocity with time. The unit is metres per second per second ($m\,s^{-2}$).

dedendum The radial distance between pitch circle and the bottom of a gear tooth.

deflection The amount of bending, compression, tension, or twisting of a part subject to load.

density The mass of a unit volume of a substance. The unit is kilograms per metre cubed ($kg\,m^{-3}$).

depth of cut The thickness of material removed from a workpiece in a machine tool during one pass.

dial gauge A sensitive mechanical instrument in which a small displacement, e.g. 0.01 mm, is indicated on a dial.

diametral clearance The difference in diameter between a shaft and the hole into which it fits or runs, e.g. in plain journal bearings.

diamond dust The hardest substance used for abrasive wheels.

diamond pyramid hardness An indentation hardness test for materials using a 136° diamond pyramidal indenter and various loads.

diamond tool A diamond shaped to the contour of a single-point cutting tool for precision machining of non-ferrous metals and plastics.

diamond wheel A grinding wheel with crushed diamonds embedded in resin or metal.

die A tool used to impart shape in many processes, e.g. blanking, cutting, drawing, forging, punching, etc.

die casting A casting made in a die. A process where molten metal is forced by high pressure into a metal mould.

differential pressure gauge A gauge which measures the difference between two pressures, e.g. across an orifice in fluid flow.

diode Thermionic or semiconductor device with unidirectional properties used as a rectifier.

direct current (d.c.) An electric current which flows in one direction only.

direct current machines Generators or motors operating on d.c.

discharge coefficient The rate of actual to theoretical flow of a fluid through an orifice, nozzle, Venturi meter, etc.

disk stresses Radial and hoop stresses in a rotating disk.

dowel A pin located in mating holes in two or more parts used to locate them relative to one another.

draft tube Discharge pipe at a water turbine outlet which reduces the water velocity and improves efficiency.

drag The resistance to motion of a body moving through a fluid.

drag coefficient A non-dimensional quantity relating drag to projected area, velocity and fluid density.

drawing Forming recessed parts by the plastic flow of metal in dies. Reducing the diameter or wire by pulling through dies of decreasing diameter.

drill A rotating end cutting tool with one or more cutting lips used for the production of holes.

drop forging A forging made using a 'drop hammer'.

dry flue gas Gaseous products of combustion excluding water vapour.

dryness fraction The proportion by mass of dry steam in a mixture of steam and water, i.e. in 'wet steam'.

ductility The ability of a material to deform plastically without fracture.

Dunkerley's method A method for determining the natural frequency of transverse vibrations of a shaft or its whirling speed when carrying several masses.

dynamic balancing The technique of eliminating the centrifugal forces in a rotor in order to eliminate vibration.

dynamic pressure Pressure in a moving fluid resulting from its instantaneous arrest equal to $\rho v^2/2$, where $\rho =$ fluid density, $V =$ velocity.

dynamics A study of the way in which forces produce motion.

dynamic viscosity (coefficient of viscosity, absolute viscosity) In a fluid the ratio of shear stress to velocity gradient. Units are newton seconds per square metre ($N\text{-}s\,m^{-2}$).

dynamo An electromagnetic machine which converts mechanical to electrical energy.

dynamometer A device for measuring the power output from a prime mover or electric motor.

effectiveness of a heat exchanger The ratio of the 'heat received by the cold fluid' to the 'maximum possible heat available in the hot fluid'.

efficiency A non-dimensional measure of the perfection of a piece of equipment, e.g. for an engine, the ratio of power produced to the energy rate of the fuel consumed, expressed as a fraction or as a percentage.

elastic constants The moduli of elasticity for direct stress, shear stress and hydrostatic stress and also Poisson's ratio.

elastic deformation Change of dimensions in a material due to stress in the elastic range.

elasticity The property of a material by virtue of which it recovers its original size and shape after deformation.

elastic limit The greatest stress that can be applied to a material without permanent deformation.

electrical resistance The real part of impedance which involves dissipation of energy. The ratio of voltage drop to current in a conductor.

electrical discharge machining (EDM) Machining process in which metal is removed by erosion due to an electric spark in a dielectric fluid using a shaped electrode.

electric potential Potential measured by the energy of a unit positive charge at a point expressed relative to zero potential.

electric strength The maximum voltage that can be applied to a piece of insulation before breakdown occurs.

electrochemical corrosion Corrosion due to the flow of current between anodic and cathodic areas on metal surfaces.

electrochemical machining (ECM) The removal of metal by electrolytic action, masks being used to

obtain the required shape. The process is the reverse of electroplating.

elongation In tensile testing the increase in length of a specimen at fracture as a percentage of the original length.

emissivity Ratio of the emissive power of a surface to that of a 'black body' at the same temperature and with the same surroundings.

end milling Machining with a rotating peripheral and end cutting tool (see face milling).

endurance limit Same as 'fatigue limit'.

energy The capacity of a body for doing work. Types are: kinetic, potential, pressure, chemical, electric, etc.

energy fluctuation coefficient The ratio of the variation in kinetic energy in a flywheel due to speed fluctuation, to the average energy stored.

enthalpy Thermodynamic property of a working substance equal to the sum of its 'internal energy' and the 'flow work' (pressure multiplied by volume). Used in the study of 'flow processes'.

enthalpy–entropy diagram (h–s or Mollier chart) A diagram used for substances on which heat and work are represented by the length of a line. Used extensively for calculations on steam cycles and refrigeration.

entropy In thermodynamics, entropy is concerned with the probability of a given distribution of momentum among molecules. In a free system entropy will tend to increase and the available energy decrease. If, in a substance undergoing a reversible change, a quantity of heat dQ at temperature T is taken in, then its entropy S is increased by an amount dQ/T. Thus the area under a curve on a T–S graph represents the heat transferred. Units: joules per kelvin ($J\,K^{-1}$).

epicyclic gear A system of gears in which one or more wheels travel round the outside or inside of another wheel the axis of which is fixed.

equilibrium The state of a body at rest or in uniform motion. A body on which the resultant force is zero.

erosion The destruction of metals, etc., by abrasive action of fluids usually accelerated by the presence of solids.

Euler strut formula A theoretical formula for determining the collapsing load for a strut.

excess air The proportion of air used in excess of the theoretical quantity for complete combustion of a fuel.

expansion The increase in volume of a working fluid, e.g. in a cylinder with moving piston. The opposite is 'compression'. In mathematics the expression of a function as an infinite series of terms.

expansion coefficient (coefficient of expansion) The

expansion per unit length, area, or volume, per unit increase in temperature.

explosive forming Shaping metal parts confined in dies using the pressure from an explosive charge.

extensometer A sensitive instrument for measuring the change in the length of a stressed body.

extrusion The conversion of a 'billet' of metal into lengths of uniform cross-section by forcing it through a die, usually when heated.

face mill A rotating milling cutter with cutting edges on the face to mill a surface perpendicular to the cutting axis.

facing Generating a flat surface on a rotating workpiece by traversing a tool perpendicular to the axis of rotation.

factor of safety The ratio between ultimate (or yield) stress for a material and the permissible stress. (Abbreviation FS or FOS).

failure The breakdown of a member due to excessive load. Several 'theories of failure' are used.

fan A device for delivering or exhausting large quantities of air or other gas at low pressure. It consists basically of a rotating axial or centrifugal impeller running in a casing.

fatigue Phenomenon leading to the failure of a part under repeated or fluctuating stress below the tensile strength of the material.

fatigue life The number of cycles of fluctuating stress required to produce failure in a fatigue test.

fatigue limit (endurance limit) The maximum stress below which a material can endure an infinite number of stress fluctuation cycles. This only applies to a specially made specimen with a high degree of surface finish.

feed The rate of advance of a cutting tool along the surface of the workpiece.

fibres In 'composites', fine threads of a long length of glass, carbon, metal, etc., used to reinforce a material (e.g. plastics, metals), known as the 'matrix'.

filler metal Metal added in soldering, brazing and welding processes, usually in the form of a rod or stick.

fillet weld A weld of approximately triangular section joining two surfaces usually at right angles to one another in a lap, T or corner joint.

film lubrication Lubrication where the shaft is separated from the bearing by a thin film of lubricant which is under pressure and supports the load.

fin One of usually a number of thin projections integral with a body (e.g. engine cylinder block, gearbox, cooler) which increase the cooling area.

finish The surface condition, quality and appearance of a metal, etc., surface.

finish machining The final machining of a component where the objectives are surface finish and accuracy of dimension.

fit The clearance or interference between mating parts. Also the term for a range of clearance suggested by standards such as British Standards.

fitting loss The pressure or head loss incurred by fittings in a pipe or duct such as valves, bends, branch, etc.

flame cutting The cutting of metal plate to a desired shape by melting with an oxygen–gas flame.

flame hardening Quench hardening where the heat is supplied by a flame.

flange A projecting annular rim around the end of a cylinder or shaft used for strengthening, fastening or locating.

flat-plate theory A study of the stresses and deflection of loaded flat plates. It is assumed that the plate is relatively thin and the deflections small.

flexible coupling A coupling usually joining rotating shafts to accommodate lateral or angular misalignment.

flowmeter An instrument for measuring the volumetric or mass flow of a fluid.

flow rate The rate of flow of a fluid. Units: cubic metres per second ($m^3 s^{-1}$) or kilograms per second ($kg s^{-1}$).

flux Material used in soldering, brazing and welding to prevent the formation of, dissolve, or facilitate the removal of, oxides, etc.

flywheel A heavy wheel on a shaft used either to reduce speed fluctuation due to uneven torque, or to store energy for punching, shearing, forming, etc.

force That quantity which produces acceleration in a body measured by the rate of change of momentum. Unit: newton (N).

forging Plastic deformation of metal, usually hot, into the desired shape using a compressive force with or without dies.

form cutter A cutter profile sharpened to produce a specified form of work.

four-stroke cycle An engine cycle of 4 strokes (2 revolutions) consisting of induction, compression, expansion (power) and exhaust strokes; e.g. in the Otto and Diesel cycles.

Francis turbine A reaction water turbine in which

water flows radially inwards through guide vanes and a runner which it leaves axially.

frequency The rate of repetition of a periodic disturbance. Units: hertz (Hz) or cycles per second. Also called 'periodicity'.

fretting corrosion Surface damage between surfaces in contact under pressure due to slight relative motion, especially in a corrosive environment.

friction The resistance to motion which takes place when attempting to move one surface over another with contact pressure.

friction coefficient The ratio of the friction force to the normal force at the point of slipping. The 'static coefficient of friction' is the value just before slipping takes place, the 'dynamic coefficient of friction' being the value just after.

friction factor in pipes A dimensionless quantity from which the pressure loss due to pipe-wall friction can be calculated. It is usually plotted against the Reynold's number for various degrees of relative pipe roughness.

friction laws These state that the coefficient of friction is independent of surface area of contact and pressure between surfaces. These laws are not strictly true.

Froude number A dimensionless number used in the study of the motion of ships through water. It is the ratio of velocity to the square root of the product of length and acceleration due to gravity, $\dfrac{V}{\sqrt{Lg}}$.

gas constant For a 'perfect gas', gas constant $R = pV/mT$, where p = pressure, V = volume, m = mass, T = temperature.

gasket A layer of usually soft material between two mating surfaces which prevents leakage of fluids.

gas processes Changes in the properties of a substance, e.g. isothermal, isentropic, constant volume, etc.

gas refrigeration cycle A cycle using a reversed constant pressure cycle in which the working substance is always a gas.

gas shielded arc welding Arc welding with a shield of inert gas, e.g. argon, helium, to prevent oxidation.

gas turbine set A prime mover consisting of one or more axial or centrifugal compressors, combustion chamber(s) (or gas heater), and one or more axial or radial flow turbines. The compressor(s) are driven by one turbine and a turbine delivers useful power.

Additional components are intercoolers between compressors, reheat between turbines and a heat exchanger.

gas welding Welding using the heat of an oxygen–gas flame.

gauge blocks (slip gauges) Accurate rectangular hard steel blocks used singly or in combination with others, the distance between them forming a gauging length.

gear ratio The speed ratio for a pair or train of gears determined by the number of teeth on each gear.

gear wheel A toothed rotating wheel used in conjunction with another wheel of the same or different diameter, to transmit motion to another shaft. The main types are spur, bevel, worm and epicyclic.

geometric factor A factor dependent on the shapes of bodies between which heat or light is radiated. This factor affects the heat-transfer coefficient.

geometric progression A series of numbers in which each number is derived by multiplying the previous number by a constant multiplier called the 'ratio'.

governor A speed regulator on variable-speed electric motors and prime movers, etc.

gravitation The attractive force between two masses. The force is proportional to the product of the masses and inversely proportional to the square of the distance between their centres of mass.

gravitational constant The gravitational force between two masses m_1 and m_2, their centres of mass a distance d apart, is given by $F = Gm_1m_2/d^2$ where G = gravitation constant = $6.67 \times 10^{-11}\,\mathrm{N\,m^2kg^{-2}}$.

grinding The removal of metal, etc., using an abrasive 'grinding wheel'.

hardness The resistance of metals to plastic deformation, usually by indentation. Measured by tests such as Brinell, Rockwell, and Vickers pyramid.

head The height of a liquid above a datum in a gravity field.

heat engine A system operating on a complete cycle developing net work from a supply of heat.

heat flow rate Heat flow per unit time in a process. Unit: watt (W).

heat transfer The study of heat flow by conduction, convection and radiation.

heat transfer coefficient A coefficient h relating, heat flow q, area of flow path A and temperature difference ΔT for heat transfer between two phases: $q = hA\Delta T$.

heat treatment Heating and cooling of solid metals to obtain the desired properties.

helical gear A gear in which the teeth are not parallel to the axis but on a helix.

helix A line, thread or wire curved into a shape it would assume if wrapped around a cylinder with even spacing.

helix angle In screw threads, etc., the angle of the helix to a plane at right angles to the axis.

honing The removal of metal, usually from a cylinder bore, by means of abrasive sticks on a rotating holder.

Hooke's law States that stress is proportional to strain up to the limit of proportionality.

hoop stress The circumferential stress in a cylinder wall under pressure or in a rotating wheel.

hot forming Forming operations such as bending, drawing, forging, pressing, etc., performed above the recrystallization temperature of a metal.

hot wire anemometer An instrument for measuring the flow of air (or other fluids) from the cooling effect on an electrically heated sensor, in the fluid stream, the resistance of which changes with temperature.

hydraulic cylinder A cylinder with piston and piston rod supplied by a liquid under pressure to provide a force with linear motion. The cylinder may be single or double acting.

hydraulic jack A device for lifting heavy loads a short distance using a hydraulic cylinder supplied by a pump, often hand operated.

hydraulic motor A motor operated by high-pressure liquids. Types: radial piston, axial piston, vane, etc.

hydraulic press A press using a hydraulic cylinder.

hydraulic pump A machine which delivers fluids at high pressure. Types: radial piston, axial piston, reciprocating, vane, gear pump.

hydraulics The science relating to the flow of fluids.

hydrocarbon fuels Solid, liquid and gaseous fuels composed primarily of hydrogen and carbon.

hydrodynamic lubrication Thick film lubrication in which the surfaces are separated and the pressurized film supports the load.

hydrodynamics The branch of dynamics which relates to fluids in motion.

hyperbola A conic section of the form $(x^2/a^2) - (y^2/b^2) = 1$.

hyperbolic functions A set of six functions, particularly useful in electrical engineering, involving the terms e^x and e^{-x}. Analogous to the trigonometrical functions sin, cos, tan, etc., they are sinh, cosh, tanh, cosech, sech, cotanh.

illuminance The quantity of light or luminous flux on unit surface area. Unit: lux (lx) = 1 lumen per square metre (lm m^{-2}).

impact extrusion A high speed cold working process for producing tubular components by a single impact by a punch. A slug of material placed in a die flows up and around the punch into the die clearance.

impact test A test to determine the behaviour of materials subjected to high rates of loading in bending, torsion and tension. The quantity measured is the 'impact energy' required to cause breakage of a specimen.

impulse When two bodies collide the impulse of the force during impact is $\int F dt$. Defined as the change of momentum produced in either body.

impulse reaction turbine A steam turbine with impulse stage(s) followed by reaction stages.

impulse turbine A steam, gas or water turbine in which the working fluid is accelerated through nozzles and impinges on blades or buckets in which there is no pressure drop.

inclined plane For a smooth plane at an angle θ to the horizontal, the force parallel to the plane to move a mass m up it is $mg \sin \theta$. It is equivalent to a 'machine' having a velocity ratio of $\cot \theta$.

inductance The property of an electric circuit carrying a current is characterized by the formation of a magnetic field and the storage of magnetic energy. Unit: henry (H).

induction hardening The use of induction heating for hardening metals.

induction heating The heating of conducting materials by inducing electric currents in the material, usually by a high-frequency source.

induction motor An a.c. motor in which the primary winding current sets up a magnetic flux which induces a current in the secondary winding, usually the rotor.

inductor An electric-circuit component which has the property of inductance. Usually a coil with air or magnetic core.

inertia The property of a body proportional to mass, but independant of gravity. Inertia opposes the state of motion of a body.

insulation *1. Heat* Material of low thermal conductivity used to limit heat gain or loss, e.g. pipe lagging. *2. Electricity* A material with very high resistivity through which there is virtually no flow of current, e.g. plastic covering on wires.

interchange factor When two bodies are involved in the interchange of heat radiation, the radiation depends upon the emissivities of both bodies. Interchange factor is a function of the emissivities which allows for this.

intercooler A cooler, usually using water, interposed between air compressor stages.

internal combustion engine (I.C. engine) An engine in which combustion takes place within a chamber, e.g. a cylinder, and the products of combustion form the working fluid, e.g. petrol engine, diesel engine, gas engine.

internal energy The difference between the heat energy supplied to a system and the work taken out. The energy is in the form of heat as measured by the temperature of the substance or its change of state.

inverse square law The intensity of a field of radiation (light, heat, radio waves) is inversely proportional to the square of the distance from the source.

investment casting Casting of metal in a mould produced by coating an expendable pattern made of wax, plastic, etc., which is removed by heating. Also 'lost wax process'.

involute gear teeth Gear-wheel teeth the flank profile of which consists of an involute curve. The commonest form of gear teeth.

isenthalpic process A process taking place at constant enthalpy, e.g. a 'throttling' process.

isentropic efficiency Defined as the actual work from the expansion of a gas, vapour, etc., divided by the work done in an isentropic expansion.

isentropic expansion The expansion of a fluid at constant entropy.

isentropic process A thermodynamic process taking place at constant entropy.

isobaric process A thermodynamic process taking place at constant pressure.

isothermal process A constant-temperature process.

Izod test A pendulum type of single blow impact test using notched test pieces.

jet A fluid stream issuing from an orifice, nozzle, etc.

jet engine An engine incorporating rotary compressor and turbine which produces a high-velocity jet for the propulsion of aircraft.

jet propulsion The propulsion of vehicles, e.g. boat, aircraft, by means of a fluid jet.

jig A device to hold a workpiece and guide a tool in cutting operations.

jig boring Boring carried out on a 'jig borer' on which the positions of holes can be positioned to a high degree of accuracy.

journal The portion of a rotating shaft which is supported in a bearing.

journal bearing A bearing which supports a journal.

Kaplan turbine A propeller water turbine with adjustable runner blades which are altered to suit the load.

key A piece of material inserted between usually a shaft and a hub to prevent relative rotation and fitting into a 'keyway'.

K factor A factor giving the proportion of, or number of, velocity head(s) lost in a pipe or in pipe fittings.

kinematic viscosity The coefficient of viscosity divided by the fluid density.

kinetic energy The energy of a body arising from its velocity. For a mass m at velocity v the kinetic energy is $\frac{1}{2}mv^2$.

labyrinth gland A gland used on steam turbines, gas turbines, etc., with radial fins on a shaft or surrounding casing, with small radial or axial clearance to limit fluid leakage.

lagging Thermal insulation on the surface of a pipe, tank, etc.

laminar flow (viscous flow) Fluid flow in which adjacent layers do not mix. It occurs at relatively low velocity and high viscosity.

lapping The finishing of spindles, bores, etc., to fine limits using a 'lap' of lead, brass, etc., in conjunction with an abrasive.

latent heat The heat required to change the 'state' of a substance without temperature change, e.g. solid to liquid, liquid to gas. The latent heat per unit mass is the 'specific latent heat'.

lathe A versatile machine tool for producing cylindrical work by turning, facing, boring, screw cutting, etc., using (usually) a single-point tool.

lead The axial advance of a helix in one revolution, e.g. in screw thread or worm.

lift The component of force on a body in a fluid stream which is at right angles to the direction of flow. The force which supports the weight of an aircraft.

lift coefficient A non-dimensional quantity relating lift to the velocity and density of the fluid and the size of the body.

limit The maximum or minimum size of a component as determined by a specified tolerance.

linear bearing A bearing in which the relative motion is linear, as opposed to rotary.

lock nut An auxiliary nut used in conjunction with a normal nut to lock the latter.

lock washer A name for many types of washer used with nuts, etc., to prevent loosening.

logarithmic mean temperature difference In heat exchangers the 'effective' difference in temperature of the fluids used in calculating heat transfer.

logarithms The logarithm of a number N to a base b is the power to which the base must be raised to produce that number. This is written $\log_b N$ or $\log N$ if the base is implied. Common logarithms are to the base 10. Natural logarithms (Naperian logarithms) are to the base e ($e = 2.7183\ldots$).

lubricant Any substance, solid, liquid or gaseous, which may be used to reduce friction between parts.

luminous flux The flux emitted in a unit solid angle of 1 steradian by a point source of uniform intensity of 1 candela. Unit: lumen (lm).

luminous intensity Unit: candela (cd). The luminance of 'black body' radiation at the temperature of solidification of platinum (2042 K) is $60\,\text{cd}\,\text{cm}^{-2}$.

machinability The relative ease of machining a particular material.

machine In mechanics, a device which overcomes a resistance at one point known as the 'load', by the application of a force called the 'effort' at another point; e.g. inclined plane, lever, pulleys, screw.

machining Removal of metal in the form of chips, etc., from work, usually by means of a 'machine tool'.

Mach number The ratio of velocity of a fluid relative to a body and the velocity of sound in the fluid. Symbol M.

magnetism The science of magnetic fields and their effect on materials due to unbalanced spin of electrons in atoms.

malleability The property of metals and alloys by which they can easily be deformed by hammering, rolling, extruding, etc.

mandrel An accurately turned spindle on which work, already bored, is mounted for further machining.

manometer An instrument used to measure the pressure of a fluid. The simplest form is the 'U tube' containing a liquid. See: pressure, Bourdon gauge.

mass The quantity of matter in a body. Equal to the inertia or resistance to acceleration under an applied force. Unit: kilogram (kg). Symbol: m.

mass flow rate The rate at which mass passes a fixed point in a fluid stream. Unit: kilograms per second $(\text{kg}\,\text{s}^{-1})$.

matrix The material in a composite in which fibres, whiskers, etc., are embedded.

mean effective pressure (m.e.p.) The average absolute pressure during an engine cycle. It gives a measure of the work done per swept volume.

mechanical advantage In a 'machine', the ratio of load to effort.

mechanical efficiency In an engine, the ratio of useful power delivered to the 'indicated power', i.e. the efficiency regarded as a machine.

Merchant's circle A diagram showing the forces on a single-point machine tool.

metal forming The shaping of metals by processes such as bending, drawing, extrusion, pressing, etc.

micrometer gauge A hand held, U-shaped length gauge in which the gap between measuring faces is adjusted by means of an accurate screw.

mild steel Carbon steel with a maximum carbon content of about 0.25%.

milling The removal of metal by a 'milling cutter' with rotating teeth on a 'milling machine'.

mixed-flow heat exchanger A heat exchanger in which the flow of one fluid is a mixture of types, e.g. alternatively counterflow and cross-flow.

mixed-flow pump A rotodynamic pump in which the general flow is a combination of axial and radial.

mixture strength The ratio of 'stoichiometric' air/fuel ratio, to the 'actual' air/fuel ratio, used for engines. 0.8 is 'weak' and 1.2 is 'rich'.

modulus of elasticity A measure of the rigidity of a material. The ratio of stress to strain in the elastic region.

modulus of section A property of plane sections used in bending-stress calculations. It is equal to the ratio of bending moment to maximum bending stress.

molecular weight The mass of a molecule referred to that of a carbon atom (12.000). The sum of the relative atomic masses in a molecule.

Mollier diagram See: enthalpy–entropy diagram.

moment The moment of a force (or other vector quantity) about a point is the product of the force and the perpendicular distance from the line of action of the force to the point.

moment of inertia The moment of inertia of a body of

mass m about a point P is equal to mk^2 where k is the 'radius of gyration' from P at which the whole mass may be assumed to be concentrated as a ring.

momentum   The product of mass and velocity of a body, i.e. mv.

multi-pass heat exchanger   A heat exchanger in which one of the fluids makes a series of passes in alternate directions.

natural vibrations   Free vibrations in an oscillatory system.

nitriding   Introducing nitrogen into solid ferrous alloys by heating in contact with nitrogenous material, e.g. ammonia, cyanide.

non-destructive testing   Inspection by methods which do not destroy a part, to determine its suitability for use.

non-flow energy equation   The equation in thermodynamics for a non-flow process such as compressing a gas in a cylinder. It states that the change in 'internal energy' of a substance is equal to heat supplied minus the work done.

non-Newtonian fluid   A fluid which does not obey the viscosity law. See: coefficient of viscosity.

notch   A vee or rectangular cut-out in a plate restricting the flow of water in a channel. The height of water above the bottom of the cut-out gives a measure of the flow.

nozzle   A convergent or convergent–divergent tube through which a fluid flows. Used to produce a high-velocity jet.

Nusselt number   A dimensionless quantity used extensively in the study of heat transfer. Defined as $\mathrm{Nu} = Qd/k\theta$, where Q = heat flow to or from a body per unit area, θ = temperature difference between the body and its surroundings, k = thermal conductivity, d = characteristic dimension of the body.

nut   A metal (or other material) collar internally screwed to fit a bolt usually of hexagonal shape but sometimes round or square.

oil seal   A device used to prevent leakage of oil, e.g. from a bearing in a gearbox.

orifice   A small opening for the passage of a fluid. Types: rounded entry, sharp edged, re-entrant.

orifice plate   A circular plate, with a central orifice, inserted in between pipe flanges or in a tank wall to measure fluid flow from the resulting pressure drop.

O ring   A toroidal O section ring of a material such as Neoprene used as a seal.

parabola   A conic section of the form $y^2 = 4ax$.

parallel-flow heat exchanger   A heat exchanger in which the two fluids flow parallel to one another and in the same direction.

pattern   A form made in wood or other material around which a mould is made.

peak value   For a waveform the maximum value of a half-wave. For a sine wave it is $r = \sqrt{2}$ times the r.m.s. (root mean square) value.

pendulum   The 'simple pendulum' consists of a small heavy mass or 'bob' suspended from a fixed point by a string of negligible weight. Its periodic time for small oscillations is $2\pi\sqrt{L/g}$, where L = length of string, g = acceleration due to gravity. The 'compound pendulum' is any body which oscillates about a fixed point a distance h from the centre of gravity with radius of gyration k. It has an equivalent simple pendulum length of $(h^2 + k^2)/h$.

perfect gas   A gas which obeys the 'gas laws'. A gas behaves as a perfect gas as the pressure is reduced.

permanent set   Plastic deformation in a material that remains after the load is removed.

Perry–Robertson formula   A practical formula for the buckling load for a strut.

p–h chart   A pressure–enthalpy chart used for refrigeration calculations.

pH value   Negative logarithm of hydrogen ion activity denoting the degree of acidity or alkalinity of a solution. At 25 °C: 7 is neutral, a lower number indicates acidity; a higher number indicates alkalinity.

pitch   The linear distance between similar features arranged in a pattern, e.g. turns of a screw thread, distance between rivets in a row.

pitch circle   An imaginary circle on gear wheels on which the teeth are constructed, a circle on which bolt holes, etc., are pitched, etc.

plain bearing   A bearing consisting of a plain bush or sleeve, as opposed to a ball or roller bearing.

plastic deformation   Deformation that remains after a load is removed.

plasticity   The ability of a metal to deform non-elastically without rupture.

Poiseuille's equation   An expression for laminar flow of a fluid through a circular pipe.

Poisson distribution   A statistical distribution characterized by a small probability of a specific event occurring during observations over a continuous interval. A limiting form of 'binomial distribution'.

Poisson's ratio The ratio of transverse to axial strain in a body subject to axial load.

polar modulus The polar second moment of area about an axis perpendicular to the area through the centroid divided by the maximum radius.

polar second moment of area The second moment for an axis through the centroid perpendicular to the plane. It is equal to the sum of any two second moments of area about perpendicular axes in the plane.

polymer A material built up of a series of smaller units (monomers) which may be relatively simple, e.g. ethane, or complex, e.g. methylmethacrylate. The mechanical properties are determined by molecular size ranging from a few hundred to hundreds of thousands.

polynomial An algebraic expression of the form $ax^n + bx^{n-1} + cx^{n-2} \ldots px + q$.

polyphase Said of a.c. power supply circuits, usually 3 phase, carrying current of equal frequency with uniformly spaced phase differences.

polytropic process A gas process obeying the law $pv^n = $ constant, where $p = $ pressure, $v = $ volume, $n = $ index of expansion not equal to 1 or γ, the ratio of specific heat capacities.

positive displacement pump A pump which displaces a 'positive' quantity of fluid each stroke or revolution, e.g. piston pump, gear pump, vane pump.

powder metallurgy The production of shaped objects by the compressing of metal powders ranging in size from 0.1 to 1000 μm.

power The rate of doing work. Unit: watt (W).

power cycle A thermodynamic cycle in which net power is produced, e.g. Otto cycle.

power factor The ratio of total power dissipation in an electrical circuit to the total equivalent volt-amperes applied to the circuit.

press A machine tool with a fixed bed and a guided reciprocating, usually vertical, ram.

press fit An interference or force fit made through the use of a press. The process is called 'pressing'.

pressure At a point in a fluid, pressure is the force per unit area acting in all directions. That is, it is a scalar quantity; e.g. in a cylinder with a piston, pressure p is the force on the piston divided by the cylinder area.

pressure transducer A device which produces a, usually electrical, signal proportional to the pressure.

prime number A natural number other than 1 divis-

ible only by itself and 1, e.g. 2, 3, 5, 7, 11, 13, . . ., 37, . . ., 5521, etc.

principal stresses Normal stresses on three mutually perpendicular planes on which there are no shear stresses.

probability The number of ways in which an event can happen divided by the total possibilities. Symbol: p.

proof stress The stress to cause a small specified permanent set in a material.

proportional limit The maximum stress at which strain is directly proportional to stress.

pump A machine driven by a prime mover which delivers a fluid, pumping it to a greater height, increasing its pressure, or increasing its kinetic energy. Main types: rotodynamic, positive displacement.

punch A tool that forces metal into a die during blanking, coining, drawing, etc. The process is called 'punching'.

push fit A fit similar to a 'snug' or 'slip' fit defined by several classes of clearance in British and other standards.

pyrometer Device for measuring temperatures above the range of liquid thermometers.

quenching The rapid cooling of heated metal to anneal, harden, etc.

rack and pinion gear A device for changing linear to rotary motion, and vice versa, in which a circular gear, or pinion, engages with a straight toothed bar or rack.

radial clearance Half the diametral clearance. The difference between the radius of a circular hole and a rod or shaft fitting into it.

radial stress The component of stress in a radial direction in pressurized cylinders, rotating disks, etc.

radiation of heat A process by which heat is transferred without the aid of an intervening medium.

radius of gyration The imaginary radius at which the mass of a rotating body is assumed to be concentrated when determining its moment of inertia.

rake The angle of relief given to faces of a cutting tool to obtain the most efficient cutting angle.

Rankine cycle An idealized steam cycle consisting of: pumping water to boiler pressure, evaporation, adiabatic expansion to condenser pressure, and complete condensation to initial point.

Rankine efficiency The thermal efficiency of a Rankine cycle under given steam conditions.

Rankine–Gordon formula An empirical formula for the buckling load of a strut.

reaction The equal opposing force to a force applied to a system. The load on a bearing or beam support.

reaction turbine A water, steam or gas turbine in which the pressure drop is distributed between fixed and moving blades. Strictly an impulse-reaction turbine.

reamer Rotary cutter with teeth on its cylindrical surface used for enlarging a drilled hole to an accurate dimension.

recess A groove or depression in a surface.

rectifier A device for converting a.c. to d.c. by inversion or suppression of alternate half-waves, e.g. diodes, mercury arc rectifier, rotary converter.

refining The removal of impurities from a metal after crude extraction from ore.

refractory Material with very high melting point used for furnace and kiln linings.

refrigerant The working fluid in a refrigerator. It may be a gas or a vapour.

refrigerator A machine in which mechanical or heat energy is used to maintain a low temperature.

regenerative heat exchanger A heat exchanger in which hot and cold fluids, usually gases, occupy the same space alternately.

reheat The process of reheating steam or gas between turbines to obtain higher efficiency. Also the injection of fuel into the jet pipe of a turbojet to obtain greater thrust.

residual stress Stress existing in a body free from external forces or thermal gradient.

resistance In electricity, the real part of impedance of a current-carrying circuit characterized by the dissipation of heat. Unit: ohm (Ω). In physics, the opposition to motion tending to a loss of energy.

resistance thermometer A thermometer using the change of resistance with temperature of a conductor. Platinum is used, as are semiconductors (thermistor).

resistance welding and brazing A process in which the resistance of a pressurized joint causes melting of the parts in contact.

resistivity A property of electric conductors which gives resistance in terms of dimensions. Resistance $R = \rho L/A$, where ρ = resistivity, L = length, A = area of conductor.

resistor An electrical component designed to give a specified resistance in a circuit.

resistor colour code A method for marking the resistance value on resistors using coloured spots or bands.

Reynold's number A dimensionless quantity used in the study of fluid flow, particularly in a pipe. If v = velocity, d = pipe diameter, ρ = density of fluid, μ = viscosity of fluid, the Reynold's number $\mathrm{Re} = (\rho v d)/\mu$.

riveting Joining two or more members by means of rivets, the unheaded end being 'upset' after the rivet is in place.

rivets A permanent fastener for connecting plates in which the unheaded end is upset, or closed, to make the joint. There are many types, e.g. snap head, pan head, pop, explosive.

roller bearing A journal or thrust bearing with straight or tapered rollers running between two 'races'.

rolling Reducing the cross-section of metal stock or the shaping of metal products using 'rolls' in a 'rolling mill'.

rolling bearings The general name given to low-friction bearings using balls and rollers running in 'races'.

root mean square (r.m.s.) A measure of the effective mean current of an alternating current. That is, with the same heating effect as a direct current. The square root of the mean of the squares of continuous ordinates for one cycle.

Roots blower An air compressor for delivering large quantities of air at relatively low pressure. It has two hour-glass shaped intermeshing rotors running with small clearances in a casing.

rotodynamic pump See: 'pump'.

roughness In machining, surface irregularities, the dimensions and direction of which establish the surface pattern. In fluid flow, the height of irregularities in pipes, etc.

runner The rotating part of a water turbine carrying vanes.

running fit Any clearance fit in the range used for relative motion.

screw A general name for fasteners with a screwed shank and a head. Also any section of bar with an external thread.

screw jack A portable lifting machine for raising heavy objects a small height. It uses a nut which carries the load rotated, usually by hand, through a lever system.

screw thread A helical ridge of vee, square, or rounded section formed on or inside a cylinder the form and pitch being standardized under various systems.

second moment of area The second moment of area of a plane figure about any axis XX is $I_{XX} = \Sigma ar^2$, where a = an element of area, r = perpendicular distance of a from XX.

seizing The stopping of a moving part by a mating surface due to excessive friction caused by 'galling'.

sets In mathematics, any collection of 'entities' (elements) defined by specifying the elements. See: 'Venn diagram'.

shaft A circular section solid or hollow bar used for the transmission of motion and/or power.

shaft coupling A solid or flexible device for connecting, usually coaxial, shafts.

shear A force causing or tending to cause adjacent parts of a body to slide relative to one another in the direction of the force.

shearing process A machine process in which shapes are produced from plate by shearing through the material.

shear modulus (modulus of rigidity, torsional modulus) The ratio of shear stress to shear strain within the elastic limit.

shear strain and stress See: 'strain' and 'stress'.

shell moulding A mould of thermosetting resin bonded with sand formed on a heated metal pattern to give a 'shell'.

shim A thin piece of metal used between two mating surfaces to obtain a correct fit, alignment or adjustment.

shrink fit An 'interference fit' between a hub and shaft, for example obtained by heating an under-sized hub to give a clearance and allowing it to cool on the shaft. Alternatively, the shaft may be cooled, e.g. by using 'dry ice'.

silver solder A brazing alloy of low melting point containing silver.

simple harmonic motion Oscillatory motion of sinusoidal form, e.g. simple pendulum, mass and spring, electric current in a tuned circuit. It follows the law $\mathrm{d}^2x/\mathrm{d}t^2 = -\omega x^2$. Abbreviation: s.h.m.

sine bar A hardened steel bar carrying two plugs of standard diameter accurately spaced to a standard distance. Used in setting out angles to a close tolerance.

single-point tool A machine tool which has a single cutting point as opposed to a number of points, e.g. a lathe tool.

sintering The bonding of particles by heating to form shapes.

slotting Cutting a groove with a reciprocating tool in a vertical shaper, broach or grinding wheel.

S–N curve A graph of stress to cause fracture against number of stress fluctuations in fatigue tests.

soldering A similar process to brazing, but with a low-melting-point filler, e.g. alloy of lead, tin, antimony.

solenoid A current-carrying coil often with an iron core used to produce a mechanical force.

solution heat treatment Heating an alloy and allowing one or more constituents to enter into solid solution.

spark erosion machining The removal of metal by means of a high-energy spark between the workpiece and a specially shaped electrode, all immersed in a bath of electrolyte.

specific fuel consumption The mass of fuel used in an engine per unit of energy delivered. Unit: kilograms per megajoule ($\mathrm{kg\,MJ^{-1}}$).

specific heat capacity The quantity of heat required to raise the temperature of unit mass of a substance by one degree. Unit: $\mathrm{J\,kg^{-1}K^{-1}}$.

specific speed A dimensionless quantity used in the study of rotodynamic pumps and turbines. It is the same for geometrically similar machines.

specific volume The volume per unit mass of substance. Unit: cubic metres per kilogram ($\mathrm{m^3\,kg^{-1}}$).

spinning Shaping of hollow metal sheet parts by rotating and applying a force.

splines Narrow keys integral with a shaft engaging with similarly shaped grooves in a hub used instead of keys.

spot facing Machining flat circular faces for the seating of nuts, bolts, etc.

spring A device capable of elastic deflection for the purpose of storing energy, absorbing shock, maintaining a pressure, measuring a force, etc.

spring washer A name for many types of washer which deflect when compressed and prevent a nut, etc., from slackening.

stagnation temperature The temperature which would be reached by a stream of fluid if it were brought to rest adiabatically.

standard deviation The root of the average of the squares of the differences from their mean $\bar{x}$ of a number n of observations x: standard deviation $\sigma = \sqrt{[(x-\bar{x})^2]/n}$

static balancing Balancing of a rotating mass in one plane only. See: 'dynamic balancing'.

static pressure The pressure normal to the surface of a body moving through a fluid.

statics The branch of applied mathematics dealing with the combination of forces so as to produce equilibrium.

steady flow energy equation For a flow process this states that $h_1 + (C_1^2/2) + Q = h_2 + (C_2^2/2) + W$, where h_1, h_2 = inlet and outlet enthalpies, C_1, C_2 = inlet and outlet velocities, Q = heat supplied, W = work out.

steam plant A power plant operating on a steam cycle, e.g. steam power station.

steam turbine A turbine using steam as a working substance. See: 'turbines'.

steel Iron based alloy containing manganese, carbon and other alloying elements.

stiffness The ability of a metal, etc., to resist elastic deformation. It is proportional to the appropriate modulus of elasticity.

stoichiometric air/fuel ratio The mixture of air and fuel for engines and boiler furnaces which contains just sufficient oxygen for complete combustion.

strain The change in shape or size of a stressed body divided by its original shape or size, e.g. 'linear strain', 'shear strain', 'volumetric strain'.

strain energy The work done in deforming a body elastically.

strain gauge A metal grid or semiconductor rod on a backing sheet which is cemented to a strained body. The increase in length alters the electrical resistance of the grid or rod from which the strain may be deduced.

strain-gauge bridge A form of Wheatstone bridge in which strain gauges are connected to give a sensitive reading of resistance change.

strain-gauge rosette A combination of three strain gauges which give the principal strains in two-dimensional stress situations.

strain hardening The increase in hardness caused by plastic deformation.

strain rate The time rate of stress application used in testing.

stress Force per unit area in a solid. The area is perpendicular to the force for tensile stress and parallel to it for shear stress. Unit: newtons per square metre $(N\,m^{-2})$.

stress concentration factor The ratio of the greatest stress at a 'stress raiser' to the nominal stress in a component.

stress raiser A local change in contour in a part, e.g. a hole, notch, change of section, etc., which gives rise to an increase in stress.

stress relieving Heating a material to a suitable temperature and holding it long enough to remove residual stresses, then slowly cooling.

stroboscope A flashing lamp of precisely variable periodicity which can be synchronized with a moving object to give a stationary appearance.

sudden contraction A sudden decrease in the cross-sectional area of a conduit, involving a loss of energy.

sudden enlargement A sudden increase in the cross-sectional area of a conduit, involving an energy loss.

superheated steam Steam heated at constant pressure out of contact with the water from which it was formed, i.e. at a temperature above saturation temperature.

surface finish The condition of a surface after final treatment.

surface grinder A grinding machine which produces a flat surface on the workpiece which is mounted on a reciprocating table.

surface hardening Heat treatment such as nitriding, cyaniding, etc., which increases the surface hardness of a metal.

surface tension Interfacial tension between two phases, one of which is a gas.

swaging Forming a reduction in a metal part by forging, squeezing or hammering, sometimes when rotating.

swarf Chips removed from a workpiece during cutting operations.

tachogenerator An electric generator producing a voltage proportional to the speed of a shaft to which it is connected. Connected to a voltmeter calibrated in speed of rotation.

tachometer An electrical or mechanical instrument which measures the rotational speed of a shaft, etc.

tap A cylindrical cutter used to produce an internal screw thread.

temperature The degree of hotness or coldness with reference to an arbitrary zero, e.g. the melting point of ice, absolute zero.

temperature coefficient of resistance A coefficient giving the change in resistance of a piece of material per degree change in temperature.

tempering The reheating of hardened steel or cast iron to a temperature below the eutectoid value to decrease hardness and increase toughness.

tensile strength Ratio of maximum load to original cross-sectional area of a component. Also called 'ultimate strength'.

tensile stress Tensile load divided by cross-sectional area.

tension The state of stress in a part which tends to increase its length in the direction of the load.

thermal shock The development of a steep temperature gradient in a component and accompanying high stress.

thermal stress Stress in a body due to a temperature gradient.

thermistor A semiconductor mixture of cobalt, nickel and manganese oxides and finely divided copper in the form of a bead with leads. The device has a high temperature coefficient of resistance and is used for temperature measurement.

thermocouple A device consisting of a junction of dissimilar metals which produce an e.m.f. approximately proportional to the temperature difference between the hot and cold junctions at the ends.

thermodynamic process A gas process involving changes in pressure, volume, temperature or state.

thermoelectricity The interchange of heat and electric energy, e.g. as in a thermocouple.

thermometer An instrument for measuring temperature.

thermoplastic Any plastic which can be melted by heat and resolidified, the process being repeatable any number of times.

thermosetting resin Compositions in which a chemical reaction takes place while being moulded under heat and pressure. The properties are changed and the product is resistant to further change.

thick cylinder A cylinder in which the thickness of wall is large compared with the bore. Stress analysis is more complicated than for a 'thin' cylinder subject to internal pressure.

thin cylinder A cylinder with a wall thickness relatively small compared with the bore. Under internal pressure a uniform hoop stress may be assumed with no radial stress.

three phase An electric supply system in which the alternating potentials on the three wires differ in phase by 120°.

throttling process The process involving the flow of a fluid through a small tortuous passage destroying all kinetic energy; there is no change in enthalpy.

thrust bearing A shaft bearing designed to take axial load through a collar on the shaft. It may be a flat surface or have balls or rollers.

thyristor A semiconductor device used for switching heavy currents.

tie rod A rod or bar which takes a tensile load.

timing belt A drive belt between two pulleys having teeth which engage with grooves in the pulleys.

timing diagram A circular diagram showing the angular positions of valve opening and closing in two-

and four-stroke engines.

tolerance The specified permissible deviation from a dimension or permissible variation in the size of a component.

toroid (torus) A solid generated by rotating a circle about an external point in its plane.

torque The algebraic sum of couples, or moments of external forces, about the axis of twist. Also called 'torsional moment'.

torsion A twisting action resulting in shear stress.

torsional oscillation Oscillations, e.g. in a shaft in which it is twisted periodically in opposite directions.

total head pressure The sum of dynamic pressure and static pressure in fluid flow.

toughness The ability of a metal to absorb energy and deform plastically before fracturing. Determined by impact tests.

transducer A device which converts a physical magnitude of one form of energy to another form according to a specified formula, e.g. mechanical to electrical energy as in a microphone.

transformer An electrical device without moving parts which transfers alternating current energy, usually with a change in voltage.

transistor A three-electrode semiconductor device used to give a voltage, current or power gain.

triaxial stress A state of stress where none of the three principal stresses is zero. Three-dimensional stress.

turbine A prime mover running on steam, gas or water, in which energy is imparted to rows of moving blades on a rotor.

turbulent flow Fluid flow in which particle motion varies rapidly in velocity and direction; characterized by a high Reynold's number.

turning Removing material from a rotating workpiece using a single-point tool as in a lathe.

twisting moment See: 'torque'.

two-dimensional stress A stress situation where two stresses act at right angles.

two-stroke cycle An engine cycle of two piston strokes, i.e. one revolution.

ultimate strength (ultimate tensile strength, UTS) The maximum tensile stress a material will withstand before failure.

ultrasonics Relating to sound with a frequency above the audible range, i.e. above about 15 kHz.

universal gas constant This is equal to the gas constant for any gas multiplied by its molecular weight, i.e. $R_0 = MR$.

upthrust The force on a floating body due to fluid pressure. Equal to the weight of fluid displaced.

U tube A simple type of pressure-measuring device, or manometer, consisting of a glass (or perspex, etc.) U-shaped tube partially filled with a liquid, e.g. water, mercury, and provided with a scale. A pressure difference across the U tube causes a difference in liquid levels.

vacuum forming A shaping process applied to a sheet of thermoplastic which is heated and sucked into a mould by vacuum.

vacuum pump General name for a pump which displaces a gas against atmospheric pressure.

vane A curved metal plate used in pumps and turbines for directing flow. Same as 'blade'.

vane anemometer A type of anemometer with a vaned rotor which rotates at a speed proportional to a fluid velocity passing through the rotor. A mechanical counter or magnetic transducer counts the revolutions which are expressed as velocity.

vane pump A type of positive-displacement pump with sliding radial vanes in slots in a rotor running eccentrically in a fixed casing.

vapour compression cycle A reversed Carnot cycle used in refrigerators.

vapour cycle A thermodynamic cycle using a vapour as the working substance, e.g. steam.

vapour process A thermodynamic process using a vapour, e.g. steam.

vector A vector, or vector quantity, has magnitude, sense and direction, e.g. velocity, force.

vee belt A power-transmission belt with a truncated vee cross-section running in a vee-groove pulley.

velocity The rate of change of position of a point with respect to time. Unit: metres per second ($m s^{-1}$).

velocity head The head equivalent of the kinetic energy of a fluid equal to $v^2/(2g)$.

velocity pressure Velocity head expressed as a pressure equal to $(\rho v^2)/2$. The pressure realized by suddenly stopping a fluid stream.

velocity ratio In a 'machine' the ratio of distance moved by the 'effort' to that moved by the 'load'.

Venn diagram In logic and mathematics, a diagram consisting of shapes, e.g. circles and rectangles, that show by their inclusion, exclusion or intersection the relationship between 'classes' and 'sets'.

Venturi A convergent–divergent duct in which pressure energy is converted to kinetic energy at the throat.

Venturi meter A flowmeter in which the pressure drop in a Venturi is used to give an indication of flow.

Vernier In instruments, such as the Vernier caliper gauge, a small movable auxiliary scale attached to a slide in contact with a main scale. It enables readings to be taken to, usually, a tenth of a division.

vibration damper A device fitted to a reciprocating engine crankshaft to minimize torsional oscillations.

Vickers' hardness test A hardness test using the indentation from a pyramidal diamond.

viscosity The resistance of a fluid to shear force. The shear force per unit area is a constant times the velocity gradient, the constant being the coefficient of viscosity. Units: newton-seconds per square metre ($Ns m^{-2}$). Symbol: μ.

viscous flow The same as 'laminar flow'.

volute The snail-shell-shaped casing into which the impeller of a centrifugal pump discharges, terminating in a circular pipe. A similar casing is used at the inlet of water turbines.

vortex flow Rotational flow. In a 'forced vortex' the fluid rotates as a solid cylinder. In a 'free vortex' (such as an eddy in a water surface) the velocity of rotation decreases with radius.

washer An annular, usually flat, piece of metal, etc., used under a nut to distribute the load.

weir A dam in a water channel sometimes used in flow measurement.

weld A union made by welding.

weld group A group of welds used to make a joint.

welding The joining of two or more pieces of material by applying heat and/or pressure, with or without a filler material, to produce local fusion.

welding rod Filler in rod or wire form used in welding.

weldment An assembly of several parts joined by welds.

wet steam A steam–water mixture such as results from partial condensation of dry saturated steam.

whirling speed (critical speed) The speed at which excessive deflection of a shaft occurs being numerically the same as the natural frequency of transverse vibration or harmonics.

white metal General term for low-melting-point alloys of lead, tin, bismuth, zinc and antimony used for plain bearings.

work A type of energy involving mechanical effort, e.g. the output from an engine.

work hardening See: 'strain hardening'.

workpiece A part upon which work is done in process operations.

worm A part of a worm gear with helical single or multi-start thread.

worm gear A high speed-ratio gear in which a single or multi-start worm engages with a worm wheel with circumferential teeth. The axes are at right angles and non-intersecting.

wrought iron Iron containing fibres of slag (iron silicate) in a ferrite matrix.

yield stress (yield point) The stress at which a material exhibits a deviation from proportionality of stress and strain. Steels tend to have a definite yield point, for ductile metals an offset of typically 0.2% is used.

Young's modulus See: 'modulus of elasticity'.

Index